Bioremediation

Bioremediation

A Sustainable Approach to Preserving Earth's Water

Edited by
Sanjay K. Sharma
JECRC University, Jaipur

CRC Press is an imprint of the
Taylor & Francis Group, an **informa** business

CRC Press
Taylor & Francis Group
6000 Broken Sound Parkway NW, Suite 300
Boca Raton, FL 33487-2742

First issued in paperback 2021

CRC Press is an imprint of Taylor & Francis Group, an Informa business

ISBN-13: 978-1-138-59307-7 (hbk)
ISBN-13: 978-1-03-217679-6 (pbk)
DOI: 10.1201/9780429489655

Publisher's Note

The publisher has gone to great lengths to ensure the quality of this reprint but points out that some imperfections in the original copies may be apparent.

Library of Congress Cataloging-in-Publication Data

Names: Sharma, Sanjay K., editor.
Title: Bioremediation : a sustainable approach to preserving earth's water / [edited by] Sanjay K. Sharma, JECRC University, Jaipur.
Description: Boca Raton, Florida : CRC Press, 2019. | Includes bibliographical references and index. |
Summary: "Bioremediation: A Sustainable Approach to Preserving Earth's Water discusses the latest research in green chemistry practices and principles that are involved in water remediation and the quality improvement of water. The presence of heavy metals, dyes, fluoride, dissolved solids and many other pollutants are responsible for water pollution and poor water quality. The removal of these pollutants in water resources is necessary, yet challenging. Water preservation is of great importance globally and researchers are making significant progress in ensuring this precious commodity is safe and potable. This volume illustrates how bioremediation in particular is a promising green technique globally"– Provided by publisher.
Identifiers: LCCN 2019030559 | ISBN 9781138593077 (hardback) | ISBN 9780429489655 (ebook)
Subjects: LCSH: Water–Purification–Biological treatment. | Bioremediation. | Green chemistry.
Classification: LCC TD475 .B576 2019 | DDC 628.1/683–dc23
LC record available at https://lccn.loc.gov/2019030559

Visit the Taylor & Francis Web site at
www.taylorandfrancis.com

and the CRC Press Web site at
www.crcpress.com

This book is dedicated to the memories of my late parents
Prof. (Dr.) M. P. Sharma
and
Smt. Parmeshwari Devi
… .who, I feel always there around me and blessing me.

Contents

Preface

'Thousands have lived without love, not one without water.'
– W. H. Auden, a British Poet

Water, the reason for life on Earth… and the Water Cycle is actually the Life Cycle on earth…continuously flows endlessly between the ocean, atmosphere, and land.

Everyone knows that, while nearly 70% of the world is covered by water, only 2.5% of it is fresh. The rest is saline and of no use practically. Even then, just 1% of our freshwater is easily accessible, with much of it trapped in glaciers and snowfields. Also, we have only the amount of water in, on, and above our planet does not increase or decrease, means Earth's water is finite and if we do not value it, we do not value life as well, seriously. Preserving Earth's water is a big challenge before all of us.

And the available less than 1% water has so many quality concerns, which make it unsafe and unhealthy, causes water pollution. This widespread problem of water pollution is a serious concern to the health of one and all around the world. It is shocking to know that water-borne diseases kill more people each year than war and all other forms of violence combined. Therefore, water remediation becomes important as well as unavoidable. Treatment of water for ensuring its drinking quality is a big challenge for the researchers all over the world. Through this book, we are also contributing our bit to the service of the mankind, which will be not more than a drop in this whole effort, but we believe that every drop counts.

Bioremediation: A Sustainable Approach to Preserving Earth's Water consists of 10 comprehensive chapters on various topics contributed by internationally acclaimed scientists and researchers.

Green chemistry considers the designing of processes and products while using minimum or no hazardous substances. Green chemistry is involved in the life cycle (design, production, usage, and disposal). It provides the most effective way to remove or reduce the harmful effects of treatment techniques, by considering energy, raw materials, and production of hazardous secondary material. Chapter 1 provides an overview of commonly used remediation techniques. Some examples and case studies on the water remediation techniques including green chemistry will be presented.

What is the share of water remediation in the scientific research publications, globally? It is a quite interesting question and we decided to give a holistic view to the readers and researchers through Chapter 2, which comprehends bibliometric features and characteristics of the research papers related to water remediation. For this purpose, research papers on water bioremediation indexed in Science Citation Index, Science Citation Index Expanded, Journal Citation Reports and Engineering Index were surveyed. The analysis of bibliometric data mainly considered scientific papers on; distribution of research papers by time perspective, geographical wise distribution of articles, analysis of authorship pattern, degree of authorship collaboration, page, citation and reference distributions and analysis of funding and publishers. It is really a nice chapter, which may encourage readers to take up similar studies in other aspects of scientific and technological researches.

Chapter 3 focuses on biofunctionalized adsorbents for treatment of industrial effluents. This study aims to identify the potentials of an integrated adsorption and biodegradation approach for commercial effluent treatment and thereby serve as a holistic source of information for hydrologists, environmentalists, and industrialists involved in conservation of water resources and eradication of water scarcity.

Comparing to the available treatment processes, biosorption offers many advantages such as cheaper, very efficient for low concentration of pollutants, sludge production is small, does not require any nutrients, easy metal recovery, and the regeneration of the adsorbent is possible. Because of the complexity of biosorption systems, relevant information about the biosorption systems is still needed. Chapter 4 provides a comprehensive review of recent advances in the biosorption area. Various aspects of biosorption such as the types of adsorption equations, modification of biosorbent, adsorption mechanisms, kinetic, and thermodynamic of biosorption are covered in this chapter.

Chapter 5 focuses on the microbial biodegradation of dye-containing wastewaters through enzymatic biotransformation. It further highlights and discusses current advances in this area such as purified enzymes treatments, pathways of dye degradation, bioreactor configurations, immobilization and treatment with mixed microbial cultures.

Olive oil mill wastewaters (OOMW) are among the most concerning contaminants, particularly in the Mediterranean region. Several treatments have been proposed to optimize the OOMW removal from aquatic matrices. The most conventional methods used are physicochemical processes but bioremediation has been arising as an environmentally friendlier alternative. The integration of both approaches allows an improvement in both discoloration, chemical oxygen demand removal and detoxification of OOMW, ensuring better results than the application of these technologies isolated. Biosorption has also been gaining attention as a surrogate approach to treat effluents, with advantages over conventional methods. Besides revisiting the state of the art on bioremediation and biosorption approaches applied to OOMW, valuation opportunities as a complement toward an improved OOMW management are revised. These include the use of OOMW as fertilizers and the application of olive oil wastes as sorbents, as well as the use of these matrices as sources of added value compounds. The development of these processes and new ideas to improve already existent solutions are crucial to mitigate the environmental impacts caused by these wastewaters while valuing the by-products resulting from the olive oil production industry, toward a circular economy approach. Chapter 6 discusses all about it.

Dyes are in general aromatic compounds with functional groups of easy ionization, able to interact with materials of opposite charge to provide color.

They may be from natural sources or synthetic. Dyes are employed in different industries as leather, textile, pharmaceutical, food and paper. Most dyes are water soluble and their application is usually in liquid medium. Therefore, high quantities of colored wastewater are generated. Physical–chemical processes and biological treatment are not always capable to remove dyes from wastewater due to their recalcitrance nature. Adsorption is a promising technique for that purpose. Hence, the search for environment-friendly and low-cost adsorbents has accelerated the development of biosorbents. Alternative materials from microorganisms, agro-industrial and agricultural wastes are used for biosorption. Mechanisms such as ion exchange, adsorption, absorption, complexation and precipitation are involved in the biosorption process, and when take place concurrently achieve high sorption capacities for dyes removal. Biosorbent materials as microalgae biomass, industrial wastes from tanneries, grape bagasse and others have been studied and show that biosorption can be a powerful technique for dye removal. Chapter 7 will be addressed the effects of dyes in the environment, and their removal from wastewater by biosorption process.

Industrial wastes classified as toxic, flammable, reactive or corrosive nonhazardous wastes and hazardous wastes are generally produced as solid, liquid and gas wastes from industrial processes. The industrial effluents from different manufacturing processes and various industrial facilities are characterized considering volume or rate of flow, their physical conditions, chemical and toxic constituents, and bacteriologic levels. Waste treatment technologies developed to minimize waste toxicity and volume use various methods that are categorized as chemical physical and biological methods.

Bioremediation is a popular, low-cost, and efficient pollution control technology that uses biological systems to ensure the deterioration of various toxic chemicals or rapid conversion to less harmful forms. For bioremediation processes, a wide variety of microorganisms are extensively used to modify, immobilize, or detoxify pollutants in contaminated water or soil.

Chapter 8 deals with the classification of industrial wastewater and the principles of bioremediation technology and its use for waste treatment.

In Chapter 9, the methodology, advantages, and disadvantages of various technologies for dye removal are discussed. Special emphasis is made on the removal of azo dyes using the biosorption method by reviewing the points on biosorption research considering the benefits, disadvantages, and future potential of biosorption as an industrial process. Adsorption, kinetic, equilibrium, and thermodynamic studies on the removal of dyes are also presented.

Green synthesis of carbonaceous adsorbents from waste materials is environment friendly when compared with procuring the same commercially. It is also cost-effective and simultaneously reduces waste. In the last few decades, varied agricultural and household wastes have been used to prepare adsorbents. A range of items such as saw dust, coconut shell, and rice husk have been used for the purpose. Immobilization of microorganisms onto carbonaceous adsorbents is another technique which combines adsorption with biodegradation, augmenting both the process, concurrently facilitating total removal of the pollutant from the environment. Chapter 10 reviews different green sources for synthesis of carbonaceous adsorbents reported in recent studies and successive application of these formulated adsorbents for treatment of PAHs. It also highlights emerging technology of combining carbonaceous adsorbent with various nanomaterials and microorganisms to enhance the efficiency of the process.

The main outcome of reading this book will be that the reader is going to have a holistic view of the immense potential and ongoing research in Bioremediation and its close connection with modern research applications. Furthermore, this book can be used as an important platform to inspire researchers in any related fields to develop greener processes for important techniques for use in several fields.

I gratefully acknowledge the efforts of all the contributors of this book, without whom these valuable chapters could not have been completed and published. I express my highest gratitude and thankfulness to all of them.

Gurudev Sri Sri Ravishankar once said 'When you care for nature, be assured that nature cares for you, more than a mother would do for her child.'

Remember, every drop counts.

Jai Gurudev!

Sanjay K. Sharma
FRSC
13 May 2019
Jaipur, India

Acknowledgments

It is time to express my gratitude to my family, friends, colleagues, and well-wishers for extending their never-ending cooperation and support during the journey of this book.

First of all, I want to express my special thanks to all the contributors of this book, without whom this book would not have been published.

I would like to thank my wife Dr. Pratima Sharma for standing beside me throughout my career and during the writing of this book. She has been my strength and motivation for continuing to improve my knowledge and move my career forward and deserves the highest level of appreciation.

I also thank my son Kunal and daughter Kritika for always making me smile and for understanding me on those Sundays when I was writing this book instead of watching television or playing games with them. I know, they are mature now, studying in a university, pursuing computer engineering, and understand why I spent so much time in front of my computer. Love you kids.

I would really like to thank Hilary Lafoe for providing me with the opportunity to become the editor for this book. I appreciate that she believed in me to make this book a reality.

I thank every person of the JECRC University. They have been nice to me and even when criticizing, they know how to be constructive. I am grateful to work with this beautiful group of people.

Last but not least, I am thankful to my students and readers because they are the actual strength behind every teacher, every author, and every book. I am sure they will find this book useful.

Jai Gurudev!

About the Editor

Prof. (Dr.) Sanjay K. Sharma is a **Fellow of Royal Society of Chemistry (UK)** and a very well-known Author and Editor of many books, research journals, and hundreds of articles from last two decades. He has also been appointed as ***Series Editor*** by **Springer, UK,** for their prestigious book series ***SpringerBriefs in Green Chemistry for Sustainability***, where he has been involved in editing of 38 different titles of various international contributors so far.

Presently, Prof. Sharma works as Professor of Chemistry and Associate Dean (Research), JECRC University, Jaipur, India, where he teaches chemistry, environmental chemistry, and green chemistry to UG and PG students, pursuing his research interest in the domain area of green chemistry with special reference of **water pollution, corrosion inhibition,** and **biopolymers** and taking care of PhD program as well as research activities of the university.

He has produced 5 PhDs so far and 3 more research scholars are pursuing PhD under his supervision.

Dr. Sharma has 18 books of chemistry published by international publishers, including Springer, Wiley, and Royal Society of Chemistry, and over 100 research papers of national and international repute to his credit.

He is also a member of American Chemical Society (USA), International Society for Environmental Information Sciences (ISEIS, Canada), and also life member of various international professional societies, including International Society of Analytical Scientists, Indian Council of Chemists, International Congress of Chemistry and Environment, and Indian Chemical Society.

He has been conferred the **Life Time Achievement Award** by the Association of Chemistry Teachers, Mumbai, and SRM University, Chennai, in 2018 for his contribution in Chemistry Education and Research.

He is a **RESOURCE PERSON** (For Green Chemistry, Water Sustainability and Manuscript Writings and Scientific Ethics) for various prestigious national and international organizations.

His popular Books are as follows:

1. *Green Chemistry for Dyes Removal from Waste Water: Research Trends and Applications* (from Wiley-Scrivener, USA)
2. *Heavy Metals in Water: Presence, Removal and Safety* (from Royal Society of Chemistry, UK)
3. *Bio-surfactants: Research Trends and Applications* (from CRC Taylor & Francis, USA)
4. *Waste Water Reuse and Management* (from Springer, UK)
5. *Advances in Water Treatment and Pollution Prevention* (from Springer, UK),
6. *Green Corrosion Chemistry and Engineering* (from Wiley, Germany)

7. *Handbook of Applied Biopolymer Technology: Synthesis, Degradation and Applications* (from Royal Society of Chemistry, UK)
8. *Handbook on Applications of Ultrasound: Sonochemistry and Sustainability* (from CRC Taylor & Francis, USA)
9. *Green Chemistry for Environmental Sustainability* (from CRC Taylor & Francis, USA)

1 Green Chemistry and Its Applications in Water Remediation

S. Atalay and G. Ersöz
Department of Chemical Engineering, Ege University, İzmir, Turkey
Sanjay K. Sharma
Department of Chemistry, JECRC University, Jaipur, India

1.1 INTRODUCTION

In the last decades, huge increases in the world population, economic development, and agricultural activities lead to increase in water demand. In addition to these, inappropriate use and degradation of natural resources have drastically accelerated the consumption of freshwater and also the contamination of water. Hence, the major environmental problems, especially in emerging countries, are the contamination of water and letting largely uncontrolled discharge of pollutants to the environment.

There are various classes of chemicals that have been targeted by the United States Environmental Protection Agency (US EPA) as priority pollutants because of their toxic and hazardous effects on the environment and public health. Hazardous compounds such as polycyclic aromatic hydrocarbon, pentachlorophenol, polychlorinated biphenyl, benzene, toluene, ethylbenzene, xylene, and aniline are persistent and may stay in the environment for a long time. They do not biodegrade easily and are known to have carcinogenic effects.

Cleaning up these compounds is one of the important goals of the green economy. Places that are polluted because of industrial activities, use of various pesticides and fertilizers, or release of other priority pollutants must be cleaned up in order for reusing or returning them to their natural state.

Several conventional processes are commonly applied for the remediation of water. But these traditional methods are not rapid, efficient, cost-effective enough, or sufficient to satisfy the discharge limitations. Therefore, there is an urgent need for a shift in the development of new technologies. The drawbacks of the traditional wastewater treatment methods can be overcome by using green technologies, taking into consideration all environmental effects of remedy implementation to maximize the net environmental benefit of cleanup applications.

This chapter aims to provide a better knowledge of the principles and processes focusing on green chemistry in water remediation. Examples and case studies shared point the ways in which green chemistry eliminates or minimizes the effect of chemical processes on the environment and human health.

1.2 GREEN CHEMISTRY: AN OVERVIEW

Green chemistry can be defined as the technologies of the invention, design, and production of chemical products by finding creative and alternate routes in a way that is sustainable, safe, and requires lowest amounts of resources and energy, nevertheless generating a small amount of waste.

Green chemistry aims to minimize or eliminate hazards of chemical feedstocks, reagents, solvents, and products from a chemical process while maximizing process efficiency. It also involves the use of sustainable materials and renewable energy sources for the processes (Anastas et al., 2001, Anastas and Lankey, 2000, 2002, Anastas and Warner, 1998).

Green chemistry lays down approaches for the production and use of chemicals that aim to decrease risks to both public health and the environment. It deals with environmental effects of both products and processes by which they are synthesized.

There are several important aspects, such as economic, materials, and waste, that make green chemistry sustainable chemistry. Application of green chemistry is generally cheaper than chemistry as it is normally applied. With respect to materials, green chemistry becomes sustainable by using materials efficiently, ensuring maximum recycling, and using minimum raw materials. Green chemistry is sustainable in terms of waste by reducing as much as possible, or even eliminating, their production, which is totally a more environmentally friendly green alternative to traditional chemistry applications (Manahan Stanley, 2006).

As it is well known, green chemistry is based on 12 principles proposed by Anastas and Warner (1998). These principles address key issues in the design and production of chemicals. Application and interpretation of these principles are becoming the main concern in academic, industrial, and governmental environments.

These 12 principles of green chemistry can be described in detail as follows:

1. Prevention

   It is better to prevent contamination than to treat or clean up after it has been generated. It is more appropriate to carry out a process by following a pathway so that formation of waste is minimized or eliminated.

2. Atom economy

   With atom economy, the aim is to design reactions in such a way that the quantity of reactants in the desired final product is the maximum possible. Methods must be designed and used to maximize the usage of all materials in the process for the final product.

Atom economy also takes into consideration the substances that are used in the process but are not directly part of the reactions. By green chemistry approach, researchers aim to either reduce the amount of or eliminate completely as much as possible. Those that cannot be eliminated are reused or recycled when possible.

3. Less hazardous chemical synthesis

   Wherever possible, synthetic methods must be used to produce substances that possess small or almost no toxicity so that they cause no hazard to public health and the environment.

4. Designing safer chemicals

   Products must be designed to affect their required function while minimizing their toxicity.

5. Safer solvents and auxiliaries

   Use of auxiliary substances such as solvents should be minimized wherever possible and innocuous when used.

6. Design for energy efficiency

   Energy needs of processes should be determined for their environmental and economic effects and should be minimized. If possible, synthetic methods should be performed at moderate conditions or ambient temperature and pressure.

7. Use of renewable feedstocks

   A raw material or feedstock should be renewable rather than depleting whenever technically and economically practicable. In other words, use of renewable energy techniques such as solar energy, biomass energy, and biofuels should be emphasized.

8. Reduce derivatives

   Unnecessary derivatization must be minimized or avoided if possible because such steps need extra reagents and can generate waste.

9. Catalysis

   Catalysts play an important role in green chemistry. They increase the selectivity of the desired product, decrease energy requirements, and ensure the use of less hazardous opcrating conditions. Consequently, catalysis is one of the main roots of green chemical applications. Catalytic reagents (as selective as possible) are superior to stoichiometric reagents.

10. Design for degradation

    Products must be designed so that in the end, they may turn into harmless degradation products and do not stay in the environment.

11. Real-time analysis for pollution prevention

    Analytical methods are required to be further developed to allow for real-time, in-process monitoring and control prior to the formation of hazardous substances.

12. Inherently safer chemistry for accident prevention

    Substances and their forms should be selected to minimize the possibility of accidents and explosions.

Scientists use these principles as a guideline and hence researches mainly deal with these principles. Consequently, it can be concluded that modern society has an increasing requirement for environmentally friendly processes.

1.2.1 Designing a Green Process

The progress in the green chemistry movement has challenged researchers in all areas to investigate the environmental impact of a chemistry process as a part of its development program.

While designing greener processes, consider

- efficient processes that minimize resources such as mass and energy required to produce the desired product,
- the environmental health and safety properties of the substances (toxicity and degradability) used in the process,
- environmental life cycle impacts of the process,
- the economic evaluation of the process,
- the waste of the process, and
- if the waste generated in the process can be recycled or recovered.

1.3 WATER REMEDIATION

Remediation is the process of transforming toxic substances in polluted water to below the limits. It involves the application of reactive materials for this detoxification and transformation process. Remedial applications are long-term cleaning activities designed to prevent or reduce the release of hazardous substances and to minimize the risk for the environment and public health.

Environmental remediation considers the removal of contaminants from the environment. For environmental remediation, treatment of both air and water, an extensive number of substances and methods have been evaluated to remove toxic substances from water, soil, and air.

Water remediation is a process to render water free from any contamination. Water remediation is applicable for groundwater, industrial wastewater, and for several other types (Mishra and Clark, 2013). Water remediation studies on the treatment of both groundwater and surface water, whereas soil remediation considers top and sub soil sediments and involves the application of reactive materials for this detoxification and transformation process.

Depending on the type and extent of the contamination, soil and water remediation approaches may be applied separately or together.

1.3.1 Remediation Technologies

There are many remediation technologies, but they can be categorized into two types based on the physical location of the remedial action:

1. ex situ remediation
2. in situ remediation

Ex situ remediation

Ex situ remediation considers physically extracting media from a polluted site and moving it to a different place for subsequent treatment. This can take place in either an above ground treatment facility or by treatment or disposal elsewhere. This remediation aims at excavating soil or sediment and treating it before coming back to its original state. After ex situ treatment, decontaminated soils may be considered for landscape goals (Kuppusamy et al., 2016).

Some of the ex situ remedial methods for polluted soil and groundwater are listed (Kuppusamy et al., 2016) as follows:

- Dig and dump
- Incineration
- Oxidation
- Adsorption
- Ion exchange
- Pyrolysis
- Soil washing
- Dehalogenation
- Solid-phase bioremediation
- Slurry-phase bioremediation: bioreactors
- Solidification/stabilization.
- Constructed wetlands
- Ultrasonic technology
- Microbial fuel cells
- Nanoremediation

In situ remediation

In situ remediation deals with the treatment of contaminants in its actual location in the subsurface. These methods treat the contamination on the site without the removal of the soil.

1.3.2 Green Remediation

Green remediation is defined by EPA as follows: "The practice of considering all environmental effects of remedy implementation and incorporating options to minimize the environmental footprint of cleanup activities."

Green remediation, which is a series of best management practices, mainly aims at effective cleaning while minimizing the environmental and energy footprints of site remediation. It is the practice of considering all environmental impacts at every stage of the remedial process and incorporating options with the objective of maximizing the net environmental benefit of a cleanup (Spellman, 2017).

Although green remediation is an intended approach with relatively low costs, with small or none secondary waste generation, it has some challenges (Tahir et al., 2015). Green remediation might need considerable time to be established and fully operational when compared to remediation that is based on engineering (Kopittke and Menzies, 2005). But these challenges of green remediation can be overcome by optimizing processes.

Green remediation objectives (EPA, 2008)
Obtain remedial action goals
Encourage the using and reusing the remediated substances
Maximize operational efficiencies
Minimize total energy use
Increase the percentage from renewable energy
Reduce the total pollutant
Minimize degradation
Enhance ecology of the site and other affected areas
Reduce air emissions and production of greenhouse gas
Minimize effects to water quality and water cycles
Preserve natural resources

Green remediation addresses mainly the following points (Green Remediation, 2008, Regulatory Council Green and Sustainable Remediation Team, 2011):

Energy requirements of the treatment system

With green remediation, the aim is to use optimized energy, with little or if possible no external utility power. It is also possible to consider using the renewable energy systems. Solar and wind power are the most frequently used sources of alternative renewable energy.

Air emissions

With green remediation, the aim is to reduce air emissions of dust, volatile organic compounds (VOCs), nitrogen oxides, sulfur oxides, and greenhouse gases.

Water requirements and effects on water resources

The consumption of the fresh water is aimed to be minimized by maximizing water reuse during treatment processes reclaiming treated water.

Land and ecosystem impacts

The aim is to minimize soil and habitat disturbances and bioavailability of contaminants through adequate contaminant source.

Material consumption and waste generation using environmentally friendly products

Opportunities should be investigated to minimize material consumption and evaluate recycle materials.

Traditional methods for the water remediation (Vitas et al., 2018) include

1. flocculation, precipitation
2. ion exchange
3. filtration by membrane
4. flotation
5. electrochemical treatment
6. coagulation and flocculation
7. reverse osmosis
8. adsorption
9. evaporation

Even though most of these methods work well, some of them have bottlenecks. In many cases, generally the chemicals used have high costs, manufacturing and operation are rather hard and energy-intensive, and they may end up with waste. As a result, it is inevitable to develop simple methods to design sustainable and innovative solutions with a small effect on the environment (Vitas et al., 2018).

Innovative treatment processes using both physical and chemical techniques such as adsorption on innovative, green adsorbents, membrane filtration, electrodialysis, and photocatalysis for treatment of industrial wastewater often involve technologies for minimizing the toxicity to meet green technology-based treatment standards.

Alternate processes based on adsorption present good performances in wastewater remediation, for example, removal of heavy metals (Barakat, 2011). Moreover, it is well known that due to the discharge of huge amounts of metal-contaminated wastewater, industries dealing with heavy metals such as Cd, Cr, Cu, Ni, As, Pb, and Zn are the most hazardous.

Using agricultural waste as biosorbents has been a promising solution. In the literature, many examples of low-cost biosorbents prepared from various biomass are reported, including sugar beet pulp, peanut hulls, and sawdust (Ajmal et al., 2000, Babel and Kurniawan, 2003, Bilal et al., 2013, Periasamy and Namasivayam, 1996, Reddad et al., 2002, Yu et al., 2000). Besides these, chemical modification of such biomass has led to improvements in their adsorption capacity, reaching results comparable with activated carbon (Vitas et al., 2018). Hence, the use of microorganisms for the removal of heavy metals and other potentially harmful, toxic inorganics in wastewater becomes a promising, sustainable alternative (Orandi et al., 2012).

1.4 GREEN WATER REMEDIATION IN THE LITERATURE

Physical and chemical approaches used in water remediation are very effective; however, they may be costly and may also complicate the actual processes. Green chemistry solutions must be performed if possible because an effective way to solve the water pollution problem is considering the least harmful ways. The review and research strategies and the regulatory environment should be encouraged to recognize the innovative green remediation technologies.

Some studies in the literature covering the green wastewater remediation technologies are summarized as follows.

Wastewater remediation by biowaste and biowaste-derived substances

In recent years, in the concept of green remediation, most of the studies have been dedicated to the biowaste and biowaste-derived materials valorization to prepare innovative, sustainable, and environmental-friendly materials for using in water remediation. In this way, the residues (industrial, urban, etc.) are to be reused, and waste generation is minimized.

In the literature, Bhomick et al. mentioned that biomass materials are promising, eco-friendly, and economical resources for the synthesis of activated carbon and emphasized on using them in water remediation. They reported that there are various kinds of activated carbons that have been prepared from different biowaste sources such as olive stones, bamboo dust, coconut and groundnut shell, rice husk and straw, banana peel, rice husk, and corn cob. Bhomick et al. concluded that activated carbon synthesized from biowaste sources will be an economical, renewable, and greener source of adsorbent for wastewater treatment applications. They recommended the preparation of activated carbons instead of using directly the expensive commercial activated carbons for efficient water remediation (Bhomick et al., 2017).

Al-Ghouti and Salih also aimed to use a biomass in their water remediation study. Eggshell, which is one of the agricultural wastes, was studied as it is cheap and environmentally friendly. They also pointed out that the presence of calcium carbonate in the eggshell makes it an effective sorbent. They tested the performance of waste eggshells as an adsorbent for the remediation of boron from desalinated seawater. They concluded that removal efficiency of boron was 96.3% at pH 6. The adsorption processes were found to be exothermic and spontaneous for both waste eggshell and eggshell that was calcined (Al-Ghouti and Salih, 2018).

Another green biomass source, gastropod shell, was used as the reactive material by Oladoja et al. for phosphorus capture and remediation of aquaculture wastewater. They obtained the gastropod shell after the snail had been removed in boiled water. The shell was rinsed thoroughly with deionized water, dried in the oven, and was ground. The green source, gastropod shell was used to catch and recover phosphorus to reach acceptable discharge standards. The equilibrium isotherm parameters and the mechanism of this process were also derived and investigated (Oladoja et al., 2017).

Nisticò et al. used composted urban biowaste-derived substances as carbon sources for the synthesis of carbon-coated magnet-sensitive nanostructured particles obtained by co-precipitation method and an additional thermal treatment. The adsorption capacity of the magnetic materials which are derived from biowaste for the removal of hydrophobic contaminants was evaluated to see if they might be used in wastewater remediation process. They concluded that they may be used as innovative, recoverable adsorbents for water remediation (Nisticò et al., 2018).

Magnacca et al. in their study prepared Fe_3O_4 nanoparticles coated with bio-based products. They used them as low-cost nanoadsorbents in the treatment of wastewater. They also realized the physical and chemical characterizations of the prepared bio-based products-coated Fe_3O_4 nanoparticles. They used crystal violet as a model contaminant to analyze the adsorption capacity of the nanoparticles. They found out that the medium pH and nanoparticle amount were important conditions to increase removal efficiency. They concluded in their study that by the industrial exploitation of the waste's chemical value, they may become a source for the revenue (Magnacca et al., 2014).

In another study, Tontiwachwuthikul et al. also used a green material for water remediation. They used potato peels for removal of water from waste oil spill. This waste biomass material was obtained from the fast-food restaurant. When they got the potato peels, they washed them enough to remove any dust or dirt material. Then they dried in an oven at 70°C during a day then washed again and dried at 80°C for another 2 days. The evaluation of potato peels powder showed that it can adsorb oil fast and can keep it for a long time. From the kinetic study they concluded that the adsorption process was well fitted to the second-order kinetics (Tontiwachwuthikul et al., 2016).

Wastewater remediation by algae

Algae are a well-known organism. They are found everywhere and one of the most common floras found in the biosphere. Generally, they are plant-like and have photosynthetic functions. Today, researchers have intensely focused on using algae biomass to promote the wastewater remediation (Nhat et al., 2018).

It is well known that algae are a key factor in pollutant remediation and can easily remove nutrient and then produce biomass very rapidly. The biomass is generated by mainly biosorption, precipitation, volatization, biosorption, nitrification, and denitrification. Carbon, nitrogen, and phosphorus serve as nutrient and substrate source for biomass production (Zhao et al., 2016). In the literature, some authors mentioned that there are various types of algae species having different characteristics (Ge and Champagne, 2017, Hou et al., 2016). Also, it is reported that the effect of operating parameters such as retention time, initial nutrient and temperature strongly affect efficiency (Sindelar et al., 2015). Therefore, optimization of these conditions is very important. Consequently, algae are currently associated with environmental problems, and there are many studies in the literature related to green process and wastewater remediation.

The polluted water sources by heavy metal ions are also in concern of researchers. Heavy metals (chromium, copper, lead, mercury, etc.) are known to be highly toxic, hazardous and hard or impossible to biodegrade. It is inevitable to remove the heavy metals from the contaminated water sources with a suitable technique before discharging them into the environment. Henriques et al. used algae in the remediation studies. They studied the metal removal from wastewater by living *Ulva lactuca*. The researchers realized the study by mono-metallic solutions of Hg, Pb, and Cd, in different environmental concentrations and multimetallic solutions having Hg, Pb, and Cd in the same concentration, and also at different concentrations. They concluded that in mono-metallic solutions, macroalgae were able to remove approximately 95% of metal. They also added that macroalgae performed well in multi-metallic solutions (Henriques et al., 2017).

Another system is related to a combination of freshwater microalgae *Chlorella zofingiensis* cultivation with piggery wastewater treatment. The researchers investigated the characteristics of algal growth, lipid, and biodiesel production. A tubular bubble column photobioreactor was used to examine the nutrient removal by using cultivate *C. zofingiensis.* Zhu et al. concluded that the piggery wastewater was efficiently treated (Zhu et al., 2013).

Sindelar with his group mentioned that algae scrubbers are a developing technique that can be efficiently used in nutrient removal from many different wastewaters. In their research, they investigated the methods for increasing total phosphorus removal in algae scrubbers. They mentioned that their research may provide information on the optimum operating conditions for algal total phosphorus removal (Sindelar et al., 2015).

The performation of biofilm-based remediation technologies in tailings waters coming from Queensland Nickel was investigated by Palma et al. Their objective was to investigate the growth potential and metal removal capacity of indigenous fungal/green microalgae consortia. They concluded that the biofilms had substantial capability for the remediation of heavy metals from tailings water (Palma et al., 2017).

Wastewater remediation by aerogels

Aerogels are a class of porous substances. They have various good properties such as low density, high porosity, high surface area, and adjustable surface chemistry. These properties combined with modifiable surface chemistry (wet synthesis approach) make them a candidate for environmental remediation. The preparation of aerogels consists of three steps: sol–gel, aging, and drying. But regarding carbon and carbide aerogels additional carbonization and carbothermal reduction step after drying are required. Recently, various environmental remediation processes of these innovative materials are focused and thoroughly reviewed. In most of them, aerogels have been found to have high sorption capacity toward the target components, and hence they are good sorbents (Maleki, 2016). Maleki in his review article gave a general summary on the aerogels production and their applications in remediation of water contaminated by oil and toxic organic compounds and heavy metal ions.

For example, Liu et al. concluded that hydrophobic silica aerogels have a huge adsorption performance toward poorly soluble organic compounds. They also concluded that hydrophilic silica aerogels were more effective in adsorption of soluble organic compounds from water (Liu et al., 2009). Also, the carbon aerogels such as carbon nanotube, carbon fiber, carbon microbelt, and graphene aerogel are very likely to be used in water remediation (Gui et al., 2013, Lei et al., 2013). They mentioned that these special materials also found in 3D forms with special properties, such as low density, high porosity, huge specific surface area, and good surface hydrophobicity (Gui et al., 2013, Lei et al., 2013, Liu et al., 2009).

Graphene aerogels have been deeply used in water remediation. Cao et al. reported a green process for the preparation of a 3D-columnar graphene aerogel. The prepared aerogel had good properties such as fast oil absorption, good fire-resistance, strong durability, and reusability. They concluded that due to these properties, the aerogel can be a suitable material for water mediation (Cao et al., 2018).

Wastewater remediation by metal organic frameworks

One of the classes of crystalline inorganic–organic hybrid porous materials is metal organic frameworks (MOFs). They have been accepted as highly powerful functional materials used in water remediation methods.

MOFs have shown high potential especially in removal by adsorption and photocatalytic degradation in water remediation as energy-effective and low-cost technologies. In the literature, there are many studies focusing on the improvements obtained in the adsorptive removal of inorganic metal cations, inorganic acids, nuclear wastes and other inorganic anions and pharmaceuticals, agricultural products, dyes commonly found in wastewater using MOF technologies (Mon et al., 2018). The adsorptive removal (Cychosz and Matzger, 2010, Han et al., 2015, Park et al., 2013, Qin et al., 2015, Sarker et al., 2018) and photodegradation (Masoomi et al., 2016; Xamena et al., 2007) of toxic phenolic compounds in water using MOF are reported by many researchers (Pi et al., 2018).

Overall, the studies indicate that in adsorption processes and photocatalytic reactions using MOFs for water remediation is promising (Pi et al., 2018) but Mon et al. concluded that despite the considerable advances, the number of MOFs applied in water remediation is still limited. Therefore, the development of innovative MOFs should be still in the interest of researchers (Mon et al., 2018).

Nanotechnology in wastewater remediation

Recently, nanotechnology has been determined as a promising alternative for conventional treatment processes and reacting agents to study at green conditions, at a lower cost while reaching the quality standards of water.

It is very important to design and prepare innovative and sustainable catalysts that can efficiently use multiple energy sources in solving the current problems faced in environmental remediation.

In this context, nanoparticles have been widely used for green water remediation containing diverse organic and inorganic pollutants due to their unique properties (Alabresm et al., 2018, Calderon and Fullana, 2015). Because of their extremely small size, nanoparticles show various physical, chemical, and biological properties when compared with larger, micro-, and macro-scale particles. Nanoparticles have large specific surface areas, a higher density of surface reaction surfaces per unit mass and hence have high adsorption capacities (Tesh and Scott, 2014).

Especially, nano-sized iron oxide particles have been used in different types of applications in the literature (Alabresm et al., 2018, Boyer et al., 2010, Laurent et al., 2008).

Alabresm et al. (2018) investigated the synergistic effect of nanotechnology and microbial remediation to remove oil spills in the marine environment. They used polyvinylpyrrolidone-coated magnetite nano-sized particles and oil-degrading bacteria for the removal of oil. They concluded that nanoparticles by themselves could remove nearly 70% of lower-chain alkanes and 65% of higher chain in 1 h of incubation. Experiments with the combination of nanoparticles and bacterial strains ended up with higher efficient results within a shorter time (Alabresm et al., 2018).

A green synthesis of the cerium dioxide nano-sized particles using leaf extract of *Azadirachta indica* plant in thermal and photocatalytic processes for dye remediation was realized by Sharma et al. They prepared nanomaterials by eco-friendly and cheaper methods and the activity and efficiency of these particles were tested by photodegradation of Rhodamine B dye. They concluded that the CeO_2 nanoparticles prepared by green methods showed a high degradation efficiency of 96% (Sharma et al., 2017).

The polyaniline:photocatalyst nanomaterial was prepared by the anoxidative polymerization method, and undoped polyaniline, polyaniline:photocatalyst dissolved in of *N*-methyl-2-pyrrolidone was coated over different surfaces by Gunti et al. They tested the efficiencies of the nanostructured photocatalyst nanomaterials to remove methyl orange in water. They concluded that this material can be used in remediation of organics by the green photo electrochemical catalytic method (Gunti et al., 2017).

Mushtaq and his friends prepared pure, single-crystalline $BiFeO_3$ nanosheets and nanowires. They mentioned that both have promising photocatalytic and piezocatalytic characteristics for water treatment. These nanocatalysts were able to harness photonic and mechanical energy for the removal of various organic pollutants such as rhodamine B. By using photocatalytic and piezoelectric properties, degradation of rhodamine B dye was greatly increased. They concluded that improvement of these nanostructures contributes to the use of green processes for efficient environmental applications (Mushtaq et al., 2018).

Photochemical remediation

Several physical and chemical methods have been applied for the green remediation of waters and wastewaters. As mentioned in the above sections, mostly adsorption is preferred in wastewater green remediation but photocatalytic oxidation has also attracted interest as a promising and alternative method for environmental remediation (Garcia et al., 2018, León et al., 2017, Vilar et al., 2017). This process is based on the degradation of pollutants by the help of a sequence of reactions taking place on the surfaces activated by the irradiation of light at a specific wavelength range (Borges et al., 2016).

Photocatalytic ozonation under visible light for the remediation of wastewater and coupling with an electro-membrane bioreactor was examined by Garcia et al. Photocatalytic oxidation and photocatalytic ozonation were performed in the purification and disinfection of the grey water. They concluded that 60% of total organic carbon removal by photocatalytic ozonation under visible light and using the two processes together enhanced global efficiency (Garcia et al., 2018).

Kumar and his friends prepared heterojunctions of graphene quantum dots (GQD) decorated ZnO nanorods (NR). They tested them as effective photocatalysts in the removal of colored pollutants (methylene blue (MB) dye) and a colorless pollutant (carbendazim (CZ) fungicide) under the irradiation of sunlight. If the results are compared the heterojunction with 2 wt% of GQD (ZGQD2) showed the best photocatalytic activity by effectively degrading (nearly 95%) of organic pollutants from water within a short time as 70 min. They concluded that this heterojunction is an effective photocatalyst for remediation processes (Kumar et al., 2018).

Leon et al. in their study focused on the degradation of the antibiotic cefotaxime under sunlight (which is simulated) radiation using heterogeneous catalysts with titanium dioxide (TiO_2) and zinc oxide (ZnO) in solutions. They concluded that the studied parameters, pH and initial catalyst loading have an important role in antibiotic degradation in both TiO_2 and ZnO suspensions. In addition, the role of photogenerated holes, hydroxyl, and superoxide anion radicals on cefotaxime degradation was investigated to determine the reaction mechanism. Antibiotic removal was described by the pseudo-first-order reaction kinetic model regardless of the catalyst type (León et al., 2017).

Lee et al. reported a layer-by-layer deposition approach for immobilizing TiO_2 nanoparticles on a porous support. They obtained a high catalytic efficiency for photochemical decomposition of bisphenol A. They mentioned that the bisphenol A solution processed with the multilayered TiO_2 nanofibers for 40 h, the estrogenic activity was significantly lower than that in the same solution treated with colloidal TiO_2 nanoparticles at the same reaction conditions. They concluded that water-based, electrostatic layer-by-layer deposition effectively immobilizes and stabilizes TiO_2 nano-sized particles on electrospun polymer nanofibers for good photochemical water remediation (Lee et al., 2010).

1.5 CONCLUDING REMARKS

Over the last few years, the requirement for clean water has increased worldwide due to fast industrialization and population growth; therefore, environmentally friendly techniques based on green chemistry have to be developed.

The green chemistry concept is based on 12 principles. These principles mainly cover minimizing or eliminating hazardous or toxic substances during the production of chemical products and in the application of chemical processes.

The goals of green chemistry in water remediation and economic gain may be reached through: biocatalysis, catalysis, use of renewable raw materials such as biomass, alternative reaction media, and reaction conditions.

Even though, green chemistry solutions have been applied in virtually every aspect of industrial activity and provide almost the most effective way to remove or reduce the harmful effects of water treatment techniques, much effort is still needed in this area.

REFERENCES

Ajmal, M., Rao, R. A. K., Ahmad, R., Ahmad, J. 2000. Adsorption studies on Citrus reticulata (fruit peel of orange): removal and recovery of Ni (II) from electroplating wastewater. *Journal of Hazardous Materials*, 79(1–2), 117–131.

Alabresm, A., Chen, Y. P., Decho, A. W., Lead, J. 2018. A novel method for the synergistic remediation of oil-water mixtures using nanoparticles and oil-degrading bacteria. *Science of the Total Environment*, 630, 1292–1297.

Al-Ghouti, M. A., Salih, N. R. 2018. Application of eggshell wastes for boron remediation from water. *Journal of Molecular Liquids*, 256, 599–610.

Anastas, P. T., Kirchhoff, M. M., Williamson, T. C. 2001. Catalysis as a foundational pillar of green chemistry. *Applied Catalysis A: General*, 221, 3–13.

Anastas, P. T., Lankey, R. L. 2002. Sustainability through Green Chemistry and Engineering. *ACS Syrup Series*, 823, 1–11.

Anastas, P.T. and Lankey, R.L., 2000. Life cycle assessment and green chemistry: the yin and yang of industrial ecology. *Green Chemistry*, 2(6), pp.289-295.

Anastas, P. T., Warner, J. C. 1998. *Green Chemistry: Theory and Practice*, Oxford University Press, New York.

Babel, S., Kurniawan, T. A. 2003. Low-cost adsorbents for heavy metals uptake from contaminated water: a review. *Journal of Hazardous Materials*, 97(1–3), 219–243.

Barakat, M. A. 2011. New trends in removing heavy metals from industrial wastewater. *Arabian Journal of Chemistry*, 4(4), 361–377.

Bhomick, P. C., Supong, A., Sinha, D. 2017. Organic pollutants in water and its remediation using biowaste activated carbon as greener adsorbent. *International Journal of Hydrogen Energy*, 1(3), 91–9291.

Bilal, M., Shah, J. A., Ashfaq, T., Gardazi, S. M. H., Tahir, A. A., Pervez, A., Mahmood, Q. 2013. Waste biomass adsorbents for copper removal from industrial wastewater—a review. *Journal of Hazardous Materials*, 263, 322–333.

Borges, M. E., Sierra, M., Cuevas, E., García, R. D., Esparza, P. 2016. Photocatalysis with solar energy: sunlight-responsive photocatalyst based on TiO_2 loaded on a natural material for wastewater treatment. *Solar Energy*, 135, 527–535.

Boyer, C., Whittaker, M. R., Bulmus, V., Liu, J., Davis, T. P. 2010. The design and utility of polymer-stabilized iron-oxide nanoparticles for nanomedicine applications. *NPG Asia Materials*, 2, 23–30.

Calderon, B., Fullana, A. 2015. Heavy metal release due to aging effect during zero valent iron nanoparticles remediation. *Water Research*, 83, 1–9.

Cao, J., Wang, Z., Yang, X., Tu, J., Wu, R., Wang, W. 2018. Green synthesis of amphipathic graphene aerogel constructed by using the framework of polymer-surfactant complex for water remediation. *Applied Surface Science*, 444, 399–406.

Cychosz, K. A., Matzger, A. J. 2010. Water stability of microporous coordination polymers and the adsorption of pharmaceuticals from water. *Langmuir*, 26(22), 17198–17202.

Garcia, D. T., Ozer, L. Y., Parrino, F., Ahmed, M., Brudecki, G. P., Hasan, S. W., Palmisano, G. 2018. Photocatalytic ozonation under visible light for the remediation of water effluents and its integration with an electro-membrane bioreactor. *Chemosphere*, 209, 534–541.

Ge, S., Champagne, P. 2017. Cultivation of the Marine Macroalgae Chaetomorpha linum in Municipal Wastewater for Nutrient Recovery and Biomass Production. *Environment and Science Technology*, 51, 3558–3566.

Green and Sustainable Remediation. 2011. State of the Science and Practice the Interstate Technology & Regulatory Council Green and Sustainable Remediation Team.

Green Remediation. 2008. Incorporating Sustainable Environmental Practices into Remediation of Contaminated Sites U.S. Environmental Protection Agency Office of Solid Waste and Emergency Response, EPA 542-R-08-002.

Gui, X., Zeng, Z., Lin, Z., Gan, Q., Xiang, R., Zhu, Y., Cao, A., Tang, Z. 2013. Magnetic and highly recyclable macroporous carbon nanotubes for spilled oil sorption and separation. *ACS Applied Materials & Interfaces*, 5(12), 5845–5850.

Gunti, S., Alamro, T., McCrory, M., Ram, M. K. 2017. The use of conducting polymer to stabilize the nanostructured photocatalyst for water remediation. *Journal of Environmental Chemical Engineering*, 5(6), 5547–5555.

Han, T., Xiao, Y., Tong, M., Huang, H., Liu, D., Wang, L., Zhong, C. 2015. Synthesis of CNT@ MIL-68 (Al) composites with improved adsorption capacity for phenol in aqueous solution. *Chemical Engineering Journal*, 275, 134–141.

Henriques, B., Rocha, L. S., Lopes, C. B., Figueira, P., Duarte, A. C., Vale, C., Pardal, M. A., Pereira, E. 2017. A macroalgae-based biotechnology for water remediation: simultaneous removal of Cd, Pb and Hg by living Ulva lactuca. *Journal of Environmental Management*, 191, 275–289.

Hou, Q., Nie, C., Pei, H., Hu, W., Jiang, L., Yang, Z. 2016. The effect of algae species on the bioelectricity and biodiesel generation through open-air cathode microbial fuel-cell with kitchen waste anaerobically digested effluent as substrate. *Bioresource Technology*, 218, 902–908.

Kopittke, P., Menzies, N. 2005. Effect of pH on Na induced Ca deficiency. *Plant Soil*, 269 (1–2), 119–129.

Kumar, S., Dhiman, A., Sudhagar, P., Krishnan, V. 2018. ZnO-graphene quantum dots heterojunctions for natural sunlight-driven photocatalytic environmental remediation. *Applied Surface Science*, 447, 802–815.

Kuppusamy, S., Palanisami, T., Megharaj, M., Venkateswarlu, K., Naidu, R. 2016. Ex-situ remediation technologies for environmental pollutants: a critical perspective. *Reviews of Environmental Contamination and Toxicology*, 236, 117–192. Springer International Publishing Switzerland.

Laurent, S., Forge, D., Port, M., Roch, A., Robic, C., Vander Elst, L., Muller, R. N. 2008. Magnetic iron oxide nanoparticles: synthesis, stabilization, vectorization, physicochemical characterizations, and biological applications. *Chemical Reviews*, 108(6), 2064–2110.

Lee, J. A., Nam, Y. S., Rutledge, G. C., Hammond, P. T. 2010. Enhanced photocatalytic activity using layer-by-layer electrospun constructs for water remediation. *Advanced Functional Materials*, 20, 2424–2429.

Lei, W., Portehault, D., Liu, D., Qin, S., Chen, Y. 2013. Porous boron nitride nanosheets for effective water cleaning. *Nature Communications*, 4, 1777.

León, D. E., Zúñiga-Benítez, H., Peñuela, G. A., Mansilla, H. D. 2017. Photocatalytic removal of the antibiotic cefotaxime on TiO_2 and ZnO suspensions under simulated sunlight radiation. *Water, Air, & Soil Pollution*, 228(9), 361.

Liu, H., Sha, W., Cooper, A. T., Fan, M. 2009. Preparation and characterization of a novel silica aerogel as adsorbent for toxic organic compounds. *Colloids and Surfaces A: Physicochemical and Engineering Aspects*, 347(1–3), 38–44.

Magnacca, G., Allera, A., Montoneri, E., Celi, L., Benito, D. E., Gagliardi, L. G., Gonzalez, M. C., Da Martire, D. O., Carlos, L. 2014. Novel magnetite nanoparticles coated with waste-sourced biobased substances as sustainable and renewable adsorbing materials. *ACS Sustainable Chemistry & Engineering*, 2, 1518−1524.

Maleki, H. 2016. Recent advances in aerogels for environmental remediation applications. *A Review Chemical Engineering Journal*, 300, 98–118.

Manahan, Stanley, E. 2006. *Green Chemistry and the Ten Commandments of Sustainability*, 2nd ed., ChemChar Research Publishers, Columbia, Missouri, USA.

Masoomi, M. Y., Bagheri, M., Morsali, A., Junk, P. C. 2016. High photodegradation efficiency of phenol by mixed-metal–organic frameworks. *Inorganic Chemistry Frontiers*, 3(7), 944–951.

Mishra, A., Clark, J. H. 2013. *Green Materials for Sustainable Water Remediation and Treatment*, The Royal Society of Chemistry, RSC Publishing.

Mon, M., Bruno, R., Ferrando-Soria, J., Armentano, D., Pardo, E. 2018. Metal–organic framework technologies for water remediation: towards a sustainable ecosystem. *Journal of Materials Chemistry A*, 6(12), 4912–4947.

Mushtaq, F., Chen, X., Hoop, M., Torlakcik, H., Pellicer, E., Sort, J., Gattinoni, C., Nelson, B. J., Pan, S. 2018. Piezoelectrically enhanced photocatalysis with $BiFeO_3$ nanostructures for efficient water remediation. *iScience*, 4, 236–246.

Nhat, P. V. H., Ngo, H. H., Guo, W. S., Chang, S. W., Nguyen, D. D., Nguyen, P. D., Bui, X. T., Zhang, X. B., Guo, J. B. 2018. Can algae-based technologies be an affordable green process for biofuel production and wastewater remediation? *Bioresource Technology*, 256, 491–501.

Nisticò, R., Cesano, F., Franzoso, F., Magnacca, G., Scarano, D., Funes, I. G., Carlos, L., Parolo, M. E. 2018. From biowaste to magnet-responsive materials for water remediation from polycyclic aromatic hydrocarbons. *Chemosphere*, 202, 686–693.

Oladoja, N. A., Adelagun, R. O. A., Ahmad, A. L., Ololade, I. A. 2017. Green reactive material for phosphorus capture and remediation of aquaculture wastewater. *Process Safety and Environmental Protection*, 105, 21–31.

Orandi, S., Lewis, D. M., Moheimani, N. R. 2012. Biofilm establishment and heavy metal removal capacity of an indigenous mining algal-microbial consortium in a photo-rotating biological contactor. *Journal of Industrial Microbiology & Biotechnology*, 39(9), 1321–1331.

Palma, H., Killoran, E., Sheehan, M., Berner, F., Heimann, K. 2017. Assessment of microalga biofilms for simultaneous remediation and biofuel generation in mine tailings water. *Bioresource Technology*, 234, 327–335.

Park, E. Y., Hasan, Z., Khan, N. A., Jhung, S. H. 2013. Adsorptive removal of bisphenol-A from water with a metal-organic framework, a porous chromium-benzenedicarboxylate. *Journal of Nanoscience and Nanotechnology*, 13(4), 2789–2794.

Periasamy, K., Namasivayam, C. 1996. Removal of copper (II) by adsorption onto peanut hull carbon from water and copper plating industry wastewater. *Chemosphere*, 32(4), 769–789.

Pi, Y., Li, X., Xia, Q., Wu, J., Li, Y., Xiao, J. 2018. Adsorptive and photocatalytic removal of Persistent Organic Pollutants (POPs) in water by metal-organic frameworks (MOFs). *Chemical Engineering Journal*, 337(1), 351–371.

Qin, F. X., Jia, S. Y., Liu, Y., Li, H. Y., Wu, S. H. 2015. Adsorptive removal of bisphenol A from aqueous solution using metal-organic frameworks. *Desalination and Water Treatment*, 54(1), 93–102.

Reddad, Z., Gérente, C., Andrès, Y., Ralet, M. C., Thibault, J. F., Le Cloirec, P. 2002. Ni (II) and Cu (II) binding properties of native and modified sugar beet pulp. *Carbohydrate Polymers*, 49(1), 23–31.

Sarker, M., Song, J. Y., Jhung, S. H. 2018. Adsorptive removal of anti-inflammatory drugs from water using graphene oxide/metal-organic framework composites. *Chemical Engineering Journal*, 335, 74–81.

Sharma, J. K., Srivastava, P., Ameen, S., Akhtar, M. S., Sengupta, S. K., Singh, G. 2017. Phytoconstituents assisted green synthesis of cerium oxide nanoparticles for thermal decomposition and dye remediation. *Materials Research Bulletin*, 91, 98–107.

Sindelar, H. R., Yap, J. N., Boyer, T. H., Brown, M. T. 2015. Algae scrubbers for phosphorus removal in impaired waters. *Ecological Engineering*, 85, 144–158.

Spellman, F. 2017. The Science of Environmental Pollution, 3rd ed., Taylor & Francis Group.

Tahir, U., Yasmin, A., Khan, U. H. 2015. Phytoremediation: potential flora for synthetic dyestuff metabolism. *Journal of King Saud University.*

Tesh, S. J., Scott, T. B. 2014. Nano-composites for water remediation: a review. *Advance Materials*, 26(35), 6056–6068.

Tontiwachwuthikul, P., Zubaidi, I. A., Rennie, E., Schubert, S., Seitz, M., Silva, C. S. 2016. Remediation of water from waste lubricating oil spill using potato peels. Proceedings of the 3rd International Conference on Fluid Flow, Heat and Mass Transfer (FFHMT'16), Ottawa, Canada.

Vilar, V. J. P., Amorim, C. C., Li Puma, G., Malato, S., Dionysiou, D. D. 2017. Intensification of photocatalytic processes for niche applications in the area of water, wastewater and air treatment. *Chemical Engineering Journal*, 310, 329–330.

Vitas, S., Keplinger, T., Reichholf, N., Figi, R., Cabane, E. 2018. Functional lignocellulosic material for the remediation of copper(II) ions from water: towards the design of a wood filter. *Journal of Hazardous Materials*, 355, 119–127.

Yu, B., Zhang, Y., Shukla, A., Shukla, S. S., Dorris, K. L. 2000. The removal of heavy metal from aqueous solutions by sawdust adsorption—removal of copper. *Journal of Hazardous Materials*, 80(1–3), 33–42.

Zhao, Z., Song, X., Wang, W., Xiao, Y., Gong, Z., Wang, Y., Zhao, Y., Chen, Y., Mei, M. 2016. Influences of iron and calcium carbonate on wastewater treatment performances of algae based reactors. *Bioresource Technology*, 216, 1–11.

Zhu, L., Wang, Z., Shu, Q., Takala, J., Hiltunen, E., Feng, P., Yuan, Z. 2013. Nutrient removal and biodiesel production by integration of freshwater algae cultivation with piggery wastewater treatment. *Water Resource*, 47, 4294–4302.

2 Share of Bioremediation in Research Journals

A Bibliometric Study

Hasan Demir
Department of Chemical Engineering, Osmaniye Korkut Ata University, Karacaoglan Campus, Turkey

Sanjay K. Sharma
Department of Chemistry, JECRC University, Jaipur, India

2.1 INTRODUCTION

It is reported that 400–500 million tons of fatal chemical materials including cyanide, sulphur and other radioactive are released per year into water (Jaganathan et al. 2014; Proverb 2011). Decreasing water quality has become an important issue of global concern, as it is causing significant confusion in water usage, ecosystem health and functioning, and the biodiversity that ecosystems underpin (IWQGES 2016). The water consumption by industrial sectors is shown in Figure 2.1. The highest water consumption was reported to be by power plants, which stands at 87.8% (Jaganathan et al. 2014). A serious problem occurs when chemicals are directly released into water resources causing contamination of water. Currently, 700 million people in 43 countries suffer from water scarcity since polluted water cannot be used for agricultural application, industry, bathing and drinking purposes (Jain and Singh 2003; United Nation Office 2013). The toxic discharges are mostly released to natural water resources in low concentrations, which can be perceived at inert and harmless levels (UNEP/MAP/RAC/CP 2004; IWQGES, United Nation Universities 2016). The quality of water is defined by EN ISO 7887:2011 and UNEP regulations.

Bioremediation is a potential technology that utilizes microorganisms (bacteria, fungi, yeast, and algae) for water treatment wasted by textile effluents, nutrients, dyes, lipids, etc. (Robinson et al. 2001; Chequer et al. 2013). Microalgae, which has almost 50,000 classes available on earth, are classified as eukaryotic and prokaryotic photosynthetic microorganisms (Fazal et al. 2017). Microalgae are used for bioremediation in wastewater treatment through two ways: consumption of effluent as carbon source and adsorbents as biosorbents. The high surface area and strong bonding of living and dead microalgae lead to high sorption (Fazal et al. 2017).

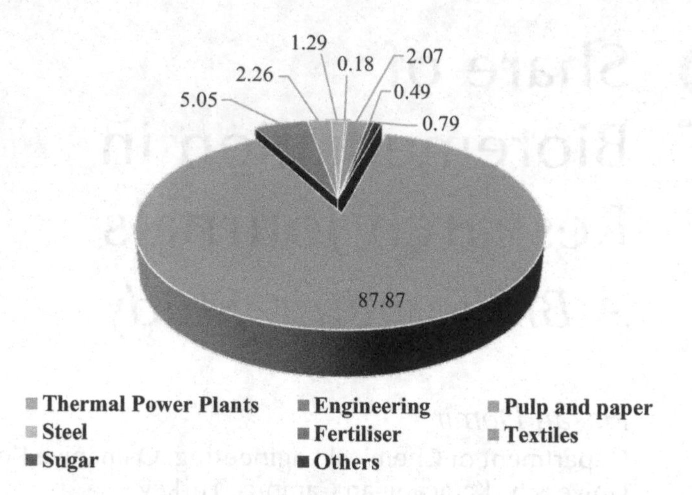

FIGURE 2.1 Consumption of water by various industrial sectors (reprinted with permission from *European Scientific Journal* (Jaganathan et al. 2014.))

The bibliometrics was identified as 'the application of mathematical and statistical methods to books and other communication medium' by Pritchard (1969) (Velmurugan and Radhakrishnan 2016). The methods of bibliometric are used to analyze, authorship, citation and publication pattern, and the relationship within scientific domains and research communities and to structure of specific fields (Vijay and Raghavan 2007; Jermann et al. 2015; Verma, Sonkar and Gupta 2015). The analysis of papers with various statistical methods can permit for investigation studies, features and behavior of published pieces of knowledge for investigation of the structures of research and scientific areas, and for assessment of administration of scientific information and research activity (Velmurugan and Radhakrishnan 2016).

The present chapter comprehends bibliometric features and characteristics of the research papers related to water remediation. The information used to build this bibliometric study was composed of scientific research papers indexed in Science Citation Index, Science Citation Index Expanded, Journal Citation Reports and Engineering Index. The analysis involves mainly discussion of papers as follows:

- Distribution of research papers on year perspective
- Geographical distribution of articles
- Authorship pattern year perspective
- Degree of authorship collaboration
- Multidisciplinary
- Page distribution
- Degree of funding
- Citation
- Reference

2.2 METHODOLOGY

The research papers were surveyed in Web of Science at all years (1969 until the first quarter of 2018). The bibliometrics analysis of 5294 scientific papers was processed according to year wise and geographical wise distribution, authorship pattern and citation, etc. All these information were transferred to MS Excel and organized, analyzed, sorted and revealed by using simple statistical methods.

2.3 RESULTS AND DISCUSSIONS

A Year-Wise Distribution of Articles

The year-wise distribution of papers is presented in Figure 2.2. The first research paper was published in 1989 by Piotrowski. In the following years, the number of published papers has increased. The number of research papers reached the maximum (548) in 2016. It was demonstrated that 90% of the research papers were published after the year 2000, simultaneously with the emerging consciousness about clean water resources.

B Geographical Wise Distribution of Contributions

Table 2.1 illustrates a number of contributions to the bioremediation papers according to countries. Some of 5294 research articles were generated by collaboration between countries. For that reason, data given in this section does not present number of papers but rather indicates number of contributions of collaborative countries. A total of 107 countries have contributed to the 5294 research papers. The United States has made maximum contribution into the

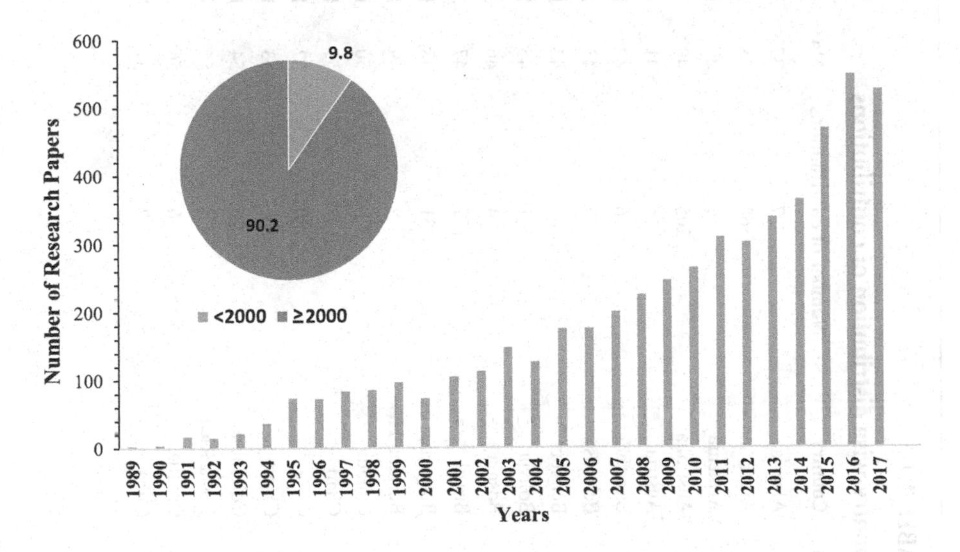

FIGURE 2.2 The number of research papers and year-wise distribution.

TABLE 2.1

Country-wise distribution of contributions

No	Country	Number of contributions	No	Country	Number of contributions	No	Country	Number of contributions
1	Algeria	7	37	Hungary	22	73	Poland	80
2	Argentina	68	38	İceland	1	74	Portugal	96
3	Armenia	1	39	İndia	562	75	Qatar	4
4	Australia	205	40	İndonesia	19	76	Romania	36
5	Austria	15	41	İran	113	77	Russia	41
6	Bangladesh	16	42	İraq	7	78	Saudi Arabia	50
7	Belarus	2	43	İreland	33	79	Scotland	48
8	Belgium	43	44	İsrael	48	80	Senegal	1
9	Bolivia	1	45	İtaly	220	81	Serbia	27
10	Bosnia	4	46	Japan	156	82	Sierra Leone	1
11	Brazil	172	47	Jordan	18	83	Singapore	32
12	Bulgaria	31	48	Kazakhstan	1	84	Slovakia	10
13	Burkina Faso	2	49	Kenya	3	85	Slovenia	4
14	Canada	270	50	Kuwait	34	86	South Africa	81
15	Chad	3	51	Laos	1	87	South Korea	132
16	Chile	30	52	Latvia	2	88	Spain	217
17	China	726	53	Lebanon	1	89	Sri Lanka	7
18	Colombia	23	54	Libya	3	90	Sweden	48
19	Costa Rica	5	55	Lithuania	3	91	Switzerland	64
20	Croatia	6	56	Malaysia	83	92	Syria	1
21	Cuba	7	57	Mexico	94	93	Taiwan	82

22	Cyprus	2	58	Monaco	1	94	Tajikistan	1
23	Czech Republic	32	59	Montenegro	2	95	Tanzania	1
24	Denmark	50	60	Morocco	7	96	Thailand	40
25	Ecuador	1	61	Namibia	1	97	Togo	1
26	Egypt	84	62	Nepal	1	98	Tunisia	33
27	Emirates	10	63	Netherlands	68	99	Turkey	56
28	England	170	64	New Zealand	19	100	Ukraine	6
29	Estonia	12	65	Nicaragua	1	101	Uruguay	4
30	Finland	40	66	Nigeria	61	102	USA	1176
31	France	178	67	Norway	20	103	Wales	25
32	Georgia	17	68	Oman	6	104	Venezuela	11
33	Germany	248	69	Pakistan	121	105	Vietnam	8
34	Ghana	1	70	Paraguay	1	106	Yemen	1
35	Greece	43	71	Peru	11	107	Zimbabwe	2
36	Hong Kong	19	72	Philippines	7			

FIGURE 2.3 Geographical distribution of contributions on the Earth map.

articles with 1176 times. China has made 726 contributions followed by India with 562 contributions.

Figure 2.3 presents geographical distribution of contributions on the Earth map. The number of contributions on published research papers was distributed all around world as shown in Figure 2.3. This situation can be interpreted as: wastewater treatment is a global problem therefore scientists around the world have focused on remediation of water with biological methods.

Figure 2.4 shows the geographical distribution of contributions over the continents. The highest contribution was provided by Asia with 34.94%. Europe

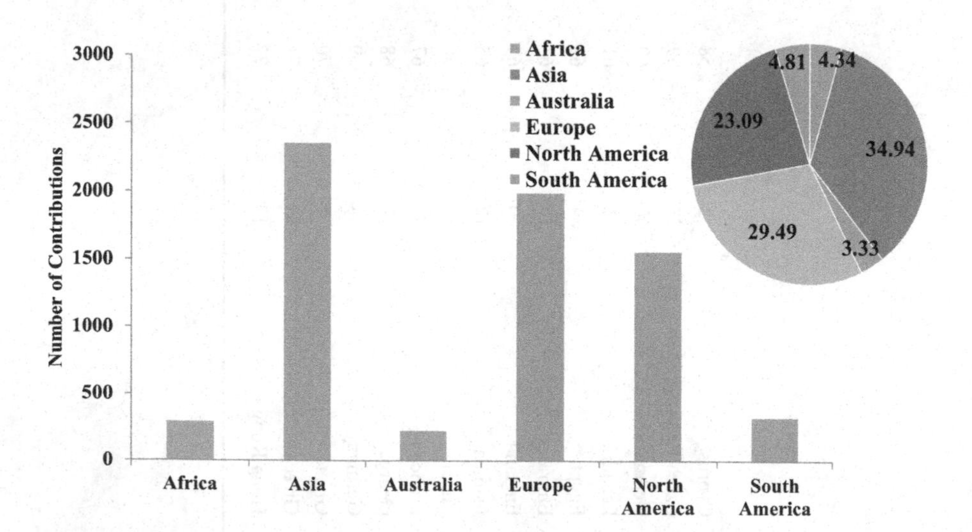

FIGURE 2.4 Geographical distribution of contributions over continents.

has made 1982 contributions with 29.49%, and North America has made 1552 contributions with 23.09%. The results can be interpreted as: the developed countries are in the need of investigating the water bioremediation due to the discharge of a huge amount of industrial wastewater.

C AUTHORSHIP PATTERN

Authorship pattern of contributions was revealed in Figure 2.5. The single authored paper has minimum percentage with 2.9%, which can be contributed to the papers tended to research in collaboration. The maximum was observed as 25.8% (1364 papers) with more than 5 authors per paper. The number of three and four authors collaborations was 21.8% (1154 papers) and 20.4% (1078 papers), respectively.

Table 2.2 tabulates the degree of collaboration. The degree of collaboration was calculated using Eq. 2.1 (Subranamayan 1983). The degree of collaboration ranges from 0.88 to 1.00. The average degree of collaboration was calculated as 0.96.

$$C = \frac{N_m}{N_m + N_s} \tag{2.1}$$

where C denotes degree of collaboration and N_m indicates number of multiple authors. The number of single author is denoted with N_s.

In terms of international collaborations, the number of international contributions to research papers is shown in Figure 2.6. Most of the published research papers (78.2%) were published without international collaboration. About 21.8% of research papers were published in collaboration with more than two countries.

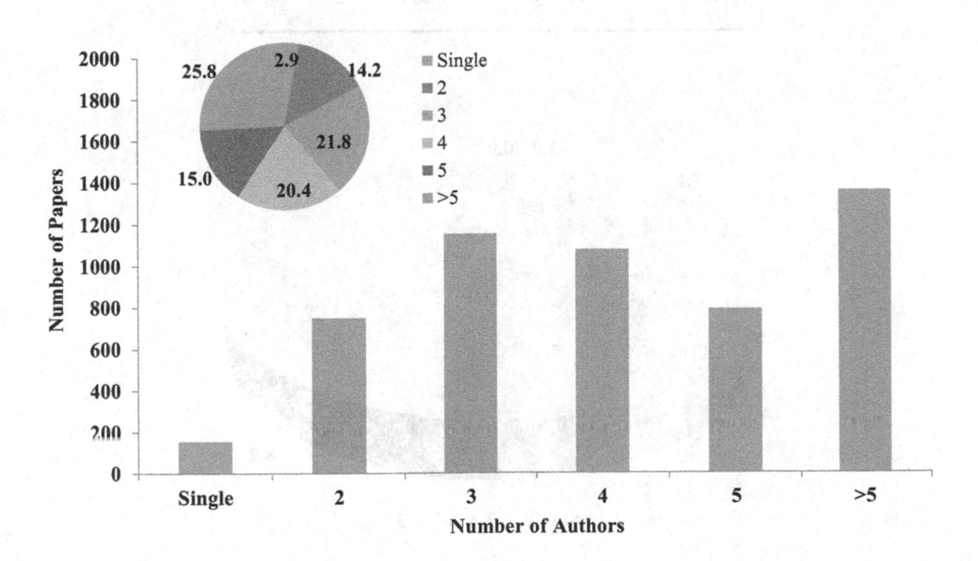

FIGURE 2.5 Authorship pattern of contributions.

TABLE 2.2
Degree of collaboration

Year	Single	Multiple	C
1995	9	65	0.88
1996	3	70	0.96
1997	0	84	1.00
1998	2	84	0.98
1999	9	88	0.91
2000	4	69	0.95
2001	10	95	0.90
2002	3	110	0.97
2003	7	141	0.95
2004	3	123	0.98
2005	6	169	0.97
2006	7	169	0.96
2007	13	187	0.94
2008	5	220	0.98
2009	4	242	0.98
2010	8	257	0.97
2011	8	302	0.97
2012	3	299	0.99
2013	6	333	0.98
2014	7	358	0.98
2015	9	460	0.98
2016	10	538	0.98
2017	11	515	0.98

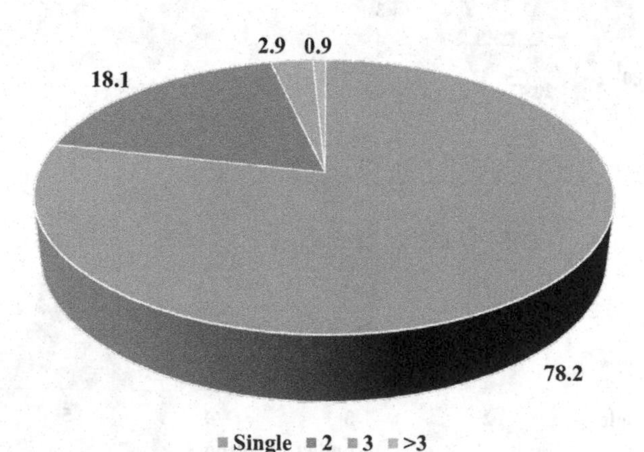

FIGURE 2.6 Number of international collaboration-wise contributions.

TABLE 2.3
Multidisciplinary collaboration-wise contributions

Multidisciplinary	Number of papers	Percentage (%)	International collaboration	Number of papers	Percentage (%)
Yes	3384	63.9	1 country	4128	78.0
No	1910	36.1	More than 1 countries	1166	22.0

Table 2.3 illustrates multidisciplinary collaboration-wise contribution. Approximately, 64% of papers were generated as a result of multidisciplinary collaborations. The topics were suitable for multidisciplinary collaborations. Chemists, biologists, chemical engineers, environmental engineers, etc., are involved in on various topics. About 98.7% of papers were written in English, as shown in Table 2.4. Also, 22% of papers were written by international collaborations, as shown in Table 2.3. Most of the papers were published in international publishers, as mentioned in Section 2e. The researchers might want to reach more researchers. These three probable reasons might affect the choice of language of the papers.

Figure 2.7 presents year-wise single and international collaboration and percentage of international collaboration. From 1989 to present, the number of papers studied with international collaboration increased progressively. In 1998, the percentage of international collaboration inclined dramatically. In 2000s, the percentage of international collaborations increased continuously

TABLE 2.4
Language of papers

Language	Number of papers
English	5225
Spanish	21
Portuguese	12
Polish	16
French	3
German	5
Croatian	4
Russian	1
Chinese	1
Japanese	2
Turkish	1
Lithuanian	1
Korean	1
Czech	1

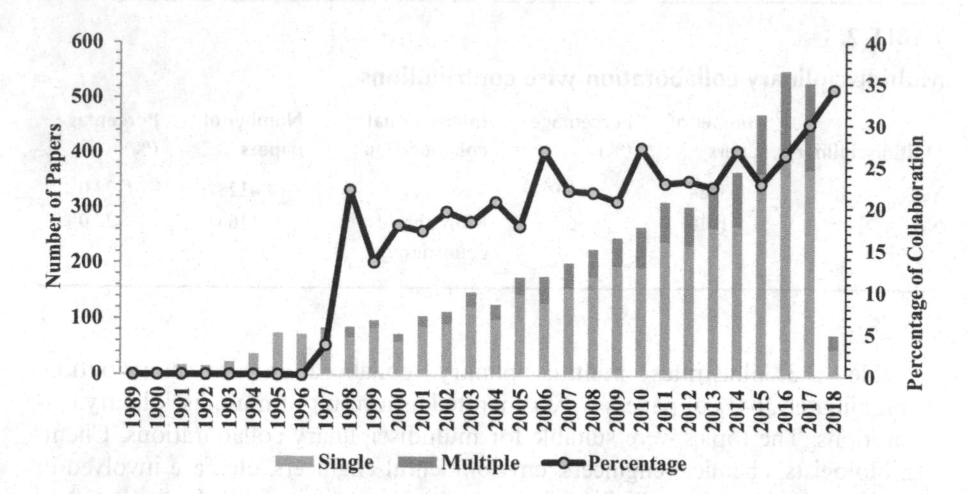

FIGURE 2.7 Year-wise single and international collaboration and percentage of international collaboration.

due to improving electronic communication systems, internet and ease of accessibility to literature.

D Page, Citation and Reference Distributions

Figure 2.8 presents the distribution of number of pages in each researched paper. Most of the research articles' length (3170 papers) varied between 6 and 11 pages. About 969 papers have length between 1 and 6 pages, and the number of papers, having length between 11 and 16 pages, was 899.

Figure 2.9 illustrates the year-wise number of citation distribution. The number of citations increased towards the year 2011. Between 2002 and 2011,

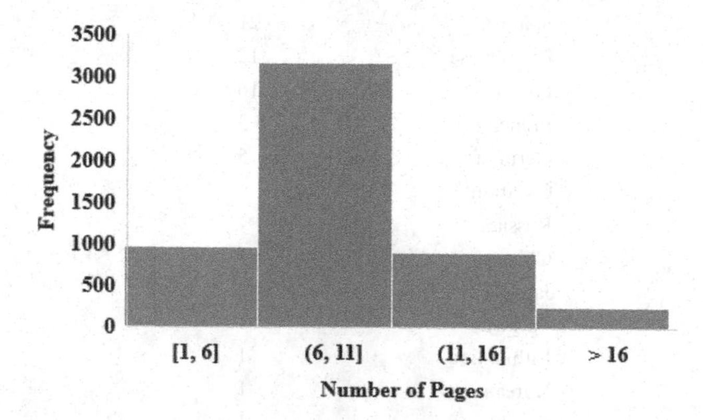

FIGURE 2.8 Number of pages in each paper.

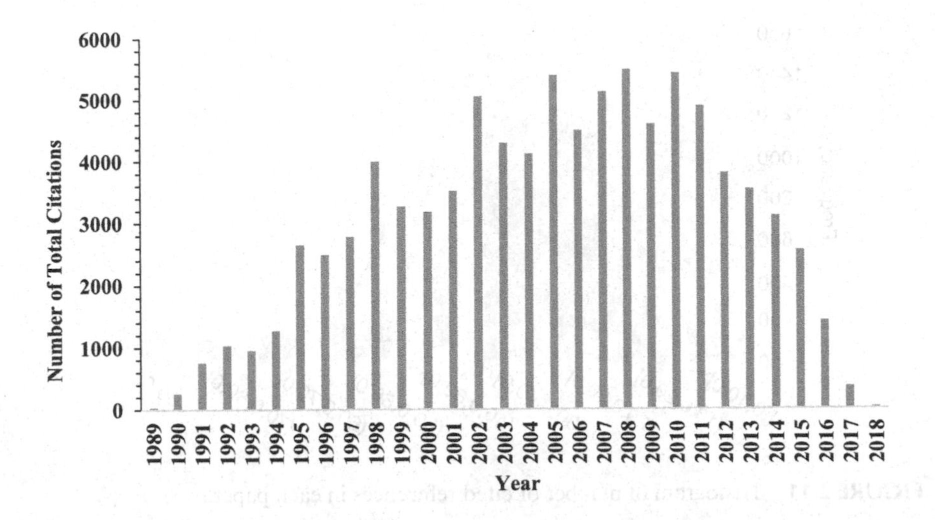

FIGURE 2.9 Year-wise number of citation distribution.

the number of citations accumulated. The number of total citations exceed 5000 times at 2002, 2005, 2007, 2008 and 2010 years. After the year 2011, the number of citations was seen to decrease gradually. The number of papers versus number of citation distribution is shown in Figure 2.10. The 2990 papers were cited less than 10 times, which can be expressed as 56% percentage. The 403 papers were cited more than 50 times – as percentage it can be defined as 8%. High citation quantity can reflect the quality of research.

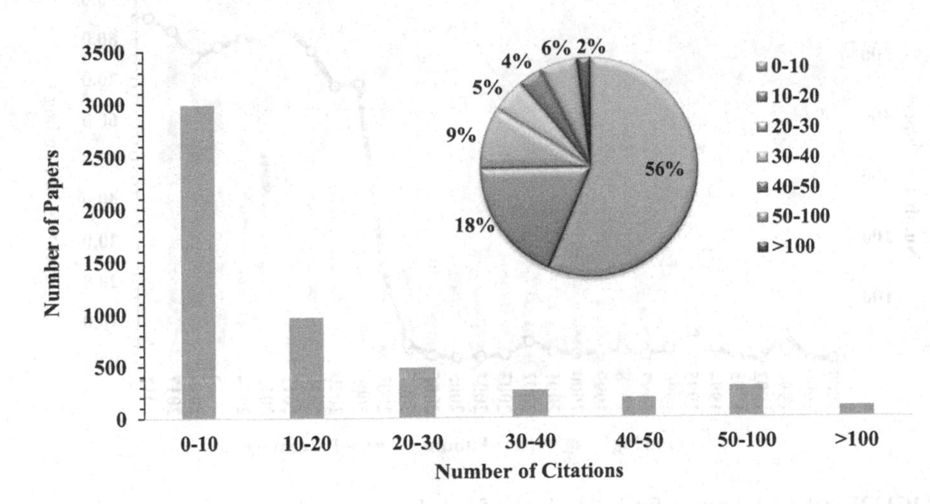

FIGURE 2.10 Number of papers against number of citation distribution.

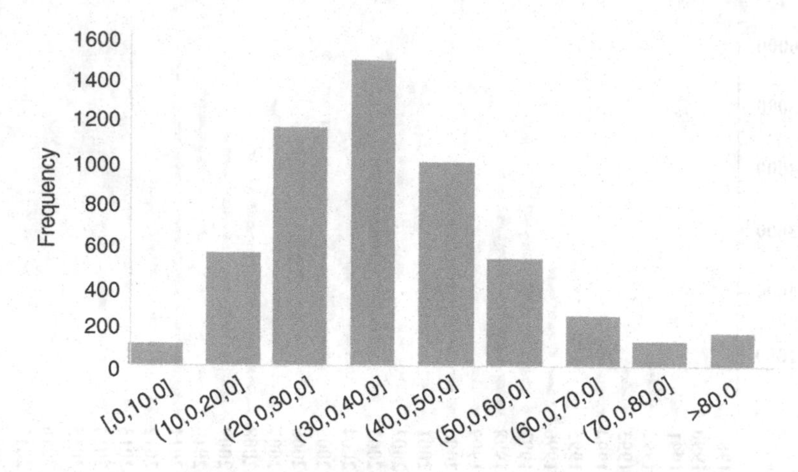

FIGURE 2.11 Histogram of number of cited references in each paper.

Figure 2.11 reveals histogram of number of cited references in each paper. The histogram of number of cited references distributed normally. The papers were studied and/or written using 30 to 40 cited references approximately.

E ANALYSIS OF FUNDING AND PUBLISHER

Figure 2.12 reveals year-wise funded and non-funded number of papers and percentage of funding. It is obvious that increment of funding caused increase in the number of papers. The percentage of funded research papers reached and exceeded

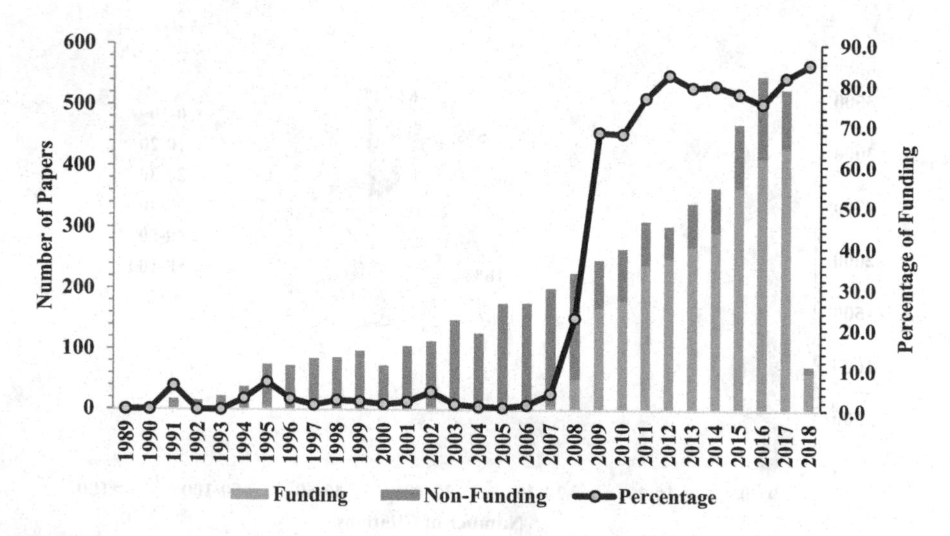

FIGURE 2.12 Year-wise funded and non-funded number of papers and percentage of funding.

TABLE 2.5
Papers' publisher names

Publisher name	Number of papers
Elsevier Sci Ltd.	1679
Springer	751
Wiley	328
Amer. Chemical Soc.	232
Taylor & Francis Inc.	227
Amer. Soc. Microbiology	84
Royal Soc. Chemistry	64
Asce-Amer. Soc. Civil Engineers	56
IWA Publishing	49
Humana Press Inc.	48
Public Library Science	42
Water Environment Federation	41
Amer Soc. Agronomy	35
MDPI Ag	35
Parlar Scientific Publications (p s p)	34
Amer. Geophysical Union	32
Nature Publishing Group	30
Frontiers Media Sa.	25
Selper Ltd. Publications Div.	20
Desalination Publ.	16
Oxford Univ. Press.	16
Triveni Enterprises	15
Gh Asachi Technical Univ. IASI	13
Soc. Bioscience Bioengineering Japan	13
Mary Ann Liebert Inc.	12

70% level after the year of 2009. Nowadays, increase of analytical equipment cost, consumable prices and labor cost reveal the importance of project grant.

Table 2.5 gives outstanding publisher names who have published number of papers more than 10 times. Approximately 61% of papers were published by widely known publishers. There may be two possible reasons for this fact. One of them is that authors want to reach readers easily. The other is these publishers incorporate wide diversity of journals.

2.4 CONCLUSION

The analysis of research paper related with water bioremediation has been carried out, and the major findings can be summarized as follows:

- Only 2.9% of papers were studied by single author.
- 107 countries have contributed on 5294 research papers.

- The United States has made maximum contribution into the articles with 1176 times.
- The highest contribution was provided by Asia with 34.94%.
- 21.8% of research papers was studied by collaboration with more than two countries.
- From 1989 to present, the number of papers studied by international collaboration has increased progressively. In 2000s, the percentage of international collaborations increased continuously due to improving electronic communication and information systems.
- The number of total citations exceed 5000 times in 2002, 2005, 2007, 2008 and 2010.
- The papers were studied and/or written using between 30 and 40 cited references approximately.
- The percentage of funded research papers reached and exceeded 70% level after 2009. Nowadays, increase of analytical equipment cost, consumable prices and labor cost reveal the importance of project grant.
- Most of the papers were published by international publishers. 98.7% of papers were written in English.

REFERENCES

Chequer, F. M. D., Oliveira, G. A. R., Ferraz, E. R. A., Cardoso, J. C., Zanoni, M. V. B., and Oliveira, D. P. 2013. Textile Dyes: Dyeing Process and Environmental impact, Chapter 6. *Eco-Friendly Textile Dyeing and Finishing*, Ed. Melih Gunay, 151–176, InTech.

Fazal, T., Mushtaq, A., Rehman, F., Khan, A. U., Rashid, N., Farooq, W., Rehman, M. S. U., and Xu, J. 2017. Bioremediation of textile wastewater and successive biodiesel production using microalgae, *Renewable and Sustainable Energy Reviews* 82:3107–3126.

IWQGES, United Nation Universities. 2016. International water quality guidelines for ecosystems (IWQGES), How to develop guidelines for healthy freshwater ecosystems, http://web.unep.org/sites/default/files/Documents/20160315_iwqges_pd_final.pdf (accessed January 18, 2018).

Jaganathan, V., Phil, M., Cherurveettil, P., Chellasamy, A., and Premapriya, M. S. 2014. Environmental pollution risk analysis and management in textile industry: A preventive mechanism, *European Scientific Journal* 2:323–329.

Jain, S. K., and Singh, V. P. 2003. Water resources systems planning and management. *Introduction to water resources systems. Developments in water science*, Eds. S.K. Jain and V.P. Singh, Volume 51, 3–858, Amsterdam: Elsevier Science B.V.

Jermann, C., Koutchma, T., Margas, E., Leadley, C., and Ros-Polski, V. 2015. Mapping trends in novel and emerging food processing technologies around the World, *Innovative Food Science and Emerging Technologies* 31:14–27.

Pritchard, A. 1969. Statistical bibliography or bibliometrics, *Journal of Documentation* 25:348-349.

Proverb, K. 2011. Policy brief: Water quality, www.unwater.org/app/uploads/2017/05/waterquality_policybrief.pdf (accessed March 6, 2018).

Robinson, T., McMullan, G., Marchant, R., and Nigam, P. 2001. Remediation of dyes in textile effluent: A critical review on current treatment technologies with a proposed alternative, *Bioresource Technology* 77:247–255.

Subranamayan, K. 1983. Bibliometric study of research collaboration: A review, *Journal of Information Science* 6:33–38.

UNEP/MAP/RAC/CP. 2004. Guidelines for the application of Best Available Techniques (BATs) and Best Environmental Practices (BEPs) in industrial sources of BOD, nutrients and suspended solids for the Mediterranean region, *MAP Technical Reports Series* 142. http://hdl.handle.net/20.500.11822/405 (accessed January 18, 2018).

United Nation Office, 2013. A 10 year story the water for life decade 2005-2015 and beyond, www.un.org/waterforlifedecade/pdf/WaterforLifeENG.pdf (accessed February 18, 2019).

Velmurugan, C., and Radhakrishnan, N. 2016. Indian Journal of Biotechnology: A bibliometric study, *Innovare Journal of Science* 4:1–7.

Verma, A., Sonkar, S. K., and Gupta, V. 2015. A Bibliometric study of the library philosophy and practice (e-journal) for the period 2005-2014, *Library Philosophy and Practice* 1292. http://digitalcommons.unl.edu/libphilprac/1292 (accessed January 19, 2018).

Vijay, K. R., and Raghavan, I. 2007. Journal of Food Science and Technology: A bibliometric study, *Annals of Library and Information Studies* 54:207–212.

3 Biofunctionalized Adsorbents for Treatment of Industrial Effluents

P. Banerjee
Department of Environmental Studies, Rabindra Bharati University, Salt Lake City, Kolkata, India
Department of Environmental Science, University of Calcutta, Kolkata, India

A. Mukhopadhyay
Department of Environmental Science, University of Calcutta, Kolkata, India

P. Das
Department of Chemical Engineering, Jadavpur University, Kolkata, India

3.1 INTRODUCTION

Consistent availability of clean and cost-effective sources of water is conceived as a fundamental humanitarian goal. Achieving this goal is presently considered as a major global challenge of the 21st century (Qu et al., 2013). As estimated at the end of 2011, approximately a global population of 2.5 billion is deprived of access to supplies of safe drinking water (WHO, 2013). Severe water crisis observed in various regions of the world has resulted in an increase in concerns regarding water resources rendering as a significant global issue (Neoh et al., 2016). Therefore, it has become imperative to carry out basic water treatment in areas facing water crisis (mostly developing countries) and having dearth of infrastructure required for water and wastewater treatment (Qu et al., 2013).

In order to meet the stringent regulations for restoration of safe environment, it has become highly essential to develop cost-effective, convenient and sustainable technologies for treatment of wastewater. In the previous decades, processes like chemical oxidation, solvent extraction, coagulation, catalytic degradation, adsorption, biodegradation (aerobic, anaerobic and electrochemical) and membrane-based separation have been widely investigated for decontamination and reclamation of polluted wastewaters (Yagub et al., 2014; Zhao et al., 2009). However, most of these processes are unsuitable for wide scale application due to their huge operational cost, quantity of sludge produced and time consumed for efficient effluent treatment (Hafez and El-Mariharawy, 2004).

Of all other processes reported till date, biological oxidation reportedly results in complete degradation of contaminants present in effluent streams (Banerjee et al., 2017a). Review of previous literature suggests that bacterial species are more capable of degrading different pollutants over a broad spectrum of environmental factors in comparison to algal or fungal species. This efficient biodegradability demonstrated by bacterial species may be attributed to their rapid growth and reproductive rate as well as simple culture techniques (Saratale et al., 2009). Recent studies have reported that the membrane bioreactor (MBR) technology for effluent treatment offers specific benefits like reduced footprint, lesser production of sludge, removal of organic micropollutants of low molecular weights and enhanced separation potential and treated effluent of superior quality (Mutamim et al., 2012; Neoh et al., 2016; Tan et al., 2015). Hence, the MBR technology is presently being preferred over the conventional activated sludge-based processes. However, the implementation of MBR-based processes on a wide-scale is limited by the long durations of treatment incurred.

Besides biodegradation, adsorption-based processes have also been widely investigated for treatment of wastewater (Roy et al., 2018a, 2018b). During adsorption, the target pollutant is simply transported between adjacent phases without degeneration of the molecular constitution of the same (Shen et al., 2010). However, adsorbents reportedly face quick exhaustion when applied in full-scale setups for treatment of huge quantities of wastewater (Shen et al., 2010). Therefore, in recent research, combined application of biodegradation and adsorption has been considered as a more efficient approach for effluent treatment (Roy et al., 2018a, 2018b). In previous studies, biomolecules have been successfully immobilized over supports prepared from natural polymers like alginates, chitosan and chitin, collagen, carrageenan, gelatin and cellulose, synthetic polymers and inorganic substrates like charcoal, ceramics, silica, celite and activated carbon (Datta et al., 2013). Biomolecules have also been reportedly immobilized on domestic, agricultural or industrial wastes like rice bran (Gurel et al., 2010), corn cobs (Genisheva et al., 2011), coconut coir, bagasse, jute gunny bag, loofah sponge (Eş et al., 2015), rice husk (He et al., 2016) and fly ash (Roy et al., 2018b). However, recent simultaneous progress achieved in fields of nanotechnology and biotechnology has paved the path for the development of various bio-functionalized nanomaterials capable of efficient degradation of toxic pollutants (Wang et al., 2011).

These advanced bio-functionalized nanomaterials have unique physicochemical properties like ultra-small dimensions as well as high surface area/mass ratio, potential for surface modification and chemical reactivity (Veerapandian and Yun, 2011). Moreover, since nanomaterials and biomolecules (microbial cells, proteins, peptides, nucleic acids etc.) have similar dimensions, the functionalized biomolecules efficiently exhibit behavior and properties similar to naturally occurring biomolecules (Veerapandian and Yun, 2011). Previous studies have reported immobilization of biomolecules on nanomaterials via non-covalent and/or covalent bonding (Knopp et al., 2009; Veerapandian and Yun, 2010), Stöber technique (Luckarift et al., 2007; Trewyn et al., 2007), deposition or coating (Bunker et al., 2007), reverse micelle and sol–gel-based processes (Yang et al., 2004) and coupling reaction (Wang et al., 2004). Biomolecule immobilization on nanomaterials has been widely achieved by the development of complex moieties interlinking complementary substrates (Veerapandian and Yun, 2011). In previous studies, nanomaterial surfaces have been modified for facilitating immobilization by treating the nano substrates with radio frequency plasma, vacuum-UV radiation as

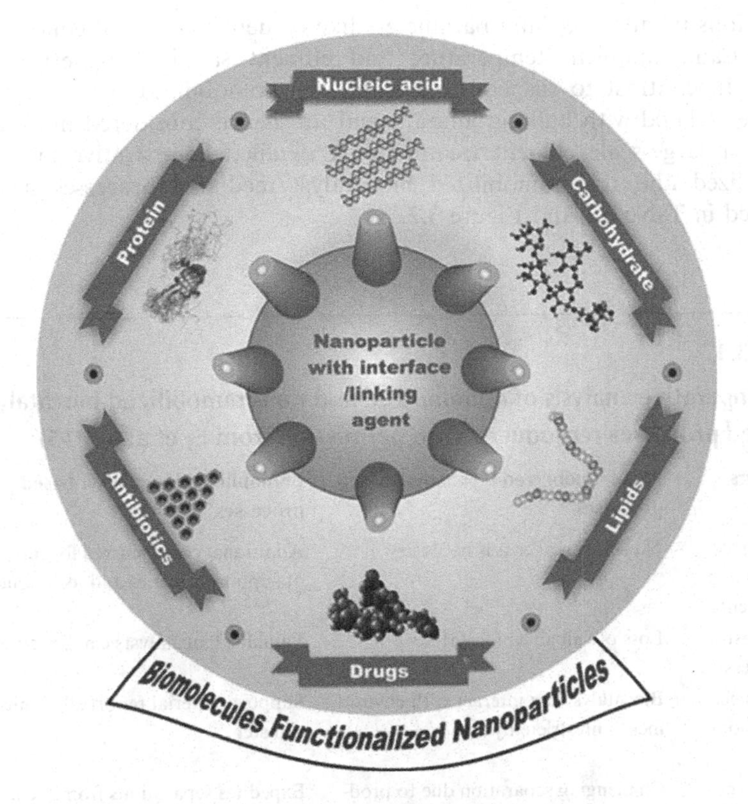

FIGURE 3.1 Schematic representation of interface/linking agent mediated biomolecule functionalization of nanoparticles (NPs) reproduced with permission from Veerapandian and Yun (2011).

well as photo-Fenton oxidation (Mazille et al., 2010; Veerapandian and Yun, 2011). A schematic representation of functionalization of nanoparticles (NPs) with biomolecules has been presented in Figure 3.1.

This review discusses different supports (made from nanomaterials) being investigated for immobilization and compares the prevalent techniques for immobilization of biomolecules. It also discusses different bioreactor configurations and optimization strategies that have been employed for facilitating the application of these immobilized biomolecules on an industrial scale. This study aims to identify the potentials of an integrated adsorption and biodegradation approach for commercial effluent treatment and thereby serve as a holistic source of information for hydrologists, environmentalists and industrialists involved in conservation of water resources and eradication of water scarcity.

3.2 COMPARISON OF NON-IMMOBILIZED AND IMMOBILIZED PROCESSES

Immobilized biomolecules (henceforth referred to as biocatalysts) reportedly exhibit improved efficiency for pollutant degradation, better resilience to fluctuations in environmental parameters like effluent pH, initial concentration of pollutants, ambient temperature and effluent salinity. (Banerjee et al., 2017b). In contrast to suspended biocatalysts, immobilized biocatalysts pose no issues related with bulking amount and are hence considered more appropriate for large-scale effluent treatment. A detailed comparative analysis of immobilized and non-immobilized biocatalyst mediated processes has been presented in Table 3.1 and Figure 3.2.

TABLE 3.1

The comparative analysis of immobilized and non-immobilized biocatalyst mediated processes reproduced with permission from Eş et al. (2015)

Parameters	Non-immobilized biocatalyst-based processes	Immobilized biocatalyst-based processes
Cost incurred for design and development	No additional cost is necessary	Additional cost incurred for fabrication of support material and technique
Overall cost-effectiveness	Loss of valuable biocatalysts	Valuable biocatalysts can be reused
Mass transfer and diffusion limitations	Biocatalyst can interact with environment unrestrictedly	Support material reportedly limits mass transfer
Downstream Process	Challenging separation due to product mixture/biocatalyst/substrate	Expedites separations from the medium of production
Pollution	Risk of pollution by reaction mixture	Minimizes or eliminates pollution of products
Growth of biomass (for cellular biocatalysts)	Control of the process is complicated as the biomass attains high concentrations in short time	Biomass growth remains same along the process
Movement (for cell biocatalysts)	Free movement—high mobility	Physical/chemical interaction with support material restricts movement
Recovery and reuse of biocatalyst	Minimal or null reuse	Conveniently recovered and reused
Stability	Low stability	Enhanced stability of activity observed over a wide range of operational conditions like solution pH, temperature, etc.)
Catalyst efficiency	Low efficiency (kg product/kg enzyme)	High efficiency (kg product/kg enzyme)

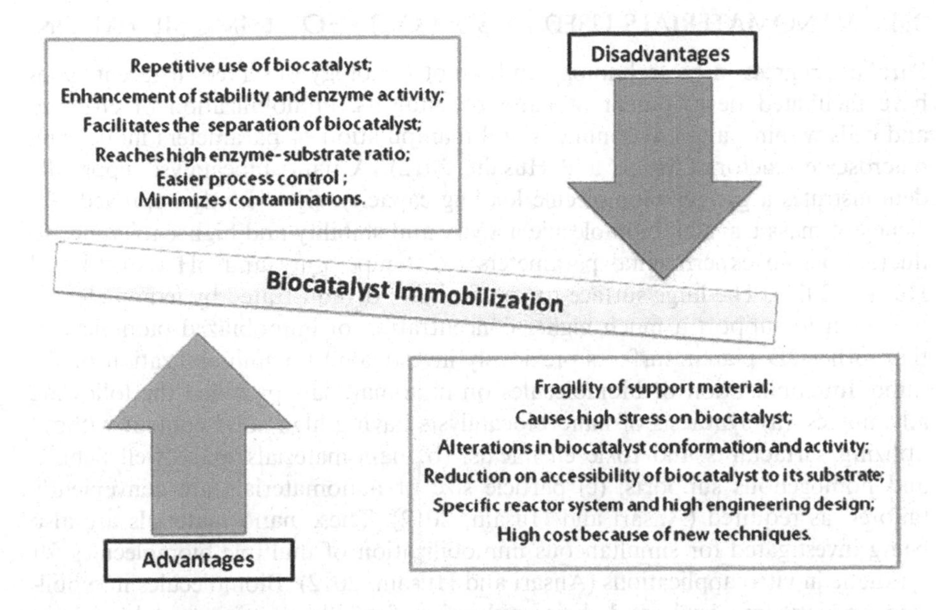

FIGURE 3.2 Schematic representation of comparison of benefits and limitations offered by immobilized biocatalysts reproduced with permission from Eş et al. (2015).

3.3 BIOMOLECULES IMMOBILIZED FOR CATALYSIS

According to Datta et al. (2013), mostly enzymes or entire microbial cells have been immobilized and investigated as biocatalysts. Enzyme moieties are immobilized on substrates after confining the same within carrier matrices or supports prepared using inert polymers or inorganic compounds (Datta et al., 2013). Besides being inert, these supports should be strong, stable, regenerable, capable of promoting enzyme activity and specificity as well as able to diminish product inhibition and microbial contamination (Singh, 2008). Application of immobilized enzymes facilitates uninterrupted operations on commercial scale, automation, large investment/capacity ratio as well as product recovery with higher purity (Datta et al., 2013).

Moreover, immobilization of the entire living cell capable of secreting these enzymes saves the cost incurred for separation and purification of enzymes, facilitates stabile and prolonged enzyme activity and provides an opportunity for multiple enzymes to act simultaneously. Immobilization of whole cells may be achieved by enabling cell attachment to support surfaces, cell entrapment in a porous matrix or behind a barrier and/or self-aggregation (Karel et al., 1985). Immobilized cell cultures reportedly facilitate improved concentration of biomass, better process stability, repeated usage of immobilized biomolecules, as well as higher hydraulic loading rates than suspended cell cultures or conventional activated sludge-based processes (Banerjee et al., 2017b).

3.4 NANOMATERIALS USED AS SUPPORTS FOR IMMOBILIZATION

Parallel progress in biotechnology and nanotechnology observed in recent times have facilitated development of nano biocatalysts, immobilization of enzymes and cells within nano environments and manipulation of parameters influencing macroscale reactors (Ansari and Husain, 2012). A nano biocatalyst reportedly demonstrates a greater biomolecule loading capacity, significantly improved efficiency of mass transfer, biomolecule activity and stability and higher tolerance to fluctuations in experimental parameters like temperature and pH (Ansari and Husain, 2012). The huge surface-to-volume ratio demonstrated by nanomaterials is known to support a much higher concentration of immobilized biomolecules than other 2D planar surfaces previously investigated for immobilization of the same. Immobilization of biomolecules on nanomaterials provided the following advantages: (a) synthesis of nano biocatalysts having high solid content without utilizing surfactants and toxic chemicals; (b) nanomaterials make well defined and homogenous supports; (c) particle size of nanomaterials are conveniently tailored as required (Ansari and Husain, 2012). These nano materials are also being investigated for simultaneous immobilization of multiple biomolecules for synthetic in vitro applications (Ansari and Husain, 2012). Biomolecules immobilized on nano materials and their application for effluent treatment have been enlisted in Table 3.2. Different groups of nano material used as supports have been discussed as follows.

3.4.1 Different Nano Carriers for Biomolecule Immobilization

The presence of specific chemical, physical, electrical and optical properties, specificity and potential of catalysis of nanomaterials have resulted in application of the same in diverse innovative biotechnological applications (Ansari and Husain, 2012). Application of nanomaterials is targeted towards reduction of limitations posed by diffusion and maximization of functional surface area for achievement of improved enzyme loading (Xie et al., 2009). In recent studies, different nanomaterials including rods, tubes, wires and rings of nano dimensions have been widely investigated (Ali and Winterer, 2009; Ni et al., 2007). Various nanomaterials reported for biomolecule immobilization have been discussed as follows.

3.4.1.1 Metal/metal Oxide NPs

Metallic NPs may have magnetic or non-magnetic properties. The large surface-to-volume ratio of magnetic NPs facilitates greater binding capacity as well as catalytic specificity of biomolecules undergoing conjugation (Johnson et al., 2008; Konwarh et al., 2009). Moreover, susceptibility to magnetic field reportedly prevented contamination of treated effluents by facilitating efficient recovery of biomolecules from mixed liquor (Ansari and Husain, 2012). Besides magnetic NPs, biomolecules have also been immobilized on non-magnetic metallic NPs. Enzymes like lipase, α-Chymotrypsin and Diastase have been immobilized on polystyrene NPs and silica-coated nickel NPs,

TABLE 3.2

Different nano materials used as supports for immobilization of biocatalysts

Biomolecule immobilized	Type of nanomaterial	Application	Major findings	Reference
Tyrosinase from edible mushrooms, *Agaricus bisporus*	Magnetic iron oxide NPs	Biotransformation of phenol	70% degradation of phenol conc. of 2500 mg L^{-1} in 4 h	Abdollahi et al., 2018
Versatile peroxidase from *Bjerkandera adusta*	Chitosan NPs	Transformation of 2,6-dimetoxy-phenol, catechol, bisphenol A, β-estradiol, 4-chlorophenol, 2,3,5,6-tetra-chlorophenol, pentachlorophenol, 2,6-dichloro-4-nitro-phenol, dichlorophen, triclosan	1 kg of immobilized enzyme was found capable of transforming 12.9 tons of phenols considered in this study in 144 h.	Alarcón-Payán et al., 2017
Laccase	Cu (II)-chelated chitosan-graft-poly glycidyl methacrylate NPs	Phenol degradation	96% of phenol removal was recorded after 12 h	Alver and Metin, 2017
Halotolerant bacterial consortium made up of *Dietzia* sp., *Pseudomonas mendocina,* and *Bacillus* sp.	Graphene oxide—Poly-acrylic acid—Gelatin composite	Treatment of hypersaline textile effluent	Optimum conditions were solution pH of 7.5, and a biocatalyst dosage of 5.0 g L^{-1} at a temperature of 30 ± 2 °C for 12 h under which a maximum of 98.82%, 99.47%, 99.12% and ~99% COD, color, surfactant and electrolyte removal respectively was achieved.	Banerjee et al., 2018
Horseradish Peroxidase	NH_2-modified Magnetic Fe_3O_4/SiO_2 nanocomposite	Removal of 2,4-dichlorophenol	80% degradation of 2,4-dichlorophenol (0.2 mmol·L^{-1}) was recorded at pH 6.4 and 25 °C in 180 mins using 1.0 mL of immobilized enzyme solution.	Chang and Tang, 2014

(*Continued*)

TABLE 3.2 (Cont.)

Biomolecule immobilized	Type of nanomaterial	Application	Major findings	Reference
Horseradish Peroxidase	Graphene oxide—Fe_3O_4 nanocomposite	Oxidation of phenol and 2,4-dichlorophenol	Easily removed 2,4-dichlorophenol facilitated the removal (94%) of hardly removed phenol from a mixture each compound was present in 0.5 mmol L^{-1}	Chang et al., 2016
Laccase	Amino-functionalized nano silica	Removal of Reactive Violet 1	96.76% dye removal was achieved after 12 h at pH 5, initial dye concentration of 100 mg L^{-1} and 30°C using biocatalyst (2000 U).	Gahlout et al., 2017
Laccase from *Coriolopsis polyzona*	Silica NPs	Transformation of ^{14}C-labeled bisphenol A (BPA).	Specific catalytic activity of the enzyme was found to decrease on immobilization but was still effective for transformation of BPA.	Galliker et al., 2010
β-Lactamase	Fe_3O_4–SiO_2–NH_2 NPs	Degradation of Penicillin G	100 mL of penicillin G (5–50 mg L^{-1}) was completely degraded in 5 min by 50 mg of biocatalyst (~128 U)	Gao et al., 2018a
Co-immobilized laccase and 2,2,6,6-tetramethyl-piperidine-1-oxyl (TEMPO)	Amino-functionalized Fe_3O_4 NPs cross-linked with glutaraldehyde	Decolorization of acid fuchsin	A maximum of 77.41% decolorization of 100 mL of acid fuchsin (50 mg L^{-1}) was recorded using 100 mg of NPs co-immobilized with TEMPO and laccase in a being 0.3 mM g^{-1}:120 U g^{-1} ratio	Gao et al., 2018b
Laccase of a *Thielavia* genus	Fumed silica NPs	Removal of bisphenol A (BPA) from effluent collected from wastewater treatment plant (WWTP).	~75% removal of BPA was achieved with 100 mg immobilized enzyme and mean BPA conc. of 1.03 ± 0.30 μg L^{-1}	Gasser et al., 2014

Laccase	CNT-coated polymer membrane support	Bisphenol-A, carbamazepine, clofibric acid, diclofenac, ibuprofen	97, 94, 85, 60 and 55 % removal of bisphenol-A, diclofenac, ibuprofen, carbamazepine and clofibric acid was obtained after 48 h with a flux of 10–15 L m^{-2} h^{-1}	Ji et al., 2016
Chloroperoxidase from *Caldariomyces fumago*	ZnO nanowire/macroporous SiO_2 composite	Removal of Acid Blue 113, Direct Black 38, Acid Black 10 BX	95.4, 92.3, and 89.1% decolorization of Acid Blue 113, Direct Black 38, and Acid Black 10 BX was observed respectively with a dye conc. of 45 mg L^{-1} and treatment of 35 min	Jin et al., 2018
Catechol 1,2-dioxygenase from *Corynebacterium glutamicum*	Ni^{2+}-magnetic Fe3O4 NPs	Elimination of toxic aromatic hydrocarbons	Immobilized systems revealed higher catalytic properties and stability at higher concs. of aromatic hydrocarbons than free enzymes.	Lee et al., 2011
Laccase	Cu_2O nanowire mesocrystal hybrid	Bioremediation of 2,4-dichlorophenol contaminated water	60 mg immobilized enzyme was capable of removing 99.17% of 2,4-dichloro-phenol	Li et al., 2018
Laccase	Poly(*p*-Phenylene-diamine)/Fe_3O_4 Nanocomposite	Removal of reactive blue 19 dye	90% dye removal was recorded in 1 h at pH 4	Liu et al., 2016
Trametes versicolor laccase	Polyamide 6/chitosan nanofibers	Removal of Bisphenol A (BPA) and 17α-ethinyl-estradiol (EE2)	Simultaneous removal of BPA (92%) and EE2 (96%) was recorded in 6 h from a mixture containing 50 μM of each pollutant. Immobilized enzyme showed activity similar to free enzyme	Maryšková et al., 2016
Laccase from *Trametes versicolor*	Silica-coated magnetic Fe_3O_4 NPs	2,2′-azino-bis(3-ethylbenzothiazoline-6-sulfonic acid) (ABTS) and bisphenol A (BPA)	A higher BPA removal from simulated effluent (85%) was recorded in comparison	Moldes-Diz et al., 2018

(*Continued*)

TABLE 3.2 (Cont.)

Biomolecule immobilized	Type of nanomaterial	Application	Major findings	Reference
		in simulated and secondary effluent of WWTP	to real effluent (80%) using 100 μgL^{-1} immobilized enzyme at pH 6 in 6 h	
Laccase	Acid treated nanobiochars	Removal of carbamazepine	83% and 86% removal of carbamazepine from simulated and secondary effluent, respectively, was recorded after treatment of 24 h	Naghdi et al., 2017
Saccharomyces cerevisiae	Chitosan-coated magnetic (Fe_3O_4) NPs	Biosorption of Cu (II)	96.8% adsorption of Cu (II) having initial conc. of 60 mg L^{-1} was recorded in 1 h	Peng et al., 2010
Laccase	ZnO and MnO_2 NPs	Degradation of alizarin red S dye	95% and 85% dye degradation was obtained using 50 mg of lac-ZnO and lac-MnO_2 at pH 7.0 and initial dye conc. of 20 mg L^{-1}	Rani et al., 2017
CotA laccase	Fe_3O_4 graphene oxide composite—Nα,Nα-Bis (carboxymethyl)-l-lysine hydrate—Ni^{2+}/Cu^{2+} nanosheets	Decolorization for Congo Red	100% dye decolorization after 5 h treatment at 60°C and pH 8	Samak et al., 2018
Laccase	Hollow mesoporous carbon nanospheres	Removal of tetracycline hydrochloride (TCH) and ciprofloxacin hydrochloride (CPH)	99.4% TCH and 96.9% CPH removal was recorded after 3 h at 150 rpm and 30°C.	Shao et al., 2019
Glucose oxidase and laccase	$MnFe_2O_4$ NP—calcium alginate composite	Decolorization of Methylene blue, Indigo and Acid red 14	Immobilized glucose oxidase exhibited higher % removal of methylene blue (93.46 % at pH 9 in 3 h), Indigo and Acid red 14 (44.03 and 46.5 %, respectively between pH 3 and 5 in 1 h) than immobilized	Shojaat et al., 2016

			laccase (82.13, 25.09, and 20.42% decolorization of methylene blue, Indigo and Acid red 14, respectively, in 1 h at pH 7, 120 rpm).	
Lipase/lysozyme	Halloysite clay nanotube/chitosan membrane	Hydrolysis of hexadecanoic acid, 4-nitrophenyl ester	Immobilization of enzymes revealed better enzyme stability and reusability.	Sun et al., 2017
Nitrilase-102 (recombinant from *Arabidopsis thaliana*)	Silica NPs	Hydrolysis of 3-cyanopyridine	Immobilized enzymes exhibited nearly similar specific activity to free enzymes, irrespective of NP size.	Swartz et al., 2009
Soybean peroxidase	Silica-coated ferroxyte NPs	Removal of ferulic acid	93% removal of ferulic acid (2.13 mM) was achieved in 30 min at pH 6.0 and 30 °C with 300 mg immobilized enzyme (demonstrating enzyme activity of 0.26 U).	Tavares et al., 2018
Horseradish Peroxidase	Silicon holders deposed with nano-gold patterns	Oxidation of 2,4,6-*tris*(dimethylaminomethyl)-phenol, 2,6-dinitrophenol, 4-metoxyphenol, β—naphtol, cathecol, hydroquinone, *o*/*p*-aminophenol and resorcine	An optimum phenol oxidation of 35.1% was recorded. Other than 4-metoxyphenol, 2,6-dinitro-phenol, 2,4,6-*tris* (dimethyl-amino- methyl)-phenol, and catechol, other compounds exhibited conversion rates similar to that of phenol	Tudorache et al., 2011
Aspergillus niger lipase	Magnetic nano-particles	Conversion of waste glycerol	Waste glycerol was successfully converted to glycerol carbonate using the biocatalyst in the presence of excess of dimethyl carbonate	Tudorache et al., 2014
Laccase	Polyvinyl alcohol/chitosan/multi-walled carbon	Diclofenac degradation	Complete degradation of diclofenac (12 mg L^{-1}) in 6 h at pH 4 and 50°C	Xu et al., 2015

(*Continued*)

TABLE 3.2 (Cont.)

Biomolecule immobilized	Type of nanomaterial	Application	Major findings	Reference
	nanotubes (CNT) composite nanofibrous membrane			
Laccase	polyacrylonitrile/polyvinylidene fluoride/nano Cu electrospun fibrous membranes	Removal of 2,4,6-trichlorophenol (2,4,6-TCP)	95.4% removal of 2,4,6-TCP (37.15 mg L^{-1}) after treatment of 270 min at 35.39°C and pH 5.66	Xu et al., 2017
Laccase	Magnetic Fe_3O_4 NPs	Removal of 4-chlorophenol (4-CP)	86% of 4-CP removal was achieved within 2 h	Zhang et al., 2017
Acinetobacter, *Pseudomonas* and *Thermophilic hydrogen bacilli*	Graphene oxide modified polyvinyl alcohol	Glucose biodegradation in hypersaline environment	The nano carrier facilitated the growth of microorganisms. COD reduction achieved with immobilized strains (62.8%) was significantly higher than that yielded by suspended cells (30.8%)	Zhou et al., 2015

respectively (Miletić et al., 2010; Prakasham et al., 2007). These immobilized enzymes have been further investigated for aminolysis, proteolysis and starch hydrolysis, respectively. Gold NPs have also been reported as excellent biocompatible supports for biomolecule immobilization due to its strong interaction with cysteine and amino groups of biomolecules (Ansari and Husain, 2012; Li et al., 2010). Nano Au microgels have been investigated for immobilization of horseradish peroxidase and urease (Xu et al., 2007). Immobilized biomolecules were found to demonstrate higher biocatalytic activity and storage stability over wide ranges of experimental conditions (solution pH, temperature etc.) in comparison to free enzymes particularly in lower concentrations (Xu et al., 2007).

3.4.1.2 Polymer Supports

Wang et al. (2009) reported electrospun polymer nano-fibers as suitable supports for immobilization of enzymes owing to their inter-fiber porosity, high specific surface area, unhindered mass transfer and large surface-to-volume ratio for enzyme loading. Sathishkumar et al. (2014) investigated electrospun nano-fibers of cellulose derivatives like carboxymethyl cellulose and cellulose nitrate for immobilization of laccase via covalent binding. These biocatalysts were further investigated for treatment of a simulated effluent bearing reactive dyes like Reactive Red 120, Remazol Brilliant Violet 5R, Remazol Black 5, Remazol Brilliant Blue R and Reactive Orange 16. Immobilization was found to increase the duration of enzyme activity and facilitate similar dye degradation efficiency.

3.4.1.3 Graphene and Graphene Derivatives

According to Wang et al. (2011), graphene and its derivatives have been reported as suitable support materials for immobilization of peptides, proteins, aptamers, bacteria and cells, small molecules, avidin–biotin and N-acryloyloxysuccinimide via chemical conjugation or physical adsorption. Biocatalysts prepared with supports containing graphene or graphene derivatives have been successfully applied for degradation of pollutants and effluents and incorporated in microchip bioreactors and biofuel cells (Pavlidis et al., 2014). Zhou et al. (2015) reported polyvinyl alcohol-modified graphene oxide as appropriate supports for immobilization of bacterial species and further investigated the biocatalyst for treatment of simulated hypersaline effluents. Banerjee et al. (2018) have also reported graphene oxide-gelatin nanocomposite for immobilization of an engineered bacterial consortium and subsequent successful implementation of the biocatalyst for removal of dyes, surfactants and electrolytes from textile effluent.

3.4.1.4 Carbon Nanotubes

The schematic representation of enzyme immobilization on CNT and the mechanism of action for pollutant sensing, monitoring and degradation using the same have been shown in Figure 3.3. CNT-based biocatalysts are primarily prepared via three major routes including adsorption (physical or chemical)

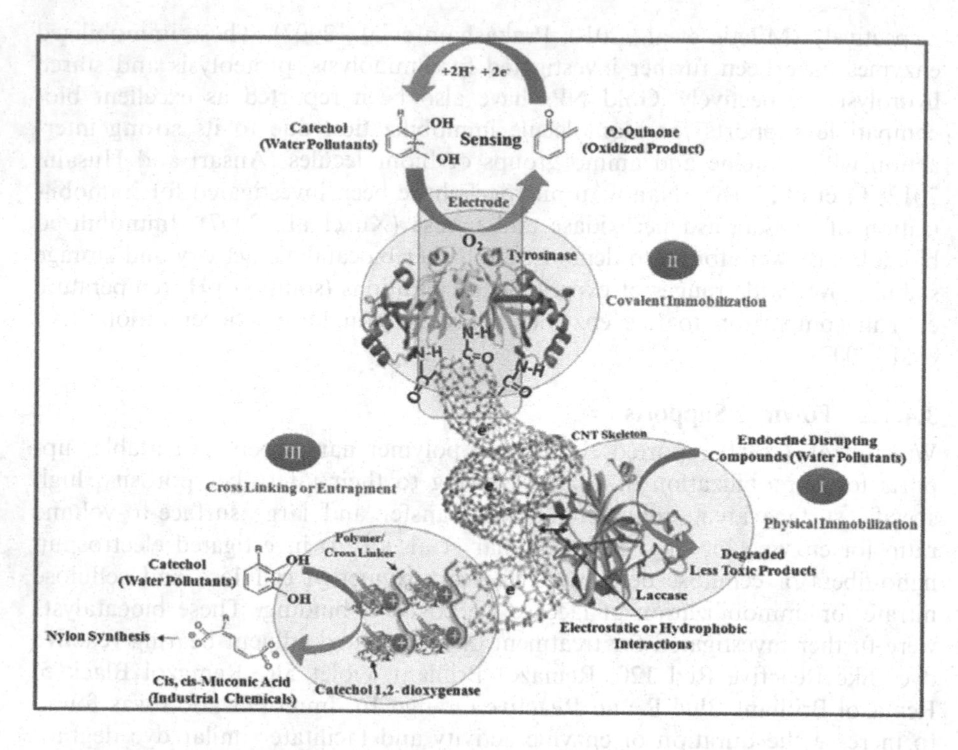

FIGURE 3.3 Different mechanisms of water purification using a potential nano-biocatalyst (reproduced with permission from Das et al., 2014).

on a support, entrapment or encapsulation and cross-linking (Hwang and Gu, 2013). Polymers like chitosan and poly (diallyldimethylammonium chloride) are used for cross-linking biomolecules on CNTs (Das et al., 2014; Subrizi et al., 2014). Moreover, biomolecule encapsulation may be attained using a layer-by-layer approach as shown in (Figure. 3.3(III)). Encapsulation of biomolecules reportedly facilitates coating of CNTs with multiple layers of biomolecule films (Feng and Ji, 2011). According to recent studies, CNT-based biocatalysts exhibit high selectivity and sensitivity for only or simultaneous sensing, monitoring, pre-concentrating, binding, removal and degradation of pollutants and thus considered highly suitable for effluent treatment on a commercial scale (Das et al., 2014).

3.5 DIFFERENT TECHNIQUES OF IMMOBILIZATION

Selection of an appropriate technique for biomolecule immobilization exerts a significant impact on biocatalyst activity. The most appropriate technique of immobilization is known to improve biocatalyst stability and rigidity by creating extremely hydrophilic microenvironments and decreasing enzyme inhibitions (Eş et al., 2015). Major techniques of immobilization have been discussed as follows.

The easiest and most widely practiced technique of immobilization is physical adsorption that generally occurs via hydrogen bonding, hydrophobic and dipole–dipole interactions as well as van der Waals forces. Another widely investigated process for biomolecule immobilization is encapsulation. Encapsulation reportedly facilitates transport of compounds having low molecular weight across the permeable matrix (Eş et al., 2015). Encapsulation or entrapment of biomolecules is usually performed within solid- or gel-based supports. Cross-linking is another process of biomolecule immobilization including both entrapment and covalent bonding (Eş et al., 2015). Cross-linking is usually achieved in the presence of agents like bisisodiacetamide and glutaraldehyde. (Chopra, 2010). Covalent binding is another route of biomolecule immobilization. In this method, the biomolecule forms a covalent bond with the support

TABLE 3.3

Comparative account of benefits and limitations of most widely investigated immobilization techniques reproduced with permission from Eş et al. (2015)

Immobilization technique	Benefits	Limitations
Physisorption	• Simple and cost-effective • Enhanced catalytic activity • No structural change of the biomolecule • No requirement of reagents • Potential for reuse	• Unstable • Probable loss of biomolecules, due to desorption
Biomolecule Encapsulation and Entrapment	• Protection of biomolecule • Facilitates transport of compounds having low molecular weight • Ensures continuous operation due to constant cell density • Ensures separation of cells and simplified process • Facilitates controlled release of product	• Restricts mass transfer • Lower capacity of enzyme loading
Formation of cross-linking	• Robust biomolecule binding • Inhibits leakage • Reduces desorption • Increases biocatalyst stability	• May alter active site • Diffusion restrictions • Reduction of enzyme activity
Formation of Covalent bonds	• Robust binding • Great heat stability • Allows the enzyme to contact with its substrate • Inhibits biocatalyst elution • Flexible design of support and method	• Restricted enzyme mobility results in reduced enzyme activity • Less effective for cell immobilization • Non-renewable support

matrix, thereby ensuring that the former is not separated from the latter in the course of utilization (Eş et al., 2015). This strong interaction between the biomolecule and the support induces high thermal stability. However, restriction of free movement of biomolecules often results in reduced enzyme activity. Nevertheless, this technique of immobilization results in better interaction between biomolecules and target pollutants as these biomolecules are localized on the support surfaces (Eş et al., 2015). A comparative account of the benefits and limitations of all processes discussed herein have been provided in Table 3.3.

3.6 BIOREACTOR DESIGNS FOR APPLICATION OF BIOCATALYSTS

Application of these biocatalysts on an industrial scale necessitates selection of appropriate reactors for effluent treatment for ensuring optimum biocatalyst efficiency (Eş et al., 2015; Rao, 2010). These reactors facilitate the monitoring of significant process parameters for optimum biocatalyst efficiency. Few reactor configurations investigated in recent studies are of continuous stirring, packed bed, plug-flow, fluidized bed types. Besides these, a separate category of reactors equipped with membranes immobilized with biomolecules (biocatalytic membrane reactors) has also been investigated. The schematic representation of these reactors have been presented in Figure 3.4 and discussed in brief as follows.

The continuous stirred tank reactor (Figure 3.4A) consists of a tank equipped with stirring facility where the biocatalysts remain in continuous flow. Gargouri et al. (2011) have investigated a similar type of reactor for treatment of hydrocarbon-rich industrial effluent. A number of such types of reactors when connected in series demonstrates better performance efficiency, as this type of reactors provides good mixing and complete utilization of reactor volume for effluent treatment (Eş et al., 2015). This type of reactors yields stable products and greater productivity with reduced energy consumption.

In a plug-flow reactor (Figure 3.4B), reactants do not undergo mixing and each particle is assumed to travel in the reactor with constant velocity (Eş et al., 2015). Rate of concentration of biocatalysts varies with the reactor length and duration of treatment. Ge et al. (2014) have reported a plug-flow reactor-based setup for removal of nitrogen from municipal wastewater. Nevertheless, it is difficult to control parameters of a plug-flow reactor.

A packed bed bioreactor (Figure 3.4C) is reportedly the most widely investigated reactor for industrial application of biocatalysts (Eş et al., 2015). The design of this reactor includes a column packed with the concerned biocatalyst that is constantly flushed by the liquid phase (Eş et al., 2015). Long et al. (2015) have described a packed bed bioreactor for heavy metals removal from industrial effluents. Owing to its configuration, this type of reactor is also highly suitable for implementation of fragile biocatalysts for effluent treatment.

In a fluidized bed reactor (Figure 3.4D), biocatalysts are stationed on a porous plate present at the bottom of the reactor. Passage of fluids (effluents) through this plate reportedly results in the suspension of the bed within

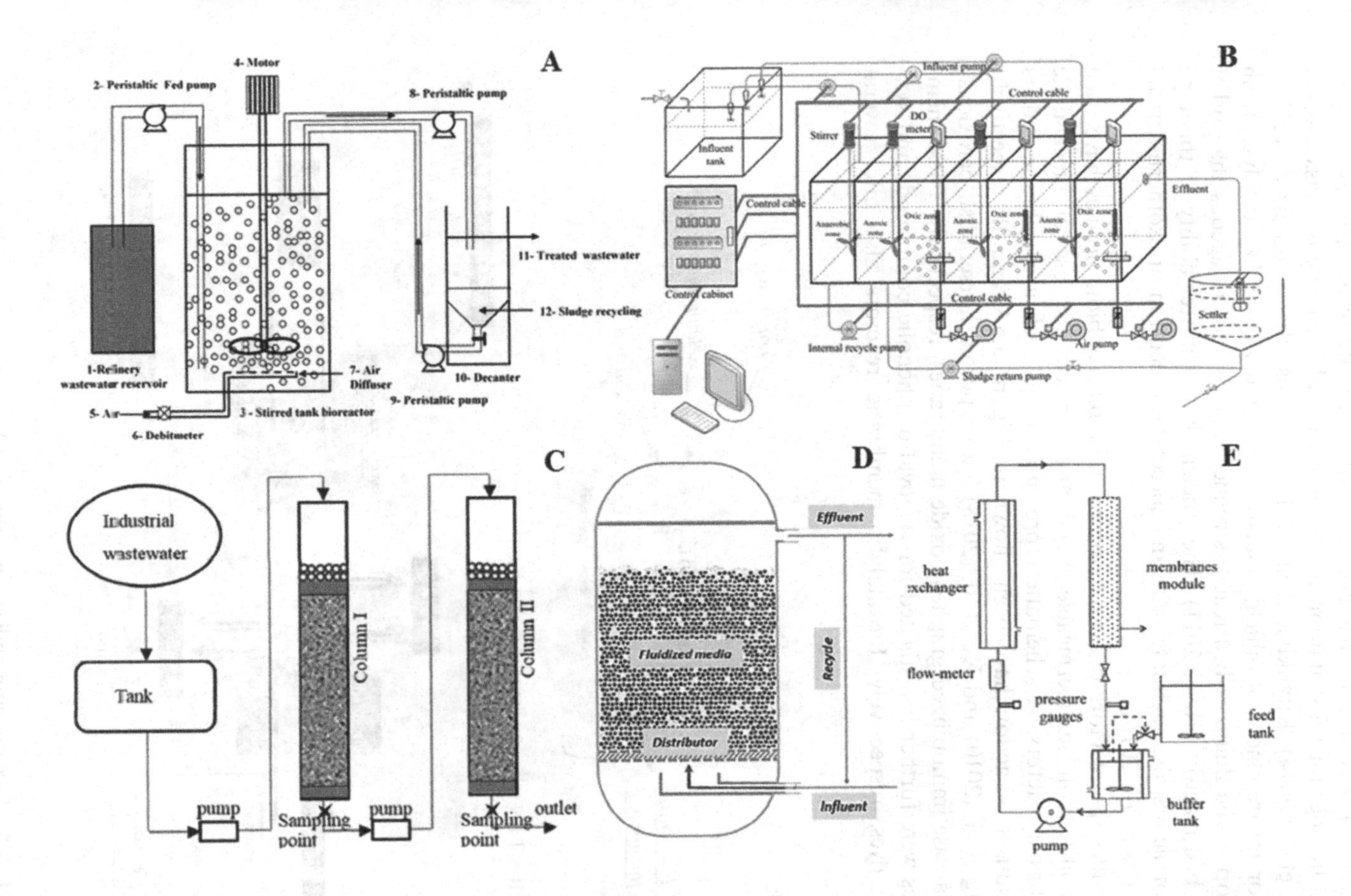

FIGURE 3.4 Schematic representation of (A) continuously stirred tank, (B) plug-flow, (C) packed bed, (D) fluidized bed and (E) enzymatic membrane bioreactors reproduced with permission from Gargouri et al. (2011), Ge et al. (2014), Long et al. (2015), Bello et al. (2017) and Rios et al. (2004) respectively.

the reactor when buoyancy force equals drag and gravitational forces (Bello et al., 2017). Fluidization has been found to facilitate brilliant mixing of particles, homogenous distribution of temperature and high rate of mass transfer (Tisa et al., 2014). Kuyukina et al. (2017) have described a fluidized bed bioreactor for treatment of oilfield effluents. The facility of adding catalysts from the top of the fluidized bed reactors renders it more advantageous than other fixed bed reactors (Ross, 2011). The efficient heat transfer ability of this type of bioreactors also facilitates safe and successful execution of exothermic reactions (Eş et al., 2015).

Besides the reactors discussed so far, immobilized biomolecules have been immobilized on semipermeable membrane surfaces for use in biocatalytic membrane reactors. A schematic representation of such a membrane and its mechanism of action have been shown in Figure 3.5A and B, respectively. Zhang et al. (2010) and Xu et al. (2018) have reported horseradish peroxidase and laccase immobilized graphene oxide membranes, respectively. These membranes were further investigated for removal of phenolic compounds and synthetic dyes, respectively. Biocatalytic membrane reactors also offer several

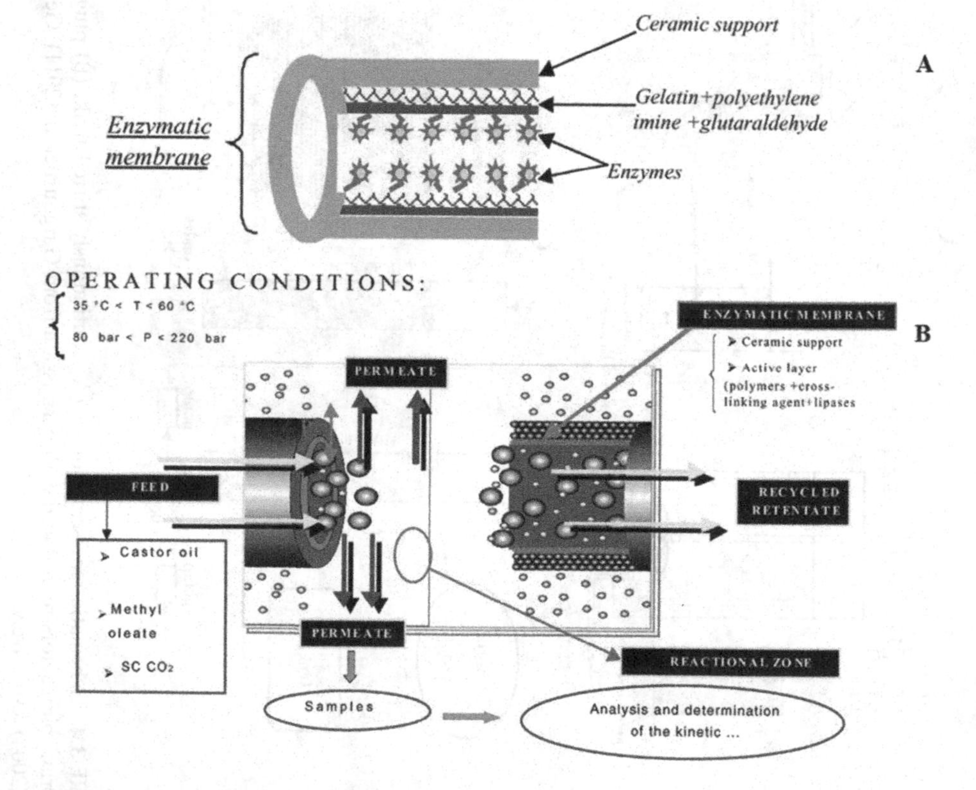

FIGURE 3.5 Enzyme immobilized membrane (A) and its principle for pollutant removal (B) as reproduced with permission from Rios et al. (2004).

advantages like prevention of product inhibition, reduction of difference in pressure experienced throughout the membrane as well as an increased rate of mass transfer between effluent and biocatalyst (Purkait and Mohanty, 2011).

3.7 PROCESS KINETICS

Process kinetics for enzymatic transformation is analyzed using the Lineweaver–Burk plots obtained from the equations guiding Michaelis–Menten kinetic (Cornish-Bowden, 2015). For determining the Michaelis constant and the optimum reaction velocity (denoted as K_m and V_{max}, respectively) initial rates of reaction of free and immobilized biomolecules are recorded in suitable buffer solutions under specific optimum experimental conditions against a standard substrate (Zhao et al., 2011). Values of K_m and V_{max} were derived from the initial rate of reaction obtained from the plot of Lineweaver–Burk. The relation between V_{max} and concentration of enzyme [E] is denoted by the equation given as follows: $V_{max} = K_{cat}[E]$, where K_{cat} represents apparent rate constant and expressed as second^{-1} (Zhao et al., 2011). K_{cat} considered all events of chemical transformation occurring between ternary enzyme conformation and product formation (DeLouise and Miller, 2005).

3.8 PROCESS OPTIMIZATION

Maximum efficiency of the immobilized biocatalysts may be achieved when implemented under optimized experimental conditions. Recent studies have reported different mathematical and statistical tools for determination of optimum values of process parameters and effect of inter-parameter interaction on the concerned process (Banerjee et al., 2017a). Two such widely employed optimization tools are artificial neural network (ANN) and response surface methodology (RSM). Implementation of RSM for determining optimum experimental and/or numerical results eradicates the probability of repetitive analysis and minimizes numerical noise. Inter-parameter interactions are revealed by RSM analysis in the format of 3D space or contour plots. The optimum experimental conditions are determined by an inbuilt Derringer's desirability function (Banerjee et al., 2017a).

ANN modeling is based upon the schema of biological nervous systems. ANN models are generally constituted of input, hidden and output layers. Appropriateness of the designed model for the concerned process is determined by the count of neurons which may be optimized by a trial and error method (Banerjee et al., 2017a). Optimum efficiency of the ANN model is achieved by maintaining a constant performance target and ramp (Banerjee et al., 2017a).

Appropriateness of RSM or ANN for optimization of the concerned process may be determined in terms of absolute average deviation, root mean squared error, mean absolute error and coefficient of regression (Banerjee et al., 2017a). Deviation between theoretically predicted and experimentally obtained results is indicated by absolute average deviation and should be as low as possible. Lower values of mean absolute and root mean squared error indicate the appropriateness of the concerned tool for optimization of the process under consideration (Banerjee et al., 2017a).

3.9 CONCLUSION

Recent studies have established nanomaterials as an excellent substrate for immobilization of biomolecules owing to their large surface areas. These novel support materials present biocompatible environments for immobilized biomolecules. Immobilization of biomolecules have been found to overcome issues related to application of free enzymes like high expense and time incurred, inactivation of enzyme, availability of purified enzyme and contamination of enzyme. Recent studies have also revealed that immobilization of whole cells is more advantageous than immobilization of enzymes. Selection of appropriate technique of immobilization and bioreactor setup as well as process optimization also plays a significant role in achieving biocatalyst efficiency. The biocatalysts discussed in this study should be investigated for their efficiency of treating industrial and domestic effluents and their feasibility for effluent treatment on a commercial scale. The present review is expected to guide efficient application of biocatalysts for wide-scale effluent treatment.

REFERENCES

Abdollahi, Kourosh, Farshad Yazdani, Reza Panahi, and Babak Mokhtarani. "Biotransformation of phenol in synthetic wastewater using the functionalized magnetic nano-biocatalyst particles carrying tyrosinase." *3 Biotech* 8, no. 10 (2018): 419.

Alarcón-Payán, Dulce A., Rina D. Koyani, and Rafael Vazquez-Duhalt. "Chitosan-based biocatalytic nanoparticles for pollutant removal from wastewater." *Enzyme and Microbial Technology* 100 (2017): 71–78.

Ali, Moazzam, and Markus Winterer. "ZnO nanocrystals: surprisingly 'alive'." *Chemistry of Materials* 22, no. 1 (2009): 85–91.

Alver, Erol, and Ayşegül Ülkü Metin. "Chitosan based metal-chelated copolymer nanoparticles: Laccase immobilization and phenol degradation studies." *International Biodeterioration & Biodegradation* 125 (2017): 235–242.

Ansari, Shakeel Ahmed, and Qayyum Husain. "Potential applications of enzymes immobilized on/in nano materials: a review." *Biotechnology Advances* 30, no. 3 (2012): 512–523.

Banerjee, Priya, Shramana Roy Barman, Dolanchapa Sikdar, Uttariya Roy, Aniruddha Mukhopadhyay, and Papita Das. "Enhanced degradation of ternary dye effluent by developed bacterial consortium with RSM optimization, ANN modeling and toxicity evaluation." *Desalination and Water Treatment* 72 (2017a): 249–265.

Banerjee, Priya, Shramana Roy Barman, Snehasikta Swarnakar, Aniruddha Mukhopadhyay, and Papita Das. "Treatment of textile effluent using bacteria-immobilized graphene oxide nanocomposites: evaluation of effluent detoxification using *Bellamya bengalensis*." *Clean Technologies and Environmental Policy* 20, no. 10 (2018): 1–12.

Banerjee, Priya, Aniruddha Mukhopadhyay, and Papita Das. "Current modifications introduced for improving bioremediation efficiency of polycyclic aromatic hydrocarbons." In *Bio-remediation: Current Research and Application*, Ed. A. K. Rathoure, pp. 231–251. IK International Publishing House, New Delhi, India, 2017b.

Bello, Mustapha Mohammed, Abdul Aziz Abdul Raman, and Monash Purushothaman. "Applications of fluidized bed reactors in wastewater treatment–a review of the major design and operational parameters." *Journal of Cleaner Production* 141 (2017): 1492–1514.

Bunker, Christopher E., Kyle C. Novak, Elena A. Guliants, Barbara A. Harruff, M. Jaouad Meziani, Yi Lin, and Ya-Ping Sun. "Formation of protein−metal oxide nanostructures by the sonochemical method: observation of nanofibers and nanoneedles." *Langmuir* 23, no. 20 (2007): 10342–10347.

Chang, Qing, Jia Huang, Yaobin Ding, and Heqing Tang. "Catalytic oxidation of phenol and 2, 4-dichlorophenol by using horseradish peroxidase immobilized on graphene oxide/Fe_3O_4." *Molecules* 21, no. 8 (2016): 1044.

Chang, Qing, and Heqing Tang. "Immobilization of horseradish peroxidase on NH_2-modified magnetic Fe_3O_4/SiO_2 particles and its application in removal of 2, 4-dichlorophenol." *Molecules* 19, no. 10 (2014): 15768–15782.

Chopra, Harish K. *Enzymes in food processing: fundamentals and potential applications.* IK International Pvt Ltd, New Delhi, India, 2010.

Cornish-Bowden, Athel. "One hundred years of Michaelis–Menten kinetics." *Perspectives in Science* 4 (2015): 3–9.

Das, Rasel, Sharifah Bee Abd Hamid, Md Eaqub Ali, Ahmad Fauzi Ismail, M. S. M. Annuar, and Seeram Ramakrishna. "Multifunctional carbon nanotubes in water treatment: the present, past and future." *Desalination* 354 (2014): 160–179.

Datta, Sumitra, L. Rene Christena, and Yamuna Rani Sriramulu Rajaram. "Enzyme immobilization: an overview on techniques and support materials." *3 Biotech* 3, no. 1 (2013): 1–9.

DeLouise, Lisa A., and Benjamin L. Miller. "Enzyme immobilization in porous silicon: quantitative analysis of the kinetic parameters for glutathione-S-transferases." *Analytical Chemistry* 77, no. 7 (2005): 1950–1956.

Eş, Ismail, José Daniel Gonçalves Vieira, and André Corrêa Amaral. "Principles, techniques, and applications of biocatalyst immobilization for industrial application." *Applied Microbiology and Biotechnology* 99, no. 5 (2015): 2065–2082.

Gahlout, Mayur, Darshan M. Rudakiya, Shilpa Gupte, and Akshaya Gupte. "Laccase-conjugated amino-functionalized nanosilica for efficient degradation of Reactive Violet 1 dye." *International Nano Letters* 7, no. 3 (2017): 195–208.

Galliker, Patrick, Gregor Hommes, Dietmar Schlosser, Philippe F-X. Corvini, and Patrick Shahgaldian. "Laccase-modified silica nanoparticles efficiently catalyze the transformation of phenolic compounds." *Journal of Colloid and Interface Science* 349, no. 1 (2010): 98–105.

Gao, X. J., X. J. Fan, X. P. Chen, and Z. Q. Ge. "Immobilized β-lactamase on Fe_3O_4 magnetic nanoparticles for degradation of β-lactam antibiotics in wastewater." *International Journal of Environmental Science and Technology* 15, no. 10 (2018a): 2203–2212.

Gao, Zhen, Yunfei Yi, Jia Zhao, Yongyang Xia, Min Jiang, Fei Cao, Hua Zhou, Ping Wei, Honghua Jia, and Xiaoyu Yong. "Co-immobilization of laccase and TEMPO onto amino-functionalized magnetic Fe_3O_4 nanoparticles and its application in acid fuchsin decolorization." *Bioresources and Bioprocessing* 5, no. 1 (2018b): 27.

Gargouri, Boutheina, Fatma Karray, Najla Mhiri, Fathi Aloui, and Sami Sayadi. "Application of a continuously stirred tank bioreactor (CSTR) for bioremediation of hydrocarbon-rich industrial wastewater effluents." *Journal of Hazardous Materials* 189, no. 1–2 (2011): 427–434.

Gasser, Christoph A., Liang Yu, Jan Svojitka, Thomas Wintgens, Erik M. Ammann, Patrick Shahgaldian, Philippe F.-X. Corvini, and Gregor Hommes. "Advanced enzymatic elimination of phenolic contaminants in wastewater: a nano approach at field scale." *Applied Microbiology and Biotechnology* 98, no. 7 (2014): 3305–3316.

Ge, Shijian, Yongzhen Peng, Shuang Qiu, Ao Zhu, and Nanqi Ren. "Complete nitrogen removal from municipal wastewater via partial nitrification by appropriately alternating anoxic/aerobic conditions in a continuous plug-flow step feed process." *Water Research* 55 (2014): 95–105.

Genisheva, Zlatina, Solange I. Mussatto, José M. Oliveira, and José A. Teixeira. "Evaluating the potential of wine-making residues and corn cobs as support materials for cell immobilization for ethanol production." *Industrial Crops and Products* 34, no. 1 (2011): 979–985.

Gurel, L., I. Senturk, T. Bahadir, and H. Buyukgungor. "Treatment of nickel plating industrial wastewater by fungus immobilized onto rice bran." *Journal of Microbial and Biochemical Technology* 2, no. 2 (2010): 1000020.

Hafez, Azza, and Samir El-Mariharawy. "Design and performance of the two-stage/two-pass RO membrane system for chromium removal from tannery wastewater. Part 3." *Desalination* 165 (2004): 141–151.

He, Mengling, Yun Li, Fuwei Pi, Jian Ji, Xingxing He, Yinzhi Zhang, and Xiulan Sun. "A novel detoxifying agent: Using rice husk carriers to immobilize zearalenone-degrading enzyme from *Aspergillus niger* FS10." *Food Control* 68 (2016): 271–279.

Hwang, Ee Taek, and Man Bock Gu. "Enzyme stabilization by nano/microsized hybrid materials." *Engineering in Life Sciences* 13, no. 1 (2013): 49–61.

Ji, Chao, Jingwei Hou, and Vicki Chen. "Cross-linked carbon nanotubes-based biocatalytic membranes for micro-pollutants degradation: performance, stability, and regeneration." *Journal of Membrane Science* 520 (2016): 869–880.

Jin, Xinyu, Saisai Li, Nengbing Long, and Ruifeng Zhang. "Improved Biodegradation of Synthetic Azo Dye by Anionic Cross-Linking of Chloroperoxidase on ZnO/SiO 2 Nanocomposite Support." *Applied Biochemistry and Biotechnology* 184, no. 3 (2018): 1009–1023.

Johnson, Andrew K., Anna M. Zawadzka, Lee A. Deobald, Ronald L. Crawford, and Andrzej J. Paszczynski. "Novel method for immobilization of enzymes to magnetic nanoparticles." *Journal of Nanoparticle Research* 10, no. 6 (2008): 1009–1025.

Karel, Steven F., Shari B. Libicki, and Channing R. Robertson. "The immobilization of whole cells: engineering principles." *Chemical Engineering Science* 40, no. 8 (1985): 1321–1354.

Knopp, Dietmar, Dianping Tang, and Reinhard Niessner. "Bioanalytical applications of biomolecule-functionalized nanometer-sized doped silica particles." *Analytica Chimica Acta* 647, no. 1 (2009): 14–30.

Konwarh, Rocktotpal, Niranjan Karak, Sudhir Kumar Rai, and Ashis Kumar Mukherjee. "Polymer-assisted iron oxide magnetic nanoparticle immobilized keratinase." *Nanotechnology* 20, no. 22 (2009): 225107.

Kuyukina, Maria S., Irena B. Ivshina, Marina K. Serebrennikova, Anastasiya V. Krivoruchko, Irina O. Korshunova, Tatyana A. Peshkur, and Colin J. Cunningham. "Oilfield wastewater biotreatment in a fluidized-bed bioreactor using co-immobilized Rhodococcus cultures." *Journal of Environmental Chemical Engineering* 5, no. 1 (2017): 1252–1260.

Lee, Soo Youn, Seonyoung Lee, Il Hwan Kho, Jin Hyung Lee, Jong Hee Kim, and Jeong Ho Chang. "Enzyme–magnetic nanoparticle conjugates as a rigid biocatalyst for the elimination of toxic aromatic hydrocarbons." *Chemical Communications* 47, no. 36 (2011): 9989–9991.

Li, Galong, Pei Ma, Yuan He, Yifan Zhang, Yane Luo, Ce Zhang, and Haiming Fan. "Enzyme–nanowire mesocrystal hybrid materials with an extremely high biocatalytic activity." *Nano letters* 18, no. 9 (2018): 5919–5926.

Li, Yuanyuang, Hermann J. Schluesener, and Shunqing Xu. "Gold nanoparticle-based biosensors." *Gold Bulletin* 43, no. 1 (2010): 29–41.

Liu, Youxun, Mingyang Yan, Yuanyuan Geng, and Juan Huang. "Laccase immobilization on poly (*p*-phenylenediamine)/Fe_3O_4 nanocomposite for reactive blue 19 dye removal." *Applied Sciences* 6, no. 8 (2016): 232.

Long, Yunchuan, Qiao Li, Jiangxia Ni, Fei Xu, and Heng Xu. "Treatment of metal wastewater in pilot-scale packed bed systems: efficiency of single-vs. mixed-mushrooms." *RSC Advances* 5, no. 37 (2015): 29145–29152.

Luckarift, Heather R., Shankar Balasubramanian, Sheetal Paliwal, Glenn R. Johnson, and Aleksandr L. Simonian. "Enzyme-encapsulated silica monolayers for rapid functionalization of a gold surface." *Colloids and Surfaces B: Biointerfaces* 58, no. 1 (2007): 28–33.

Maryšková, Milena, Inés Ardao, Carlos A. García-González, Lenka Martinová, Jana Rotková, and Alena Ševců. "Polyamide 6/chitosan nanofibers as support for the immobilization of Trametes versicolor laccase for the elimination of endocrine disrupting chemicals." *Enzyme and microbial technology* 89 (2016): 31–38.

Mazille, F., A. Moncayo-Lasso, D. Spuhler, A. Serra, J. Peral, N. L. Benítez, and C. Pulgarin. "Comparative evaluation of polymer surface functionalization techniques before iron oxide deposition. Activity of the iron oxide-coated polymer films in the photo-assisted degradation of organic pollutants and inactivation of bacteria." *Chemical Engineering Journal* 160, no. 1 (2010): 176–184.

Miletić, Nemanja, Volker Abetz, Katrin Ebert, and Katja Loos. "Immobilization of Candida antarctica lipase B on polystyrene nanoparticles." *Macromolecular Rapid Communications* 31, no. 1 (2010): 71–74.

Moldes-Diz, Y., M. Gamallo, G. Eibes, Z. Vargas-Osorio, C. Vazquez-Vazquez, G. Feijoo, J. M. Lema, and M. T. Moreira. "Development of a superparamagnetic laccase nanobiocatalyst for the enzymatic biotransformation of xenobiotics." *Journal of Environmental Engineering* 144, no. 3 (2018): 04018007.

Mutamim, Noor Sabrina Ahmad, Zainura Zainon Noor, Mohd Ariffin Abu Hassan, and Gustaf Olsson. "Application of membrane bioreactor technology in treating high strength industrial wastewater: a performance review." *Desalination* 305 (2012): 1–11.

Naghdi, Mitra, Mehrdad Taheran, Satinder K. Brar, Azadeh Kermanshahi-pour, M. Verma, and Rao Y. Surampalli. "Immobilized laccase on oxygen functionalized nanobiochars through mineral acids treatment for removal of carbamazepine." *Science of the Total Environment* 584 (2017): 393–401.

Neoh, Chin Hong, Zainura Zainon Noor, Noor Sabrina Ahmad Mutamim, and Chi Kim Lim. "Green technology in wastewater treatment technologies: integration of membrane bioreactor with various wastewater treatment systems." *Chemical Engineering Journal* 283 (2016): 582–594.

Ni, Yonghong, Xiaofeng Cao, Guogen Wu, Guangzhi Hu, Zhousheng Yang, and Xianwen Wei. "Preparation, characterization and property study of zinc oxide nanoparticles via a simple solution-combusting method." *Nanotechnology* 18, no. 15 (2007): 155603.

Pavlidis, Ioannis V., Michaela Patila, Uwe T. Bornscheuer, Dimitrios Gournis, and Haralambos Stamatis. "Graphene-based nanobiocatalytic systems: recent advances and future prospects." *Trends in Biotechnology* 32, no. 6 (2014): 312–320.

Peng, Qingqing, Yunguo Liu, Guangming Zeng, Weihua Xu, Chunping Yang, and Jingjin Zhang. "Biosorption of copper (II) by immobilizing *Saccharomyces cerevisiae* on the surface of chitosan-coated magnetic nanoparticles from aqueous solution." *Journal of Hazardous Materials* 177, no. 1–3 (2010): 676–682.

Prakasham, R. S., G. Sarala Devi, K. Rajya Laxmi, and Ch Subba Rao. "Novel synthesis of ferric impregnated silica nanoparticles and their evaluation as a matrix for enzyme immobilization." *The Journal of Physical Chemistry C* 111, no. 10 (2007): 3842–3847.

Purkait, Mihir K., and Kaustubha Mohanty. *Membrane technologies and applications.* CRC press, United States, 2011.

Qu, Xiaolei, Pedro JJ Alvarez, and Qilin Li. "Applications of nanotechnology in water and wastewater treatment." *Water Research* 47, no. 12 (2013): 3931–3946.

Rani, Manviri, Uma Shanker, and Amit K. Chaurasia. "Catalytic potential of laccase immobilized on transition metal oxides nanomaterials: degradation of alizarin red S dye." *Journal of Environmental Chemical Engineering* 5, no. 3 (2017): 2730–2739.

Rao, Dubasi Govardhana. *Introduction to biochemical engineering.* Tata McGraw-Hill Education, New York, United States, 2010.

Rios, G. M., M. P. Belleville, D. Paolucci, and J. Sanchez. "Progress in enzymatic membrane reactors–a review." *Journal of Membrane Science* 242, no. 1–2 (2004): 189–196.

Ross, Julian RH. *Heterogeneous catalysis: fundamentals and applications.* Elsevier, Amsterdam, Netherlands, 2011.

Roy, Uttariya, Shubhalakshmi Sengupta, Priya Banerjee, Papita Das, Avijit Bhowal, and Siddhartha Datta. "Assessment on the decolourization of textile dye (Reactive Yellow) using *Pseudomonas* sp. immobilized on fly ash: Response surface methodology optimization and toxicity evaluation." *Journal of Environmental Management* 223 (2018b): 185–195.

Roy, Uttariya, Shubhalakshmi Sengupta, Papita Das, Avijit Bhowal, and Siddhartha Datta. "Integral approach of sorption coupled with biodegradation for treatment of azo dye using *Pseudomonas* sp.: batch, toxicity, and artificial neural network." *3 Biotech* 8, no. 4 (2018a): 192.

Samak, Nadia A., Yeqiang Tan, Kunyan Sui, Ting-Ting Xia, Kefeng Wang, Chen Guo, and Chunzhao Liu. "CotA laccase immobilized on functionalized magnetic graphene oxide nano-sheets for efficient biocatalysis." *Molecular Catalysis* 445 (2018): 269–278.

Saratale, R. G., G. D. Saratale, D. C. Kalyani, Jo-Shu Chang, and S. P. Govindwar. "Enhanced decolorization and biodegradation of textile azo dye Scarlet R by using developed microbial consortium-GR." *Bioresource Technology* 100, no. 9 (2009): 2493–2500.

Sathishkumar, Palanivel, Seralathan Kamala-Kannan, Min Cho, Jae Su Kim, Tony Hadibarata, Mohd Razman Salim, and Byung-Taek Oh. "Laccase immobilization on cellulose nanofiber: the catalytic efficiency and recyclic application for simulated dye effluent treatment." *Journal of Molecular Catalysis B: Enzymatic* 100 (2014): 111–120.

Shao, Binbin, Zhifeng Liu, Guangming Zeng, Yang Liu, Xin Yang, Chengyun Zhou, Ming Chen, Yujie Liu, Yilin Jiang, and Ming Yan. "Immobilization of laccase on hollow mesoporous carbon nanospheres: Noteworthy immobilization, excellent stability and efficacious for antibiotic contaminants removal." *Journal of Hazardous Materials* 362 (2019): 318–326.

Shen, Liang, Yu Liu, and Hai-Lou Xu. "Treatment of ampicillin-loaded wastewater by combined adsorption and biodegradation." *Journal of Chemical Technology & Biotechnology* 85, no. 6 (2010): 814–820.

Shojaat, Rahim, Naghi Saadatjoo, Afzal Karimi, and Soheil Aber. "Simultaneous adsorption–degradation of organic dyes using MnFe2O4/calcium alginate nanocomposites coupled with GOx and laccase." *Journal of Environmental Chemical Engineering* 4, no. 2 (2016): 1722–1730.

Singh, B. D. "Biotechnology (Expanding Horizons)." (2008).

Subrizi, Fabiana, Marcello Crucianelli, Valentina Grossi, Maurizio Passacantando, Lorenzo Pesci, and Raffaele Saladino. "Carbon nanotubes as activating tyrosinase supports for the selective synthesis of catechols." *ACS Catalysis* 4, no. 3 (2014): 810–822.

Sun, Jiajia, Raghuvara Yendluri, Kai Liu, Ying Guo, Yuri Lvov, and Xuehai Yan. "Enzyme-immobilized clay nanotube–chitosan membranes with sustainable biocatalytic activities." *Physical Chemistry Chemical Physics* 19, no. 1 (2017): 562–567.

Swartz, Joshua D., Scott A. Miller, and David Wright. "Rapid production of nitrilase containing silica nanoparticles offers an effective and reusable biocatalyst for synthetic nitrile hydrolysis." *Organic Process Research & Development* 13, no. 3 (2009): 584-589.

Tan, Jia-Ming, Guanglei Qiu, and Yen-Peng Ting. "Osmotic membrane bioreactor for municipal wastewater treatment and the effects of silver nanoparticles on system performance." *Journal of Cleaner Production* 88 (2015): 146–151.

Tavares, Tássia Silva, Juliana Arriel Torres, Maria Cristina Silva, Francisco Guilherme Esteves Nogueira, Adilson C. Da Silva, and Teodorico C. Ramalho. "Soybean peroxidase immobilized on δ-FeOOH as new magnetically recyclable biocatalyst for removal of ferulic acid." *Bioprocess and Biosystems Engineering* 41, no. 1 (2018): 97–106.

Tisa, Farhana, Abdul Aziz Abdul Raman, and Wan Mohd Ashri Wan Daud. "Applicability of fluidized bed reactor in recalcitrant compound degradation through advanced oxidation processes: a review." *Journal of Environmental Management* 146 (2014): 260–275.

Trewyn, Brian G., Igor I. Slowing, Supratim Giri, Hung-Ting Chen, and Victor S-Y. Lin. "Synthesis and functionalization of a mesoporous silica nanoparticle based on the sol–gel process and applications in controlled release." *Accounts of Chemical Research* 40, no. 9 (2007): 846–853.

Tudorache, Madalina, Diana Mahalu, Cristian Teodorescu, Razvan Stan, Camelia Bala, and Vasile I. Parvulescu. "Biocatalytic microreactor incorporating HRP anchored on micro-/nano-lithographic patterns for flow oxidation of phenols." *Journal of Molecular Catalysis B: Enzymatic* 69, no. 3–4 (2011): 133–139.

Tudorache, Madalina, Alina Negoi, Bogdan Tudora, and Vasile I. Parvulescu. "Environmental-friendly strategy for biocatalytic conversion of waste glycerol to glycerol carbonate." *Applied Catalysis B: Environmental* 146 (2014): 274–278.

Veerapandian, Murugan, and Kyusik Yun. "Synthesis of silver nanoclusters and functionalization with glucosamine for glyconanoparticles." *Synthesis and Reactivity in Inorganic, Metal-Organic, and Nano-Metal Chemistry* 40, no. 1 (2010): 56–64.

Veerapandian, Murugan, and Kyusik Yun. "Functionalization of biomolecules on nanoparticles: specialized for antibacterial applications." *Applied Microbiology and Biotechnology* 90, no. 5 (2011): 1655–1667.

Wang, Hua, David G. Castner, Buddy D. Ratner, and Shaoyi Jiang. "Probing the orientation of surface-immobilized immunoglobulin G by time-of-flight secondary ion mass spectrometry." *Langmuir* 20, no. 5 (2004): 1877–1887.

Wang, Ying, Zhaohui Li, Jun Wang, Jinghong Li, and Yuehe Lin. "Graphene and graphene oxide: biofunctionalization and applications in biotechnology." *Trends in Biotechnology* 29, no. 5 (2011): 205–212.

Wang, Zhen-Gang, Ling-Shu Wan, Zhen-Mei Liu, Xiao-Jun Huang, and Zhi-Kang Xu. "Enzyme immobilization on electrospun polymer nanofibers: an overview." *Journal of Molecular Catalysis B: Enzymatic* 56, no. 4 (2009): 189–195.

World Health Organization, and UNICEF. "Progress on sanitation and drinking water-2013 update." (2013).

Xie, Tian, Anming Wang, Lifeng Huang, Haifeng Li, Zhenming Chen, Qiuyan Wang, and Xiaopu Yin. "Recent advance in the support and technology used in enzyme immobilization." *African Journal of Biotechnology* 8, no. 19 (2009): 4724–4733.

Xu, Hui-Min, Xue-Fei Sun, Si-Yu Wang, Chao Song, and Shu-Guang Wang. "Development of laccase/graphene oxide membrane for enhanced synthetic dyes separation and degradation." *Separation and Purification Technology* 204 (2018): 255–260.

Xu, Jing, Fang Zeng, Shuizhu Wu, Xinxing Liu, Chao Hou, and Zhen Tong. "Gold nanoparticles bound on microgel particles and their application as an enzyme support." *Nanotechnology* 18, no. 26 (2007): 265704.

Xu, Ran, Jingyuan Cui, Rongzhi Tang, Fengting Li, and Bingru Zhang. "Removal of 2, 4, 6-trichlorophenol by laccase immobilized on nano-copper incorporated electrospun fibrous membrane-high efficiency, stability and reusability." *Chemical Engineering Journal* 326 (2017): 647–655.

Xu, Ran, Rongzhi Tang, Qijun Zhou, Fengting Li, and Bingru Zhang. "Enhancement of catalytic activity of immobilized laccase for diclofenac biodegradation by carbon nanotubes." *Chemical Engineering Journal* 262 (2015): 88–95.

Yagub, Mustafa T., Tushar Kanti Sen, Sharmeen Afroze, and Ha Ming Ang. "Dye and its removal from aqueous solution by adsorption: a review." *Advances in Colloid and Interface Science* 209 (2014): 172–184.

Yang, Huang-Hao, Shu-Qiong Zhang, Xiao-Lan Chen, Zhi-Xia Zhuang, Jin-Gou Xu, and Xiao-Ru Wang. "Magnetite-containing spherical silica nanoparticles for biocatalysis and bioseparations." *Analytical Chemistry* 76, no. 5 (2004): 1316–1321.

Zhang, Di, Manfeng Deng, Hongbin Cao, Songping Zhang, and He Zhao. "Laccase immobilized on magnetic nanoparticles by dopamine polymerization for 4-chlorophenol removal." *Green Energy & Environment* 2, no. 4 (2017): 393–400.

Zhang, Feng, Bin Zheng, Jiali Zhang, Xuelei Huang, Hui Liu, Shouwu Guo, and Jingyan Zhang. "Horseradish peroxidase immobilized on graphene oxide: physical properties and applications in phenolic compound removal." *The Journal of Physical Chemistry C* 114, no. 18 (2010): 8469–8473.

Zhao, Guanghui, Jianzhi Wang, Yanfeng Li, Xia Chen, and Yaping Liu. "Enzymes immobilized on superparamagnetic Fe3O4@ clays nanocomposites: preparation, characterization, and a new strategy for the regeneration of supports." *The Journal of Physical Chemistry C* 115, no. 14 (2011): 6350–6359.

Zhao, Xinmei, Baohua Zhang, Kelong Ai, Guo Zhang, Linyuan Cao, Xiaojuan Liu, Hongmei Sun, Haishui Wang, and Lehui Lu. "Monitoring catalytic degradation of dye molecules on silver-coated ZnO nanowire arrays by surface-enhanced Raman spectroscopy." *Journal of Materials Chemistry* 19, no. 31 (2009): 5547–5553.

Zhou, Guizhong, Zhaofeng Wang, Wenqian Li, Qian Yao, and Dayi Zhang. "Graphene-oxide modified polyvinyl-alcohol as microbial carrier to improve high salt wastewater treatment." *Materials Letters* 156 (2015): 205–208.

4 Applications of Biosorption in Heavy Metals Removal

F.E. Soetaredjo, S.P. Santoso, and L. Laysandra
Department of Chemical Engineering, Widya Mandala Surabaya Catholic University, Indonesia

K. Foe
Faculty of Pharmacy, Widya Mandala Surabaya Catholic University, Pakuwon City, Indonesia

S. Ismadji
Department of Chemical Engineering, Widya Mandala Surabaya Catholic University, Indonesia

4.1 INTRODUCTION

Utilization of cheap, renewable, and abundantly available materials as the adsorbents for hazardous pollutants, especially heavy metals removal from water environments, is currently intensively studied. Agricultural wastes and other lignocellulosic materials possibly are very potential candidates as the alternative adsorbents. Thousands of studies have employed these materials as unconventional adsorbents to remove heavy metals from the solutions or synthetic wastewater. Other biological substances have also been studied for its adsorption capability toward heavy metals.

In general, the adsorption of heavy metals by biological materials involved complex adsorption mechanisms. Living or dead biological materials also give different adsorption mechanisms. The existence of different adsorption mechanisms in biological materials also gives some difficulties on the term of adsorption. Some authors prefer the term of biosorption for the adsorption of substances on biological materials (for both living and dead), while others prefer to use the term of bioaccumulation for the adsorption of substances on living organisms. For more details about the terminology and concepts of biosorption, the readers can refer to the review paper by Fomina and Gadd (2014). The term *biosorption* is used in this chapter for the uptake of heavy metals by biological materials.

4.2 LIGNOCELLULOSIC MATERIALS AS BIOSORBENTS

Lignocellulosic biomass is a renewable material that is available abundantly in most of the countries in the world. In the future, lignocellulosic materials are the most important raw materials for industrial (biofuels and other chemicals) and environmental applications. Table 4.1 summarizes the results of the adsorption studies that utilize lignocellulosic biomasses to remove several heavy metals from solutions.

Pristine lignocellulosic biomasses usually posses low adsorption capacity toward certain metals ions (Table 4.1). The low uptake capacity of pristine lignocellulosic biomass is possibly due to the low BET (Brunauer–Emmett–Teller) surface area. In this case, physical adsorption does not perform an important function in the uptake of heavy metal ions. The heavy metals uptake by pristine lignocellulosic biomass occurs primarily through the interaction between some surface functional groups with the metal ions (Kosasih et al., 2010).

The chemical compositions of lignocellulosic biomass strongly depend on geographic location, climate, soil condition, etc. The chemical composition of lignocellulosic biomass consists of lignin, hemicellulose, and cellulose. Both lignin and hemicellulose are natural amorphous polymers in which both of them are chemically connected through covalent bonding (Putro et al., 2016). These amorphous polymers protect the porous part of the lignocellulosic material (cellulose). Thus, it becomes inaccessible (low BET surface area). A typical value of the BET surface area of pristine lignocellulosic biomass is usually around 1 to 10 m^2/g. As indicated in Figure 4.1, low intake of nitrogen gas during the nitrogen sorption measurement is a strong indication that pristine lignocellulosic biomass has low porosity. Therefore, a low BET surface area is as estimated (in this case, a pristine cassava peel was taken as the lignocellulosic sample).

This evidence is also supported by DFT (density functional theory) pore size distribution (Figure 4.2).

From Figure 4.2, it can be seen that the pore structure of pristine cassava peels mostly in the macroporous region with some in the mesoporous region (10 to 70 nm).

Lignocellulosic materials have a complex structure and a variety of structural components. Distinct types of surface functional groups can be found in different lignocellulosic materials. Depending on the condition of the adsorption process (pH of the solution), some of the available functional groups (hydroxyl, carboxyl, etc.) can interact with heavy metals. Wang and Wang (2018) studied the removal of Cu^{2+}, Cd^{2+}, and Ni^{2+} from aqueous solution using camphor leaf. The FTIR (Fourier Transform Infra-Red) analysis indicates that carboxyl groups act as metal binding. Addition of proton to the carboxyl groups at low pH could increase the heavy metals uptake on camphor leaf. Carboxyl groups are also responsible for binding other metal ions on the durian shell (Kurniawan et al., 2011; Ngabura et al., 2018).

Ion exchange mechanism during the adsorption of Cd^{2+} with pristine rice straw was claimed by Ding et al. (2012). The presence of several cationic

TABLE 4.1
Heavy metal removal using lignocellulosic materials

Metal ions	Adsorbent	Conditions	q_{max} (mg/g)	Ref.
As^{5+}	Pine leaves	pH = 4, T = 25°C, adsorbent dosage = 1 g, shaking at 100 rpm for 35 min	3.27	Shafique et al., 2012
Cd^{2+}	Banana stalks Sunflower Corn cob	Shaker at 175 rpm at room temperature (25°C), t = 60 min	3.66 11.40 13.58	Mahmood-ul-Hassan et al., 2015
	Apple pomace	C_o = 10–200 mg/L in 50 mL solution, pH = 6.0	4.45	Chand et al., 2014
	Corn stalk	T = 25°C, adsorbent dosage = 5 g/L, pH = 7.0	4.85	Zheng and Meng, 2016
	Cicer arietinum husk	pH = 7.0, T = 25°C	8.58	Pandey, 2016
	Rice straw	pH = 5.0, T = 25°C, shaking at 150 rpm for 3 h	13.89	Ding et al., 2012
	Cashew nut shell	pH = 5.0, t = 1 h, T = 25°C	11.23	Coelho et al., 2014
	Chestnut bur	C_o = 100 ppm, T = 20–25°C, pH = 4.0, shaking at 200 rpm for 6 h	34.77	Kim et al., 2015
	Portulaca oleracea dried plant	pH = 6.0, T = 27°C, t = 100 min, shaking at 100 rpm speed	43.48	Dubey et al., 2014
Cr^{3+}	Cashew nut shell	pH = 5.0, t = 1 h, T = 25°C	8.42	Coelho et al., 2014
	Sugarcane pulp	pH = 5.0, shaking at 140 rpm for 24 h	15.85	Yang et al., 2013
	Spent grains	T = 30°C, shaking at 150 rpm	16.68	Ferraz et al., 2015
	Peanut shell	pH = 5.0, T = 20°C, and t =1 h	27.86	Witek-Krowiak et al., 2011
Cr^{6+}	Almond green hull	pH = 6, t = 30 min, at room temperature	2.04	Sahranavard et al., 2011
	Wheat bran	C_o = 200 mg/L, pH = 2, T = 23°C and shaking at 120 rpm for 200 min	4.53	Kaya et al., 2014
	Moringa pods	T = 20°C, adsorbent dosage = 1 g, t = 60 min	5.50	Matouq et al., 2015
	Banana stalks Sunflower Corn cob	Shaker at 175 rpm at room temperature (25°C), t = 60 min	6.86 12.21 18.78	Mahmood-ul-Hassan et al., 2015
	Durian peel	C_o = 75 mg/L, pH = 2, t = 30 min	10.67	Saueprasearsit, 2011
	Rice husk Rice straw Rice bran Hyacinth roots	C_o = 25 mg/L, T = 30°C, pH = 1.5 (for rice husk) and pH = 2 (for other adsorbents), t = 3 h	11.40 12.17 12.34 15.28	Singha and Das, 2011

(Continued)

TABLE 4.1 (Cont.)

Metal ions	Adsorbent	Conditions	q_{max} (mg/g)	Ref.
	Ash gourd (*Benincasa hispida*) peel waste	C_o = 125 mg/L, T = 28 ± 1°C, pH = 1, adsorbent dosage 6 g/L, shaking at 180 rpm.	18.70	Sreenivas et al., 2014
	Ficus carica	pH = 3, agitation speed = 100 rpm at 30°C	19.68	Gupta et al., 2013
	Portulaca Oleracea	C_o = 100 mg/L, pH = 2, shaking at 100 rpm	54.95	Mishra et al., 2015
	Mangrove leaves	T = 30°C, pH = 2, t = 30 min	60.24	Sathish et al., 2015
	Wheat bran	T = 50°C, pH = 1, t = 12 h	70.03	Wang et al., 2008
	Rice bran		95.35	
	Laminaria japonica		96.31	
	Durian shell waste	C_o = 200 ppm, pH = 2.5, t = 60 min, at 60°C and shaking at 100 rpm	117	Kurniawan et al., 2011
Cu^{2+}	Coir pith	Co = 1,171 mg/L, T = 30°C, pH = 2, t = 22 h	165.00	Suksabye and Thiravetyan, 2012
	Sugarcane bagasse	pH = 6, shaking at 100 rpm for 90 min	3.65	Putra et al., 2014
	Coconut tree		3.89	
	Barley straw	pH = 6–7, T = 25 ± 1°C, t = 120 min	4.64	Pehlivan et al., 2012
	Pine cone shell	T = 25°C, pH = 5, t = 100 min	6.81	Tenorio et al., 2012
	Moringa pods	T = 20°C, adsorbent dosage = 1 g, t = 60 min	6.07	Matouq et al., 2015
	Cicer arietinum husk	pH = 5.0, T = 25°C	9.70	Pandey, 2016
	Cashew nut shell	C_o = 10–50 ppm, pH = 5.0, T = 30°C, t = 30 min, adsorbent dosage = 0.3 g	20	Senthilkumar et al., 2011b
	Peanut shell	pH = 5.0, T = 20°C, and t =1 h	25.39	Witek-Krowiak et al., 2011
	Palm oil fruit shells	pH = 6 at room temperature, t = 24 h, adsorbent dosage = 0.5 g	32.46	Hossain et al., 2012
	Peganum harmala-L	C_o = 50 mg/L, T = 25°C, pH = 6.0, t = 60 min	68.02	Ghasemi et al., 2014
	Caesalpinia bonducella leaf	T = 50°C, pH = 5, t = 120 min, shaking at 180 rpm	76.92	Yuvaraja et al., 2012
Ni^{2+}	Sugarcane bagasse	T = 25°C, pH = 5	2.23	Alomá et al., 2012
	Moringa pods	T = 20°C, adsorbent dosage = 1 g, t = 60 min	5.54	Matouq et al., 2015

(*Continued*)

TABLE 4.1 (Cont.)

Metal ions	Adsorbent	Conditions	q_{max} (mg/g)	Ref.
	Pigeon peas hulls	T = 40°C, pH = 4, shaking at 300 rpm	23.63	Venkata et al., 2012
	Lycopersicum esculentum (tomato)	T = 50°C, pH = 5.5, t = 3 h	58.82	Gutha et al., 2015
	Litchi chinensis seeds	pH = 7.5, T = 25°C	66.62	Flores-Garnica et al., 2013
	Rice bran	T = 30°C, pH = 7, shaking at 130 rpm for 24 h	238.3	Zafar et al., 2015
Pb^{2+}	Bornean oil palm shell	pH = 8.0, t = 480 min	3.39	Chong et al., 2013
	Olive tree	C_o = 150 ppm, pH = 5	12.97	Ronda et al., 2013
	Cashew nut shell	T = 40°C, pH = 5, t = 30 min	20.00	SenthilKumar et al., 2011a
	Banana stalks	Shaking at 175 rpm at room temperature (25°C), t = 60 min	20.90	Mahmood-ul-Hassan et al., 2015
	Sunflower		22.64	
	Corn cob		29.17	
	Sugarcane bagasse	pH = 6, shaking at 100 rpm for 90 min	21.28	Putra et al., 2014
	Coconut tree		25.00	
	Pigeon peas hulls	T = 40°C, pH = 4, shaking at 300 rpm	23.64	Venkata et al., 2012
	Terminalia arjuna fruit powder	C_o = 10–100 mg/L, T = 30°C, pH = 4, t = 30 min	27.39	Rao et al., 2016
	Cashew nut shell	pH = 5.0, t = 1 h, T = 25°C	28.65	Coelho et al., 2014
	Hazelnut shell	Co = 100 mg/L, T = 20°C, t = 1 h, and shaking at 150 rpm	41.90	Şencan et al., 2015
	Cyclosorus interruptus	T = 25°C	45.25	Zhou et al., 2015
	Chestnut bur	C_o = 100 ppm, T = 20–25°C, pH = 4.0, shaking at 200 rpm for 6 h	74.35	Kim et al., 2015
	Corn silk	T = 40°C, pH = 5, shaking at 250 rpm	84.20	Petrović et al., 2016
	Cucumber peel	Co = 20–350 ppm, pH = 5.0, T = 30°C, t = 1 h, adsorbent dosage = 1 g	133.60	Basu ct al., 2017
Zn^{2+}	Raw bagasse	T = 27 ± 1°C, pH = 6.0, t = 3 h	13.4	Salihi et al., 2016
	Coconut tree	pH = 6, shaking at 100 rpm for 90 min	23.81	Putra et al., 2014
	Sugarcane bagasse		40.00	
	Cashew nut shell	T = 30°C, pH = 5.0, t = 30 min	24.98	Senthilkumar et al., 2012

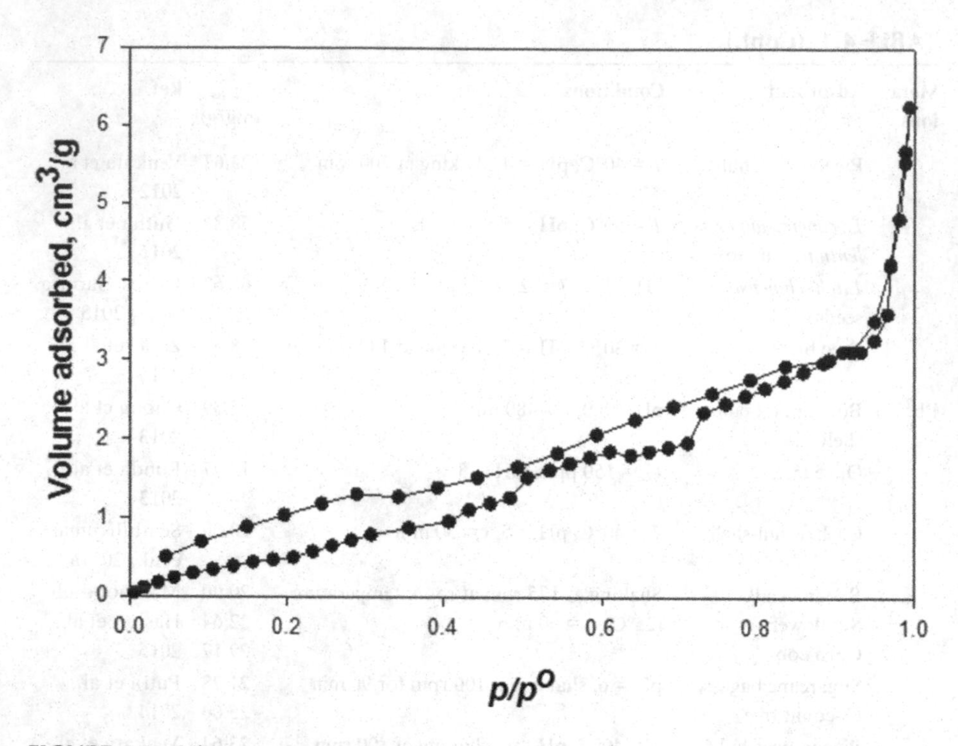

FIGURE 4.1 Nitrogen sorption isotherm of cassava peel.

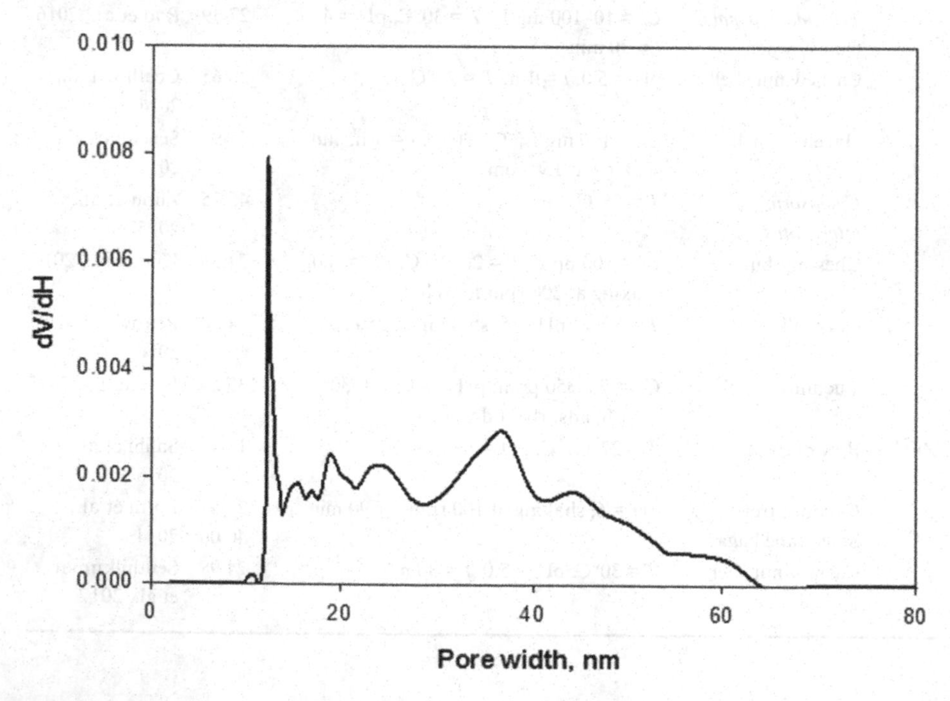

FIGURE 4.2 DFT pore size distribution of cassava peels.

metals viz. Ca^{2+}, K^{+}, Mg^{2+}, and Na^{+} facilitates the exchange with Cd^{2+}. The chelation between Cd^{2+} and some surface functional groups (hydroxyl and carboxyl groups) also contributes to the uptake of heavy metal by pristine rice straw. Another study by Reddy et al. (2011) also revealed that chelation of heavy metal (Ni^{2+}) with several surface functional groups (hydroxyl, carboxyl, and carbonyl) was the main mechanism for the biosorption process.

4.3 MODIFICATION OF LIGNOCELLULOSIC MATERIALS FOR ADSORPTION PURPOSE

Due to low adsorption capacity, modification of lignocellulosic materials often required to increase its adsorption capacity. The simplest way to increase adsorption by lignocellulosic biomass is through the delignification process. This process removes both lignin and hemicellulose, so the cellulose is readily exposed to the heavy metal solution, and it will increase the attachment of heavy metals on the lignocellulose surface. Strong acids and alkaline solutions, such as nitric acid (Calero et al., 2013; Ronda et al., 2013; Wang et al., 2018), sulphuric acid (Blazquez et al., 2014), hydrochloric acid (Ngabura et al., 2018), and sodium hydroxide (Blazquez et al., 2014) were widely used for this purpose. Thermal treatment is also a popular method to increase the capability of the lignocellulosic biomass to remove heavy metal from solution. The modification processes of several lignocellulosic biomasses for the adsorption of heavy metals are listed in Table 4.2.

Removal of Pb^{2+} from aqueous solution using pristine- and nitric acid-modified *Phytolacca americana L.* biomass was studied by Wang et al. (2018). *Phytolacca americana L.* is a herbaceous plant that is rich in cellulose, hemicellulose, and lignin. The treatment of biomass using HNO_3 solution increased the adsorption capacity of pristine biomass less than 20%, from 10.83 mg/g to 12.66 mg/g. A small increase in the adsorption capacity of biomass after acid treatment is possibly due to the low concentration of HNO_3 used in the study (0.1 N). At this concentration, lignin and hemicellulose could not be completely removed from the structure of biomass, so only small parts of the cellulose were exposed to the Pb^{2+} solution, resulting in the low uptake of Pb^{2+} by the biomass.

Physical as well as chemical characteristics of biosorbent are very crucial in the adsorption process. The adsorbents with high BET surface area and pore volume are more desirable than nonporous adsorbents. The adsorbents with high porosity provide a large number of active sites which increase its adsorption capacity toward heavy metals. In many cases, the treatment of lignocellulosic materials using strong or oxidizing acids did not improve or escalate BET quantity of the biomass as indicated in Table 4.3.

As explained earlier, the low BET quantity gave poor adsorption performance, as seen in Table 4.3.

Thermal treatment of lignocellulosic biomass at a high temperature significantly improves the adsorption performance of the material. The pyrolysis of

TABLE 4.2
The performance of biosorbent modified with another compound for heavy metal removal in aqueous solution

Metal ions	Adsorbent	Modifying agent	Conditions	q_{max} (mg/g)	Ref.
Cd^{2+}	Banana	NaOH	Shaker at 175 rpm at room temperature (25°C), t = 60 min	5.82	Mahmood-ul-Hassan et al., 2015
	Sunflower	HNO_3		6.74	
	Corn cob	H_2SO_4		5.21	
		NaOH		14.28	
		HNO_3		16.28	
		H_2SO_4		12.74	
		NaOH		19.86	
		HNO_3		18.46	
		H_2SO_4		13.89	
	Corn stalk	Alkali treatment (NaOH), crosslinking reaction, CS_2 and Mg^{2+} substitution reactions	T = 25°C, adsorbent dosage = 5 g/L, pH = 7.0	20.58	Zheng and Meng, 2016
	Wood apple shell	H_2SO_4	T = 26°C, pH = 6.5, t = 4 h, shaking at 150 rpm	28.33	Sartape et al., 2013
	Pineapple peel fiber	Succinic anhydride	C_o = 30–300 mg/L, pH = 7.5, t = 30 min	34.18	Hu et al., 2011
	Wheat (*Triticum aestivum*)	Urea solution under the effect of microwave radiation	pH = 6, amount of biomass = 100 mg, volume of solution = 50 mL, shaking speed = 150 rpm for 10 min	39.22	Farooq et al., 2011
	Grapefruit peel	0.1 M HCl, CH_3OH	C_o = 50 mg/L, pH = 5, t = 150 min	42.09	Torab-Moestadi et al., 2013
	Apple pomace	Succinic anhydride	C_o = 10–200 mg/L in 50 mL solution, pH = 5	91.75	Chand et al., 2014
	Moringa oleifera	Citric acid	T = 40°C, pH = 5, shaking at = 200 rpm	171.37	Reddy et al., 2012
	Maize straw	Succinic anhydride in xylene	pH = 5.8, T = 20°C, t = 1.5 h, adsorbent dose = 1 g/L	196.1	Guo et al., 2015
Cr^{3+}	Pineapple crown leaves	AcOH, H_2O_2, and HCl	T = 25°C, pH = 5	2.54	Gogoi et al., 2018
	Sugarcane pulp	Pyrolysis 500°C	pH = 5.0, shaking at 140 rpm for 24 h	3.43	Yang et al., 2013

(*Continued*)

TABLE 4.2 (Cont.)

Metal ions	Adsorbent	Modifying agent	Conditions	q_{max} (mg/g)	Ref.
Cr^{6+}	Pineapple crown leaves	AcOH, H_2O_2, and HCl	T = 25°C, pH = 2.5	3.91	Gogoi et al., 2018
	Wheat bran	Tartaric acid	C_o = 200 mg/L, pH = 2.2, T = 23°C and shaking at 120 rpm for 200 min	5.28	Kaya et al., 2014
	Banana	NaOH	Shaker at 175 rpm at room temperature (25° C), t = 60 min	13.35	Mahmood-ul-Hassan et al., 2015
	Sunflower	HNO_3		12.41	
	Corn cob	H_2SO_4		7.42	
		NaOH		20.36	
		HNO_3		14.41	
		H_2SO_4		12.26	
		NaOH		34.97	
		HNO_3		27.80	
		H_2SO_4		23.67	
	Neem leaves	NaOH 0.1 N and H_2SO_4 0.1 N	C = 25 mg/L, T = 30°C, pH = 2, t = 3 h	15.95	Singha and Das, 2011
	Coconut shell			18.70	
	Peanut shell	Activated by KOH and pyrolysis at 450 ± 5°C	T = 25°C, pH = 4, shaking at 200 rpm,	16.26	Al-Othman et al., 2012
	Eichhornia crassipes root	Thermal treatment 600° C, H_2SO_4 5N	T = 25 ± 2°C, pH = 4.5, t = 30 min	36.34	Giri et al., 2012
	Coir pith	50% H_2O_2	C_o = 1,171 mg/L, T = 30°C, pH = 2, t = 22 h	51.06	Suksabye and Thiravetyan, 2012
		2 M acrylamide-grafted		128.29	
		2 M acrylic acid-grafted		196.00	
	Rice husk	Carbonized at 800°C using ozone as an activating agent	Co = 100 mg/L, pH = 2.0, t = 2.5 h, adsorbent dosage 0.2 g	62.90	Sugashini and Begum, 2015
	Corn stalk	Fe_3O_4, $FeCl_3$, $NH_3.H_2O$ (25%), epichlorohydrin, N,N-dimethylformamide, ethylenediamine, and triethylamine	C_o = 50–600 mg/L, pH = 3, T = 45°C, t = 8 h, adsorbent dosage = 0.1 g	231.1	Song et al., 2015
Cu^{2+}	Pineapple peel fiber	Succinic anhydride	C_o = 40–300 mg/L, pH = 5.4, t = 30 min	27.68	Hu et al., 2011
	Barley straw	Citric acid	pH = 6–7, t = 120 min, T = 25 ± 1°C	31.71	Pehlivan et al., 2012
	Rice husk	3 M NaOH, carbonized at 400–650°C	Co = 400 mg/g, T = 35°C, pH = 7.0	55.2	Ye et al., 2010

(Continued)

TABLE 4.2 (Cont.)

Metal ions	Adsorbent	Modifying agent	Conditions	q_{max} (mg/g)	Ref.
	Pineapple peel Orange peel	DIC process and modification by sodium hydroxide and citric acid	$T = 25°C$	64.33 163.01	Romero-Cano et al., 2017
	Moringa oleifera	Citric acid	$T = 40°C$, pH = 5	167.90	Reddy et al., 2012
Ni^{2+}	Loquat bark	0.1 N NaOH	$C_o = 10–100$ mg/L, $T = 40°C$, pH = 6, $t = 30$ min	29.54	Salem and Awwad, 2014
	Grapefruit peel	0.1 M HCl, CH_3OH	$C_o = 50$ mg/L, pH = 5, $t = 150$ min	46.13	Torab-Moestadi et al., 2013
	Bamboo stem	Thermal treatment, H_3PO_4	$T = 25°C$, pH = 5, shaking at 200 rpm for 1 h	98.07	Rajesh et al., 2014
	Theobroma cacao (cocoa) shell	18 N H_2SO_4, Na_2CO_3 18 N H_2SO_4, Na_2CO_3, thermal treatment 350°C	pH = 6. $T = 60$ min	97.59 158.8	Kalaivani et al., 2015
	Sugarcane bagasse pith	Thermal treatment	$T = 30°C$, pH = 6, $t = 4$ h, shaking at 200 rpm	140.85	Anoop et al., 2011
	Moringa oleifera	Citric acid	$T = 40°C$, pH = 5	163.88	Reddy et al., 2012
	Caesalpinia bonducella seed	0.1 N NaOH and 0.1N H_2SO_4	$T = 50°C$, pH = 5	188.7	Gutha et al., 2011
	Rice bran	HCl H_2SO_4 $Ca(OH)_2$ NaOH	$T = 30°C$, pH = 7, shaking at 130 rpm for 24 h	324.7 327.9 338.3 400.0	Zafar et al., 2015
Pb^{2+}	Olive tree	HNO_3 H_2SO_4 NaOH	$C_o = 150$ ppm, pH = 5	14.15 15.36 16.04	Ronda et al., 2013
	Camellia oleifera	Citric acid	$T = 40°C$, pH = 5, $t = 30$ min	49.65	Guo et al., 2016
	Banana Sunflower Corn cob	NaOH HNO_3 H_2SO_4 NaOH HNO_3 H_2SO_4 NaOH HNO_3 H_2SO_4	Shaker at 175 rpm at room temperature (25°C), $t = 60$ min	59.39 39.91 36.45 39.23 27.87 23.60 56.67 51.75 33.39	Mahmood-ul-Hassan et al., 2015

(Continued)

TABLE 4.2 (Cont.)

Metal ions	Adsorbent	Modifying agent	Conditions	q_{max} (mg/g)	Ref.
	Potato starch	Citric acid Carbon disulfide	t = 2 h, shaker speed = 100 rpm at 30°C	57.60 109.10	Ma et al., 2015
	Hazelnut shell	Pyrolisis 250°C Pyrolisis 700°C Pyrolisis 700°C, $ZnCl_2$	C_o = 100 mg/L, T = 20°C, t = 1 h, and shaken at 150 rpm	62.00 116.20 151.90	Şencan et al., 2015
	Pineapple peel fiber	Succinic anhydride	C_o = 60–500 mg/L, pH = 5.6, t = 30 min	70.29	Hu et al., 2011
	Bamboo fibers	Citric acid	pH = 5.6, t = 12 h at room temperature	127.10	Wang et al., 2013
	Sugarcane bagasse	Dry sludge, KOH 3M, pyrolysis 800°C, 3M H_2SO_4, 60% HNO_3	C_o = 100 ppm, T =35°C, pH = 4, shaking at 180 rpm for 90 min, adsorbent dose of 2 g/L,	135.54	Tao et al., 2015
	Typha angustifolia	$SOCl_2$-activated EDTA	C_o = 20–600 mg/L, pH = 5.0, biomass dosage = 2.0 g/L, T =, 298°C, t = 300 min.	263.90	Liu et al., 2011
	Cashew nut shell	H_2SO_4	T = 60°C, t = 30 min	480.5	SenthilKumar et al., 2011a
Zn^{2+}	Bagasse	Carbonization process at 900°C and was immersed in a weak acid	pH = 6.0, T = 27 ± 1°C, t = 3 h	21.05	Salihi et al., 2016

sugarcane pulp residue at 500°C could increase the adsorption capacity of the biomass toward Cr^{3+} from 3.43 mg/g to 15.85 mg/g (Yang et al., 2013). During pyrolysis process at high temperature, the breakdown of lignin, hemicellulose, and cellulose occurs to form solid (biochar), gases (CO_2, CO, CH_4, H_2O), and organic liquids (tar). The thermal degradation of lignocellulosic material creates a new structure within the solid, and the solid becomes porous. The creation of new pores within the solid will increase the surface area of the resulting biochar. The increase of the surface area will enhance the uptake of Cr^{3+} by the biochar.

Tao et al. (2015) utilized sludge and sugarcane bagasse as the raw materials for the production of biomass-based active charcoal. The active charcoal was employed to remove Pb^{2+} in solution. Activation by using potassium hydroxide and thermal treatment at 800°C greatly enhanced the BET surface area of the solid biomass from 3.57 m^2/g to 806.57 m^2/g, as well as increased the adsorption capacity of the solid from few mg/g to 135.54 mg/g. The creation of porous structure (mostly micropore and some mesopore) during the pyrolysis process significantly increased the surface area of the resulting solid (called as activated biochar or activated carbon). In this process, potassium hydroxide

TABLE 4.3

The pore structure of lignocellulosic materials after surface modification using a chemical method

Lignocellulosic biomass	Chemical/ physical	Initial BET surface area, m^2/g	The adsorption capacity of pristine biomass, mg/g	BET surface area after modification, m^2/g	The adsorption capacity of modified biomass, mg/g	Ref.
Apricot stone	NaOH	15.4	4.24 (Cu^{2+})	20.0	8.99 (Cu^{2+})	Sostaric et al., 2018
Durian shell	HCl	0.6793	20.20 (Zn^{2+})	0.8807	34.84 (Zn^{2+})	Ngabura et al., 2018
Olive tree pruning	H_2SO_4	0.63	12.97 (Pb^{2+})	0.61	15.36	Ronda et al., 2013
	HNO_3			0.43	14.15	
	NaOH			3.53	16.04	
Wheat straw	Urea	8.17	4.25 (Cd^{2+})	15.21	39.22 (Cd^{2+})	Farooq et al., 2011

acted as the activating agent; it developed a microporous structure, widening the existing pores of biochar. With the existence of micropore along with mesopore in active charcoal structure, the adsorption of Pb^{2+} also occurred via van der Waals interaction (physisorption), so it increased the uptake of Pb^{2+} from solution.

The influence of heat treatment temperature to BET quantity of hazelnut shell active charcoal was studied by Şencan et al. (2015). The hazelnut shell was pyrolyzed at temperature 250 and 700°C. The BET surface area of pristine hazelnut shell was 5.92 m^2/g and increased to 270.2 m^2/g when heated at a temperature of 250°C, and increased to 686.7 m^2/g when heated at a temperature of 700°C. Depending on the type of lignocellulosic material, during the pyrolysis process, the degradation of lignocellulosic material occurs in several steps of processes. At a temperature between 100 and 200°C, free moisture content and bound water will evaporate from the matrix of hazelnut shell, degradation of hemicellulose occurs at a temperature range of 200 to 300°C, while cellulose will decompose at a temperature range of 300 to 360°C. Lignin, which possesses more stable bonds, will decompose at the temperature range of 400 to 500°C (Hartono et al., 2016). At a temperature of 250°C, the decomposition of hemicellulose of hazelnut shell occurred, leaving empty spaces (pores) in the matrices of the hazelnut shell. The creation of new pores in the structure of the heat-treated hazelnut shell will increase its BET surface area. If the temperature of the pyrolysis process increased to 700°C, the decomposition of hemicellulose, cellulose, and lignin occurred, more pores were created within the structure of biochar, and it increased the BET of the resulting carbon. The adsorbate uptake of pristine hazelnut shell, heat-treated at 250°C, and heat-treated at 700°C toward Pb^{2+} were 41.9, 62.0, and 116.2 mg of adsorbate per g of adsorbent, successively.

4.4 OTHER BIOLOGICAL MATERIALS AS BIOSORBENTS

Due to its potential application as a cheap material for purifying water from metals and sewage, microbial biomasses and algae have been intensively studied as renewable and nonconventional adsorbents. Thousands of studies have been conducted to searching new kinds of microbial biomass and algae that are capable of removing and recovering heavy metals from environment efficiently. Apart from the advantages as the alternative adsorbents, the use of microbial biomasses and algae also has many disadvantages, which cause these alternative adsorbents cannot compete with commercially available adsorbents. Therefore, more efforts are still needed to make these unconventional biosorbents applicable for industrial-scale application.

Some recent biosorption studies using microbial biomasses and algae to remove heavy metals from aqueous solutions are listed in Table 4.4.

From Table 4.4, it is obvious that several algae have potential application to be a substitute for heavy metal adsorbent because of their excellent performance in removing certain heavy metals. The ability of freshwater alga *Spirogyra hyalina* for removing several metal cations (Cd^{2+}, Pb^{2+}, As^{5+}, Hg^{2+}, Co^{2+}) was studied by Kumar and Oommen (2012). The uptake of heavy metal ions by the alga depends on the onset amount of metal cation; the highest uptake was recorded for Hg^{2+} (35.71 mg/g).

The effect of initial concentration, onset pH, as well as the biosorbent dose for the uptake of Cr^{3+} and Cr^{6+} by a genus of freshwater green algae (*Scenedesmus quadricauda*) was studied by Pakshirajan et al. (2013). The pH of the solution plays a significant role in the uptake of chromium ions by green algae. The maximum removal of chromium ions by *Scenedesmus quadricauda* was achieved at pH around 5, and the percent removal declines as pH increases. Chromium is a unique metal element; the aqueous forms of Cr^{6+} are CrO_4^{2-}, $HCrO_4^-$, and $Cr_2O_7^{2-}$. All of these chemical species of Cr are in the anion forms. At low pH value, the protonation of some surface functional groups in green algae occurs and the surface of green algae becomes positively charged; whereas, at high pH value, the deprotonation occurs and the surface of the green algae becomes negatively charged. Because all of the forms of Cr^{6+} in the solution are anions, at high pH, the repulsive force between the anion and negatively charged surface occurred, which reduced the uptake of chromium ions. The adsorption capacity of green algae *Scenedesmus quadricauda* toward chromium ions was 12 mg/g. Wang et al. (2008) found that the removal of Cr^{6+} from the solution using marine macroalgae (*Laminaria japonica* and *Porphyra yezoensis Ueda*) was favorable at the acidic condition.

The effect of the solution acidity in removing Pb^{2+} by a macroalgae *Caulerpa fastigiata* was investigated at acidity scale of 2–8 (Sarada et al., 2014). Deflation of acidity from 2 to 5 increased the removal of Pb^{2+}; further increase in the pH of the solution significantly decreased the removal of Pb^{2+} by

TABLE 4.4
Heavy metal removal using different types of other biosorbents

Type	Name of species	Metal ions	Conditions	q_{max} (mg/g)	Ref.
Algae	*Spirogyra hyalina*	As^{5+} Co^{2+} Cd^{2+} Pb^{2+} Hg^{2+}	At room temperature for 120 min, shaking at 120 rpm	4.81 12.82 18.18 31.25 35.71	Kumar and Oommen, 2012
	Scenedesmus quadricauda	Cr^{3+}	pH = 5, t = 2 h.	12	Pakshirajan et al., 2013
	Caulerpa fastigiata	Pb^{2+}	T = 40°C, pH = 5	17.99	Sarada et al., 2014
	Green algae (*Ulva lactuca*)	Cd^{2+} Pb^{2+}	T = 20°C, pH = 5, t = 60 min	29.2 34.7	Sari and Tuzen, 2008
	Oedogonium hatei	Ni^{2+}	T = 25°C, pH = 5, t = 80 min	40.9	Gupta et al., 2010
	Gracilaria caudata *Sargassum muticum*	Ni^{2+}	T = 20°C, pH = 5 (for Gc) and 3 (for Sm)	50.10 75.60	Bermúdez et al., 2011
	P. yezoensis Ueda	Cr^{6+}	T = 50°C, pH = 1, t = 12 h	95.81	Wang et al., 2008
	Red algae: *Jania rubens* *Pterocladiella capillacea* *Corallina mediterranea* *Galaxaura oblongata* *Jania rubens* *Pterocladiella capillacea* *Corallina mediterranea* *Galaxaura oblongata* *Jania rubens* *Pterocladiella capillacea* *Corallina mediterranea* *Galaxaura oblongata*	Cd^{2+} Cr^{3+} Pb^{2+}	T = 25°C, pH = 5, t = 60 min, biomass dosage = 10 g/L	 30.5 33.5 64.1 85.5 28.5 34.7 70.3 105.2 30.6 34.1 64.3 88.6	Ibrahim, 2011
Bacteria	*Kocuria rhizophila*	Cd^{2+} Cr^{3+}	T = 35 ± 2°C, pH = 8 (for Cd^{2+}) and 4 (for Cr^{3+}), t = 1 h	9.07 14.40	Haq et al., 2016
	Halomonas sp.	Cd^{2+}	pH = 11.5 t = 60 min, at room temperature	12.02	Rajesh et al., 2014
	Pseudomonas putida CZ1	Cu^{2+} Zn^{2+}	C_o = 63.5 (for Cu^{2+}) and 65.3 ppm (for Zn^{2+}), T = 30°C, pH = 5	29.9 27.4	Chen et al., 2005
	Bacillus megaterium *Bacillus circulans*	Cr^{6+}	T = 28 ± 3°C, pH = 4.0, t = 1 h, shaking at 150 rpm	32.0 34.5	Srinath et al., 2002

(Continued)

TABLE 4.4 (Cont.)

Type	Name of species	Metal ions	Conditions	q_{max} (mg/g)	Ref.
	Nostoc commune (cyanobacterium)	Zn^{2+} Cd^{2+}	T = 50°C, pH = 6, t = 30 min	115.41 126.32	Morsy et al., 2011
	Pseudomonas putida	Cu^{2+} Pb^{2+}	T = 30°C, pH = 5.5, shaking at 150 rpm for 24 h	96.85 159.96	Uslu and Tanyol, 2006
	Klebsiella sp.	Cd^{2+}	T = 30°C, pH = 5	170.4	Hou et al., 2015
	Acinetobacter haemolyticus	Cr^{3+}	C_o = 100 mg/g, pH = 5.0, t = 30 min, adsorbent dosage = 15 mg cell dry weight	198.80	Yahya et al., 2012
Fungi	*Ganoderma lucidum*	Cr^{3+}	T = 25°C, pH = 4.5, t = 5 h	2.16	Shoaib, 2012
	Aspergillus flavus	Cu^{2+} Pb^{2+}	T = 25 ± 1°C, pH = 5.0, t = 2 h	10.82 13.46	Akar and Tunali, 2006
	Mucor hiemalis	Ni^{2+}	C_o = 50 ppm, pH = 8.0, t = 3 h	21.49	Shroff and Vaidya, 2011
	Agaricus bisporus	Pb^{2+}	T = 25°C, pH = 5.5	86.4	Long et al., 2014
	Yarrowia lipolytica	Ni^{2+}	T = 35°C, pH = 7.5, t = 2 h	95.33	Shinde et al., 2012
	Trametes versicolor	Cd^{2+} Pb^{2+}	T = 50°C, pH = 5.0, t = 210 min, shaking at 120 rpm	166.6 208.3	Subbaiah et al., 2011
Yeast	Dried yeast *Saccharomyces* biomass	Cu^{2+}	C_o = 25–200 ppm, T = 50°C, pH = 4.0, t = 24 h	2.59	Cojocaru et al., 2009
	Brewer's yeast	Ni^{2+} Cd^{2+}	T = 25°C, pH = 6.0, shaking at 160 rpm	5.34 10.17	Cui et al., 2010
	Baker's yeast	Ni^{2+}	C_o = 10–400 ppm, T = 27°C, pH = 6.75	9.01	Padmavathy, 2008

macroalgae *Caulerpa fastigiata*. In this case, an interesting phenomenon is observed. As mentioned by Sarada et al. (2014) that macroalgae *Caulerpa fastigiata* contains an abundant amount of carboxyl groups. Carboxyl groups found in biological substances is an acidic functional group. At low pH, the protonation of carboxyl groups causes the surface of the biosorbent to be a net positive charge, while the deprotonation at high pH causes the surface to become negatively charged. In this case, the uptake of Pb^{2+} should be higher at high pH as a result of differences in charge of adsorber with positive ions of Pb. Furthermore, the formation of $Pb(OH)_2$ at the basic condition will enhance the removal of Pb^{2+} from the solution. In this situation, the carboxyl groups did not play an important role in the removal of Pb^{2+}, possibly other biosorption mechanisms also took place, since the biosorption process involves very complex adsorption mechanisms.

Bacteria are also one of the unconventional adsorbents, which widely studied as the alternative adsorbents to remove metal cations. Uptake of Cd^{2+} and Cr^{3+} from water using endophytic bacterium *Kocuria rhizophila* was studied by Haq et al. (2016). Similar to other biosorption studies, the system acidity strongly influences the withdrawal of heavy metals by *Kocuria rhizophila*. The best pH of the solution for the removal of Cd^{2+} was 8, while for Cr^{3+}, it was 4. The adsorption capacity of Cr^{2+} and Cd^{3+} by *Kocuria rhizophila* was 14.4 and 9.07 mg of the metal per g adsorber, successively. The surface functional groups that may be responsible for the biosorption of both metals were C=N, OH, CH_2, C=O, N–H, PO_2, C–O, C–H, and C–O–C.

For biosorption purpose, the bacteria will be used, as the biosorbent should be isolated from the contaminated environment to obtain the maximum uptake of the contaminant by bacteria. The isolation of the bacteria (*Klebsiella species*) from a wastewater treatment facility for biosorption of Cd^{2+} and Mn^{2+} was performed by Hou et al. (2015). From the experimental results, they found that *Klebsiella species* that isolated from the contaminated environment can persist at high concentration of heavy metal. The high amount of heavy metals adsorbed by the bacteria was also observed. The biosorption capacity of *Klebsiella species* for Cd^{2+} and Mn^{2+} was 170.4 and 114.1 mg/g, respectively. Biosorption mechanisms were controlled by the binding process between heavy metals and active sites of *Klebsiella species*. Biosorption process parameters such as pH, temperature, initial concentration, etc., gave a significant influence on the metal uptake by bacteria.

Several fungi are used as an alternative biosorbent of metals in aqueous solution. The presence of various functional groups on the cell walls of the fungi could increase the uptake of heavy metals. The fungus *Trametes versicolor* has been employed as a good biosorbent in removing Pb^{2+} as well as Cd^{2+} in water (Subbaiah et al., 2011). The lower biosorption of metal cations was observed during the acidic condition. At acidic condition, fungus surface was positively charged, so the charge refusal of biosorbent against the metal cations retarded the biosorption. At higher pH value, the deprotonation of the surface of fungus occurred, and the surface becomes a less positive charge. Hence biosorption increased. Maximum uptake of Pb^{2+} and Cd^{2+} was achieved at pH 5. The examination of functional sites suggests that the surface sites assigned to uptake Pb^{2+} and Cd^{2+} were hydroxyl, amine, and carboxyl.

4.5 ADSORPTION ISOTHERM AND KINETIC STUDIES

For the design of the adsorption process in commercial scale, the relevant information about how fast the equilibrium condition can be achieved and how much amount of the pollutant can be removed from the contaminated water or wastewater is required. The main constraints for obtaining valid adsorption data that cover various concentrations and temperature for isotherm and kinetic are tedious, time-consuming, and expensive. To overcome this problem, various isotherm and adsorption kinetics models applicable to various adsorption processes have been developed and tried in diverse processes. Some proposed systems are purely empirical with two or more adjustable parameters, while others possess a theoretical background. Several adsorption isotherm equations, which are widely used to represent the biosorption of heavy metals, are listed in Table 4.5.

TABLE 4.5
Available adsorption isotherm models for biosorption of heavy metals

Isotherm	Mathematical form	Parameters and its physical meaning
Langmuir	$q_e = q_{\max}\frac{K_L C_e}{1+K_L C_e}$	q_{max} (mg/g) represents the adsorption capacity of adsorbent, while K_L (L/mg) is adsorption affinity of Langmuir equation.
Freundlich	$q_e = K_F C_e^{1/n}$	Parameter K_F ((mg/g)(L/mg)$^{-n}$) represents Freundlich parameter; n represents the heterogeneity of the system adsorption capacity.
Dubinin-Radushkevich (DR)	$q_e = q_{\max}\exp\left(-\left(\frac{RT\ln(C_s/C_e)}{\beta E_o}\right)^2\right)$	q_{max} (mg/g) represents the adsorption capacity of adsorbent; the parameter E_o represents the characteristic energy of the biosorbent toward a reference adsorbate (benzene usually chosen as the reference). The parameter β is proportional to the molar volume of the adsorbate.
Temkin	$q_e = \frac{RT}{b}\ln(aC_e)$	The parameters a (L/mg) and b (g.kJ/mol.mg) are specific to the adsorption system (adsorbate – biosorbent).
Sips	$q_e = q_{\max}\frac{(K_s C_e)^{1/n}}{1+(K_s C_e)^{1/n}}$	q_{max} (mg/g) represents the adsorption capacity of adsorbent; K_S (L/mg) is adsorption affinity of Sips equation, and parameter n represents the heterogeneity of the system.
Toth	$q_e = q_{\max}\frac{K_T C_e}{\left(1+(K_T C_e)^t\right)^{1/t}}$	q_{max} (mg/g) represents the adsorption capacity of adsorbent; K_T (L/mg) is adsorption affinity of Toth equation, and parameter t represents the heterogeneity of the system.
Redlich – Peterson	$q_e = \frac{K_{RP} C_e}{1+\beta_{RP} C_e^{n_{RP}}}$	K_{RP} (L/mg)(mg/g) is Redlich – Peterson constant, which represents a combination between adsorption affinity and capacity of the adsorbent. Parameter n_{RP} represents the heterogeneity of the system, and β_{RP} (L/g)^n is a constant.

Langmuir, Freundlich, Dubinin-Radushkevich, and Temkin are two-parameter adsorption models. Sips, Toth, and Redlich-Peterson are three-parameter adsorption equations. Among the adsorption isotherm models listed in Table 4.5, only Langmuir model possesses a theoretical background.

Langmuir isotherm was proposed under several basic prejudgements: solid has a homogeneous surface, the adsorption localization of solute onto the surface of solid, and active group for the specific solute molecule. The advantage of using this equation is: it is valid over a wide range of concentration. Langmuir holds a restricted surfeit border at the high solute amount, and it obeys the law of Henry in a limited amount of solute.

Parameter q_{max} represents the highest uptake of solute (also called as removal/adsorption capacity of biosorbent). One of the important features of the Langmuir equation is the adsorption affinity given by the parameter K_L. It measures the strength of the attachment or attraction between the molecules of solute toward biosorbent superficies. Small K_L is an indication that the solute will fully cover the surface of biosorbent at high concentration. The high value of K_L indicates that the interaction between solute and biosorbent is very strong, and the maximum uptake of solute by the biosorbent will be obtained at low equilibrium concentration. The influence of the values of K_L on the uptake of solute by biosorbent is given in Figure 4.3.

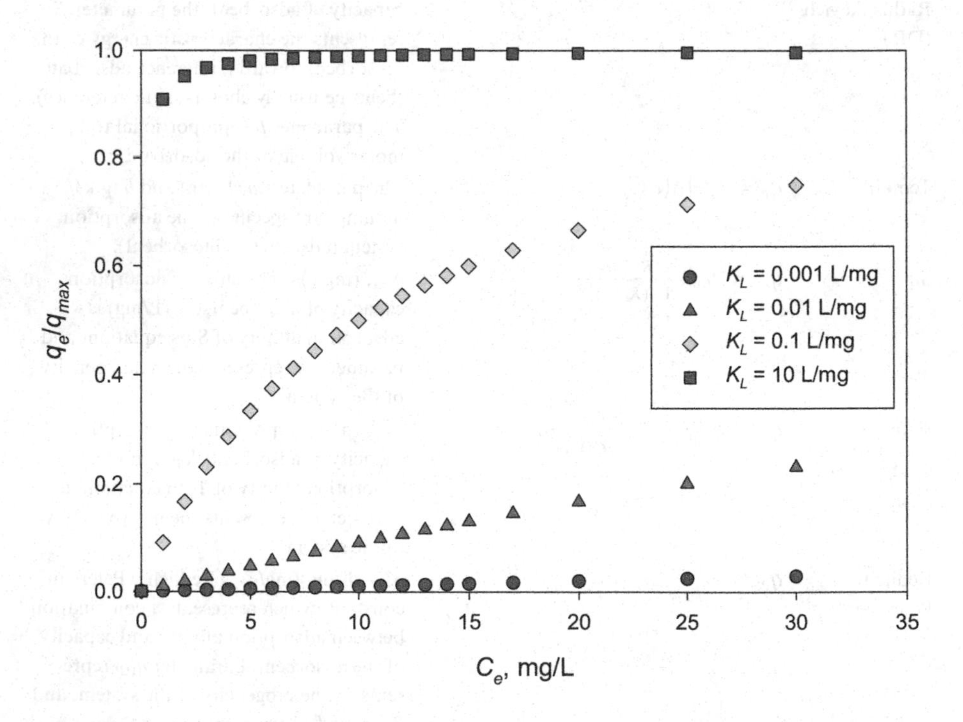

FIGURE 4.3 The influence of adsorption affinity on surface coverage.

Freundlich isotherm is the earliest empirical equation to correlate adsorption equilibrium data on the heterogeneous surface. Unlike Langmuir, the value of K_F cannot truly represent the performance of bioadsorbent because of the absence of finite saturation limit. The absence of Henry law in the model makes this equation only superior in the moderate range of concentration. The advantage of using Freundlich isotherm to correlate the experimental data is: it can extract the heterogeneity of the adsorption system through the parameter n. For most adsorption systems, the magnitude of n is 1 to 10. The system is called as homogeneous (linear) when n is equal to 1. When n diverges away from 1, adsorption occurs heterogeneously and becomes irreversible if n is approaching 10 (rectangular isothermic).

Dubinin-Radushkevich (DR) equation develops for the uptake of solute onto microporous solid (based on micropore filling theory). Although it was proposed originally for microporous solid, this equation is also often used to describe the experimental adsorption data for heavy metal–biosorbent systems (Sari and Tuzen, 2008; Farooq et al., 2011; Krishnan et al., 2011; Reddy et al., 2011; Ding et al., 2012; Pehlivan et al., 2012; Reddy et al., 2012; Senthilkumar et al., 2012; Yahya et al., 2012; Flores-Garnica et al., 2013; Ghasemi et al., 2014; Manasi et al., 2014; Rao et al., 2016; Zheng and Meng, 2016). Since the biomass materials are not microporous solids, typically, the DR equation cannot represent the data well as expected. In case this model could describe the adsorption isotherm pretty well, the characteristic curve should be given as well to confirm the validity of the DR model.

The physical meanings of the parameters of Temkin model are not well understood. Therefore, it is difficult to judge the validity of the model based on its physical phenomena. The easiest way to confirm the validity of Temkin isotherm in representing the experimental adsorption data is based on the value of R^2. However, it must be remembered before utilizing this equation. Temkin model was developed for the removal of hydrogen by using Pt at low alkaline condition (Do, 1998). Temkin fits the experimental points in a good manner when chemisorption mechanism governs the adsorption process (Hossain et al., 2012; Shafique et al., 2012; Reddy et al., 2012; Flores-Garnica et al., 2013). However, in general, for biosorption systems, Temkin model failed to represent the experimental data correctly (Gupta et al., 2010; Reddy et al., 2011; Ghasemi et al., 2014; Kalaivani et al., 2015; Song et al., 2015; Rao et al., 2016; Zheng and Meng, 2016; Basu et al., 2017). This equation has a similar limitation with the Freundlich equation; it does not have correct Henry law for low concentration and finite saturation limit for high concentration.

Sips equation, which is also known as Langmuir–Freundlich isotherm, is an empirical equation. This mathematical expression is used to define biosorption of several heavy metals over some biosorbents (Cojocaru et al., 2009; González Bermúdez et al., 2011; Witek-Krowiak et al., 2011; Hossain et al., 2012; Flores-Garnica et al., 2013; Pakshirajan et al., 2013). This equation was developed to fix the problem in the Freundlich model, especially at a high concentration of solute. However, the lack of correct Henry law in Sips equation still creates the problem at very low concentration. Quantity n denotes system heterogeneity. The system is homogeneous when $n = 1$, the Sips equation turns to Langmuir isotherm. The behavior of the system heterogeneity on the surface coverage is given in Figure 4.4.

It is obvious that the heterogeneity of the system gives a negative impact on adsorption performance. The increase of heterogeneity of the system will decrease the surface coverage of the adsorbent. The heterogeneity of the adsorption system is from both the adsorbent and the adsorbate. For the biosorption system, the heterogeneity of biosorbent is mostly from the surface functional groups and some small part from the structure of the biomass. The more heterogeneous of the biosorbent, the amount of surface functional groups responsible for the adsorption process will also increase. With the increase of surface functional groups, the uptake of solute molecules by the biosorbent should increase, and it will increase the surface coverage of the biosorbent. However, the theoretical results depicted in Figure 4.4 are contradicting with this phenomenon. In this case, the reader should be more careful to use the Sips equation to determine the system heterogeneity.

Parameter K_S in Sips represents spontaneity manner of adsorption. A theoretical representation of parameter K_S on the surface coverage is depicted in Figure 4.5.

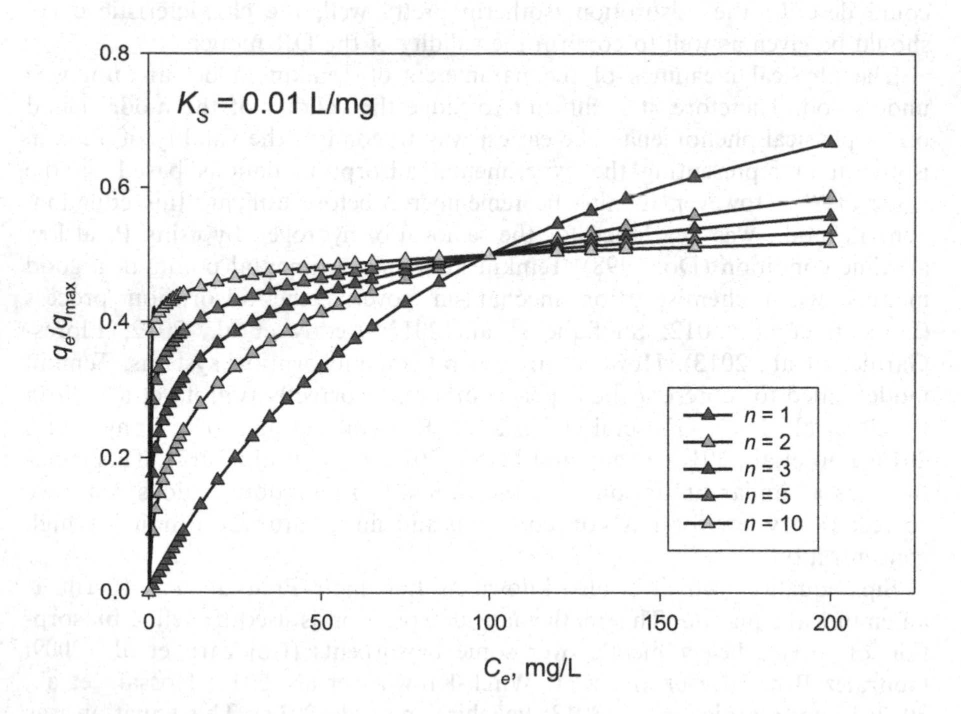

FIGURE 4.4 The influence of system heterogeneity of Sips model on the surface coverage.

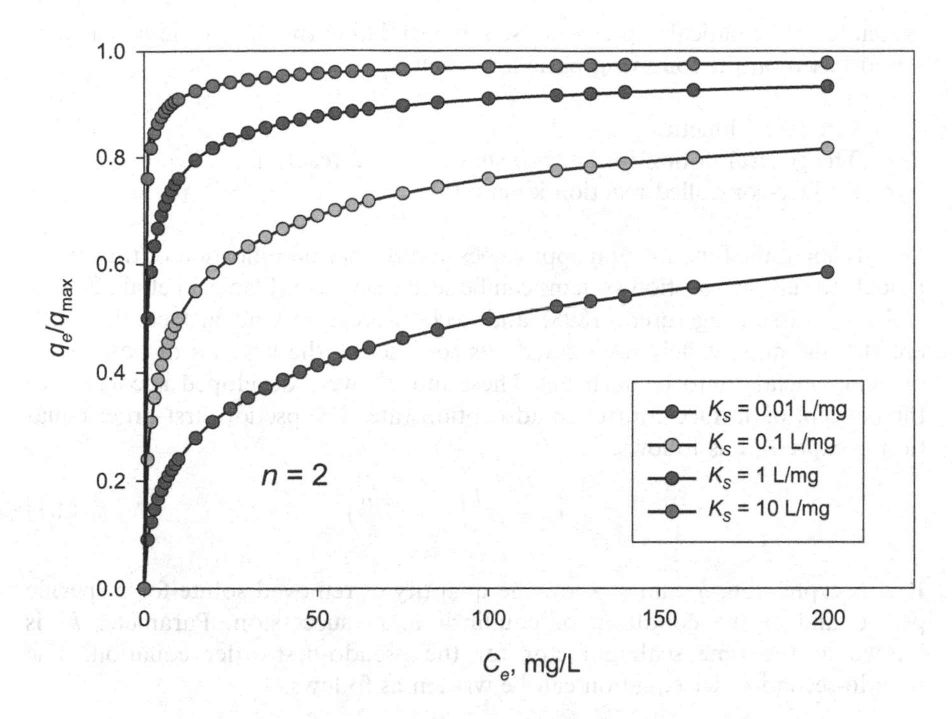

FIGURE 4.5 The influence of K_S on the surface coverage of an adsorbent.

With the increase of the adsorption affinity, the attraction of solute molecules onto the adsorbent material also becomes stronger, and the adsorbate molecules become easily adsorbed. This phenomenon supports the uptake of solute molecules, leading to the increase of surface coverage as seen in Figure 4.5.

Toth equation is another empirical equation which is also utilized for correlating the data (Hossain et al., 2012; Flores-Garnica et al., 2013). Unlike Freundlich and Sips equations, Toth isotherm possesses the correct Henry's proposition as well as finite surfeit boundary. Even Toth equation is more superior than the Freundlich or Sips model; this equation is rarely used for modeling the metal–biosorbent systems. The other three-parameter adsorption isotherm model popularly utilized to correlate the empiric and theoretic data of heavy metal–biosorbent systems is Redlich–Peterson (Padmavathy, 2008; Cojocaru et al., 2009; Witek-Krowiak et al., 2011; Hossain et al., 2012; Flores-Garnica et al., 2013; Kalaivani et al., 2015; Zheng and Meng, 2016). As the empirical model, the Redlich–Peterson equation was proposed to combine the feature of both Freundlich and Langmuir isotherm models. It has a correct Henry's law at low concentration, but it does not have a finite saturation limit at high concentration.

The ability to determine the rate in the adsorption is very important in designing the adsorption system. According to Plazinski et al. (2009), the available mathematical expression used in correlating the data usually was proposed based on the following basic approaches:

- Interfacial kinetics
- The general notion for surface sites chemical reaction
- Surface-controlled reaction kinetic rate

Details about the fundamental approaches and further modification of the kinetic models to suit the sorption systems can be seen elsewhere (Plazinski et al., 2009).

Pseudo-first (Lagergren, 1898) and second-order (Blanchard et al., 1984) are still the most widely used equations to describe the kinetics of adsorption of heavy metals onto biosorbents. These models were developed according to the concept of surface-controlled adsorption rate. The pseudo-first-order equation is expressed as follows:

$$q_t = q_e\left(1 - e^{(-k_1 t)}\right) \tag{4.1}$$

In this expression, q_t and q_e show the quantity of removed solute for a specific period and at the condition of equilibrium, in succession. Parameter k_1 is known as the time scaling factor for the pseudo-first-order equation. The pseudo-second-order equation can be written as follows:

$$q_t = \frac{q_e^2 k_2 t}{1 + q_e k_2 t} \tag{4.2}$$

In equation (4.2), the parameter k_2 poses as a time scale magnitude for the pseudo-second-order equation; it indicates the biosorption pace toward an equilibria condition. Theoretically, the higher the value of k, the more rapid the system gains equilibrium, as seen in Figure 4.6. Both k_1 and k_2 strongly depend on the onset quantity of solute. Usually, a higher initial solute concentration will give the lower value of those time scaling factors (Plazinski et al., 2009; Senthilkumar et al., 2012; Rao et al., 2016; Wen et al., 2018).

Depending on the type of adsorption (physisorption or chemisorption), the temperature also gives a significant effect on the value of k_1 and k_2. In the case of a chemisorption-controlled process, the rise in heat energy causes a synergistic impact for k_2 (Awwad and Salem, 2014; Ngabura et al., 2018). If the physisorption controls the mechanism of biosorption, the value of k_1 declines at higher heat energy.

Other kinetics models which have been used to represent the adsorption kinetic data heavy metals onto biosorbents are as follows:

- Elovich (Basha and Murthy, 2007; Farooq et al., 2011; Singha and Das, 2011; Witek-Krowiak et al., 2011; Hossain et al., 2012; Flores-Garnica et al., 2013; Ronda et al., 2013; Kalaivani et al., 2015)
- Fractional power (Basha and Murthy, 2007; Flores-Garnica et al., 2013)

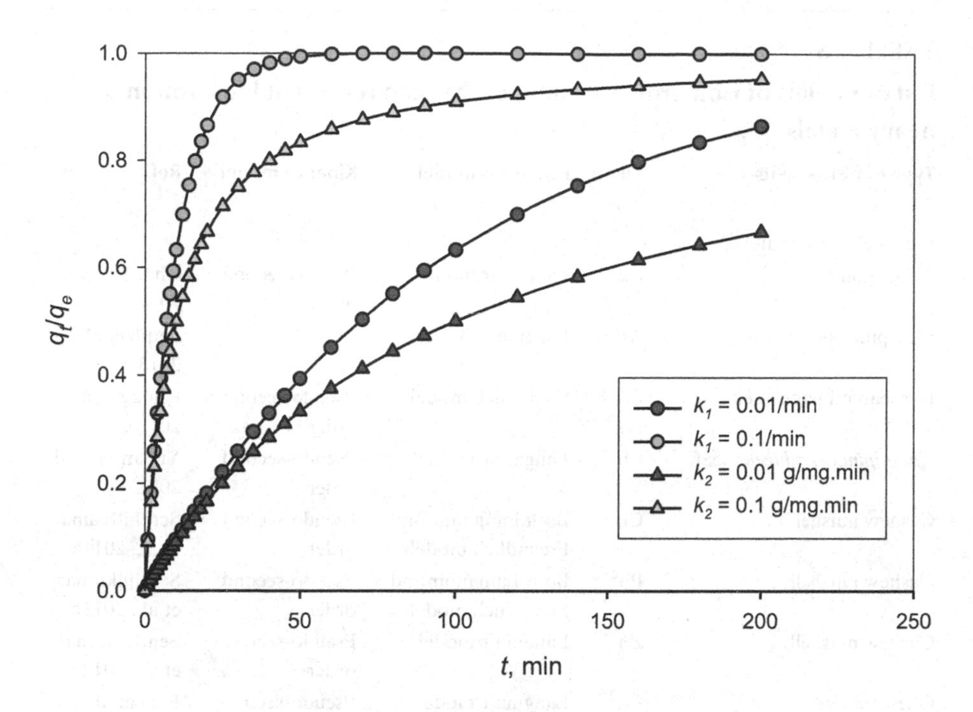

FIGURE 4.6 Effect of time scaling factor of pseudo-first-order and pseudo-second-order equations on the surface coverage of biosorbent.

- Weber and Morris intraparticle diffusion model (Farooq et al., 2011; Gutha et al., 2011; Reddy et al., 2011; Shroff and Vaidya, 2011; Singha and Das, 2011; Subbaiah et al., 2011; Ding et al., 2012; Reddy et al., 2012; Yuvaraja et al., 2012; Pakshirajan et al., 2013; Ronda et al., 2013; Sartape et al., 2013; Ghasemi et al., 2014; Kalaivani et al., 2015; Rajesh et al., 2014; Guo et al., 2015; Ma et al., 2015; Zhou et al., 2015; Haq et al., 2016; Basu et al., 2017)
- Pseudo-nth-order equation (Ronda et al., 2013)
- Double exponential equation (Ronda et al., 2013)
- Reichenberg model (Singha and Das, 2011)
- Boyd kinetic model (Kalaivani et al., 2015)
- Avrami model (Hossain et al., 2012)

For easy reference, the prospective adsorption isotherm and kinetic models are summarized in Table 4.6. Langmuir adsorption isotherm and pseudo-second-order kinetic model can describe well the equilibrium and kinetic data for most heavy metal–biosorbent systems as indicated in Table 4.6.

As referred, Langmuir is based on three prejudgements; also the adsorption process took place in monolayer. Because chelation and ion exchange process control the main mechanism of metal cations removal from water, the concept of one adsorption site only accommodate one molecule of solute is suitable

TABLE 4.6
List of models of isotherms and kinetics that can represent biosorption of heavy metals

Type of biosorbents	Metal ions	Isotherms model	Kinetics model	Ref.
Lignocellulosic materials:				
Apple pomace	Cd^{2+}	Langmuir model	Pseudo-second order	Chand et al., 2014
Blue pine and walnut	As^{5+}	Langmuir model	-	Saqib et al., 2013
Bornean oil palm shell	Cu^{2+} Pb^{2+}	Freundlich model	Pseudo-second order	Chong et al., 2013
Caesalpinia bonducella leaf	Cu^{2+}	Langmuir model	Pseudo-second order	Yuvaraja et al., 2012
Cashew nutshell	Cu^{2+}	Both Langmuir and Freundlich models	Pseudo-second order	SenthilKumar et al., 2011a
Cashew nutshell	Pb^{2+}	Both Langmuir and Freundlich models	Pseudo-second order	Senthilkumar et al., 2011b
Cashew nutshell	Zn^{2+}	Langmuir model	Pseudo-second order	Senthilkumar et al., 2012
Chestnut bur	Cd^{2+} Pb^{2+}	Langmuir model	Pseudo-second order	Kim et al., 2015
Cicer arietinum husk	Cd^{2+} Cu^{2+}	Langmuir model	Pseudo-second order	Pandey, 2016
Cucumber peel	Pb^{2+}	Langmuir model	Pseudo-second order	Basu et al., 2017
Corn silk	Pb^{2+}	Freundlich model	Pseudo-second order	Petrović et al., 2016
Durian shells waste	Cr^{6+}	Both Langmuir and Freundlich models	Pseudo-first order	Kurniawan et al., 2011
Ficus carica	Cr^{6+}	Langmuir model	Pseudo-second order	Gupta et al., 2013
Litchi chinensis seeds	Ni^{2+}	Langmuir model	Pseudo-second order	Flores-Garnica et al., 2013
Lycopersicum esculentum (tomato)	Ni^{2+}	Langmuir model	Pseudo-second order	Gutha et al., 2015
Mangrove leaves	Cr^{6+}	Langmuir model	Pseudo-second order	Sathish et al., 2015
Moringa pods	Cu^{2+} Ni^{2+} Cr^{6+} Zn^{2+}	Langmuir model	Pseudo-second order	Matouq et al., 2015
Peganum harmala-L	Ni^{2+}	Langmuir model	Pseudo-second order	Ghasemi et al., 2014

(*Continued*)

TABLE 4.6 (Cont.)

Type of biosorbents	Metal ions	Isotherms model	Kinetics model	Ref.
Pine cone shell	Cu^{2+}	Langmuir model	Pseudo-second order	Tenorio et al., 2012
Pine leaves	As^{5+}	Langmuir and Flory–Huggins models	Pseudo-second order	Shafique et al., 2012
Pineapple stem	Ni^{2+}	Freundlich model	Pseudo-second order	Rajesh et al., 2014
Peanut shell	Cr^{3+} Cu^{2+}	Langmuir model	Pseudo-second order	Witek-Krowiak et al., 2011
Portulaca oleracea	Cr^{6+}	Langmuir model	Pseudo-second order	Mishra et al., 2015
Portulaca oleracea dried plant	Cd^{2+}	Langmuir model	Pseudo-second order	Dubey et al., 2014
Rice bran	Ni^{2+}	Langmuir model	Pseudo-second order	Zafar et al., 2015
Spent grains	Cr^{3+}	Langmuir model	Pseudo-second order	Ferraz et al., 2015
Sugarcane bagasse	Ni^{2+}	Langmuir model	Pseudo-second order	Alomá et al., 2012
Terminalia arjuna fruit powder	Pb^{2+}	Freundlich model	Pseudo-second order	Rao et al., 2016
Lignocellulose modified material:				
Acid-bagasse–sludge-based adsorbent–KOH	Pb^{2+}	Langmuir model	Pseudo-second order	Tao et al., 2015
Amine-functionalized magnetic corn stalk composites	Cr^{6+}	Langmuir model	Pseudo-second order	Song et al., 2015
Apple pomace modified with succinic anhydride	Cd^{2+}	Langmuir model	Pseudo-second order	Chand et al., 2014
Bamboo stem activated charcoal	Ni^{2+}	Freundlich model	Pseudo-second order	Rajesh et al., 2014
Caesalpinia bonducella Seed modified with alkali and acid solution	Ni^{2+}	Langmuir model	Pseudo-second order	Gutha et al., 2011
Cactus fibers (activated biochar)	Cu^{2+}	Langmuir model	Pseudo-first order	Hadjittofi et al., 2014
Cashew nut shell modified with sulfuric acid	Pb^{2+}	Freundlich model	Pseudo-second order	SenthilKumar et al., 2011a
Camellia oleifera modified with citric acid	Pb^{2+}	Langmuir model	Pseudo-second order	Guo et al., 2016
Cornstalk xanthates	Cd^{2+}	Langmuir model	Pseudo-second order	Guo et al., 2016

(*Continued*)

TABLE 4.6 (Cont.)

Type of biosorbents	Metal ions	Isotherms model	Kinetics model	Ref.
Grapefruit peel modified with chloride acid	Cd^{2+} Ni^{2+}	Freundlich model	Pseudo-second order	Torab-Moestadi et al., 2013
Loquat bark modified with sodium hydroxide	Ni^{2+}	Langmuir and Temkin isotherm model	Pseudo-second order	Salem and Awwad, 2014
Rice husk biochar	Cd^{2+} Cu^{2+} Pb^{2+} Zn^{2+}	Langmuir model	-	Xu et al., 2013
Sugarcane bagasse pith	Ni^{2+}	Langmuir model	Pseudo-second order	Anoop-Krishnan et al., 2011
Maize straw modified with succinic anhydride in xylene	Cd^{2+}	Langmuir model	Pseudo-first order	Guo et al., 2015
Moringa oleifera leaves powder modified with citric acid	Cd^{2+} Cu^{2+} Ni^{2+}	Langmuir model	Pseudo-second order	Reddy et al., 2012
Pineapple crown leaves modified with AcOH, H_2O_2, and HCl	Cr^{3+} Cr^{6+}	Liu model	Pseudo-second order	Gogoi et al., 2018
Pineapple peel fiber modified with succinic anhydride	Cd^{2+} Cu^{2+} Pb^{2+}	Langmuir model	Pseudo-second order	Hu et al., 2011
Sugarcane pulp biochar	Cr^{3+}	Langmuir model	Pseudo-second order	Yang et al., 2013
Theobroma cacao (cocoa) shell modified with sulfuric acid and thermal treatment	Ni^{2+}	Langmuir model	Pseudo-second order	Kalaivani et al., 2015
Wood apple shell	Cd^{2+}	Langmuir model	Pseudo-second order	Sartape et al., 2013
Other biosorbent:				
Acinetobacter haemolyticus	Cr^{3+}	Freundlich model	Pseudo-second order	Yahya et al., 2012
Baker's yeast	Ni^{2+}	Redlich – Peterson	Pseudo-second order	Padmavathy, 2008
Caulerpa fastigiata	Pb^{2+}	Freundlich model	Pseudo-second order	Sarada et al., 2014
Dried yeast *Saccharomyces* biomass	Cu^{2+}	Langmuir model	-	Cojocaru et al., 2009
Klebsiella sp.	Cd^{2+}	Langmuir model	-	Hou et al., 2015
Kocuria rhizophila	Cd^{2+} Cr^{3+}	Langmuir model	Pseudo-second order	Haq et al., 2016

(*Continued*)

TABLE 4.6 (Cont.)

Type of biosorbents	Metal ions	Isotherms model	Kinetics model	Ref.
Halomonas sp.	Cd^{2+}	Langmuir model	Pseudo-second order	Rajesh et al., 2014
Mucor hiemalis	Ni^{2+}	Both Langmuir and Freundlich models	Pseudo-first order	Shroff and Vaidya, 2011
Oedogonium hatei	Ni^{2+}	Freundlich model	Pseudo-second order	Gupta et al., 2010
P. yezoensis Ueda	Cr^{6+}	Freundlich model	Pseudo-second order	Wang et al., 2008
Pseudomonas putida CZ1	Cu^{2+} Zn^{2+}	Langmuir model	-	Chen et al., 2005
Pseudomonas putida	Cu^{2+} Pb^{2+}	Langmuir model	-	Uslu and Tanyol, 2006
Trametes versicolor	Cd^{2+} Pb^{2+}	Langmuir model	Pseudo-second order	Subbaiah et al., 2011
Undaria pinnatifida	Cd^{2+} Zn^{2+}	Langmuir model	Pseudo-second order	Plaza-Cazón et al., 2013

with the mechanism of biosorption of heavy metal. Therefore, this model shows preferable data fitness than other isotherm equations.

In most cases, the pseudo-second-order is more superior than the pseudo-first-order equation. The actual value of amount adsorbed at equilibrium condition (q_e) also can be predicted very well by this equation. The superiority of the pseudo-second-order equation is probably due to its insensitivity for the effect of the experimental error randomly (Plazinski et al., 2009). Furthermore, to acquire the parameters of pseudo-first-order equation from the experimental data using its linear form is more sensitive to the influence of the experimental error as mentioned by Plazinski et al. (2009).

4.6 THERMODYNAMIC STUDIES

The thermodynamic properties of the biosorption system are one of the fundamental requirement for determining the feasibility of the process and the design of good biosorption system. The thermodynamic properties of the biosorption system include enthalpy-change (ΔH°), entropy-change (ΔS°), and Gibb's energy change (ΔG°). All of these thermodynamic properties can easily be determined from the biosorption isotherm at various temperatures.

ΔG° measures spontaneous manner of biosorption. This thermodynamic property can be determined using the following equation:

$$\Delta G^o = -RT \ln K_D \tag{4.3}$$

where R is a gas constant, T is the temperature of the biosorption process, and K_D is a linear distribution coefficient of the biosorption process. The value of K_D represents the ratio between the equilibrium concentration of the solute on the surface of biosorbent with the equilibrium concentration of the solute in the solution (q_e/C_e). The value of K_D that represents the whole adsorption isotherm can be obtained through the plotting of C_e as the x-axis and $\ln(q_e/C_e)$ as the y-axis; a straight line is drawn in the plot and extrapolated it to zero C_e (Khan and Singh, 1987) as shown in Figure 4.7.

Classical van't Hoff equation can be employed to calculate other thermodynamic properties of biosorption system (ΔH° and ΔS°):

$$\ln K_D = \frac{\Delta S^o}{R} - \frac{\Delta H^o}{RT} \tag{4.4}$$

The nature of biosorption system can be determined based on the values of ΔG°, ΔH°, and ΔS°. If the value of ΔG° is negative, it indicates that biosorption is spontaneous and thermodynamically feasible (Padmavathy, 2008; Sari and Tuzen, 2008; Farooq et al., 2011; Gutha et al., 2011; Reddy et al., 2011; Singha and Das, 2011; Subbaiah et al., 2011; Al-Othman et al., 2012; Senthilkumar et al., 2012; Suksabye and Thiravetyan, 2012; Torab-Moestadi et al., 2013; Manasi et al., 2014). The positive value of ΔG_o reveals that the biosorption of heavy metal onto biosorbent is not a spontaneous process.

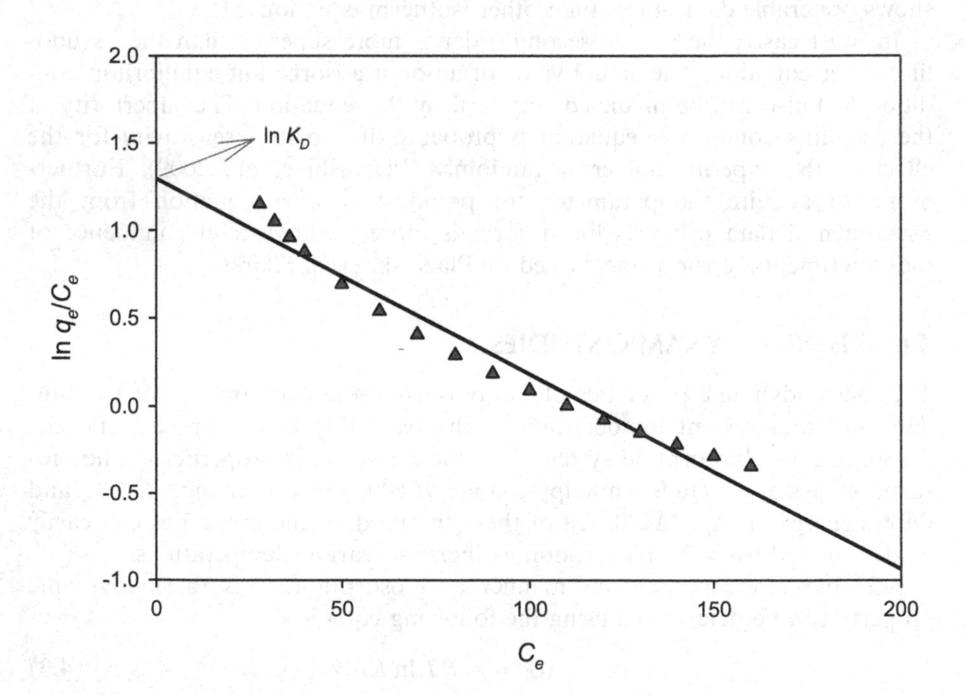

FIGURE 4.7 The plot to determine the value of K_D.

Thermodynamic equilibrium for biosorption and desorption process is never achieved, and desorption of heavy metal from the biosorbent occurs naturally (Flores-Garnica et al., 2013).

$\Delta H°$ less than zero means that the biosorption produces heat energy. If biosorption is exothermic, the temperature gives antagonistic behavior for the uptake of heavy metals by biosorbent; metal uptake goes down with the increase of temperature (Padmavathy, 2008; Sari and Tuzen, 2008; Gutha et al., 2011; Senthilkumar et al., 2012; Manasi et al., 2014). In this case, the physical adsorption controls the mechanism of the biosorption process. The positive value of ΔH_o is an indication that the biosorption process is endothermic; the temperature gives a positive effect on the uptake of heavy metal. In some of the studies, the interaction between surface functional groups of biosorbents and heavy metals were through chemisorption (Farooq et al., 2011; Reddy et al., 2011; Singha and Das, 2011; Subbaiah et al., 2011; Al-Othman et al., 2012; Suksabye and Thiravetyan, 2012; Flores-Garnica et al., 2013; Torab-Moestadi et al., 2013; Manasi et al., 2014).

The standard of entropy change, ΔS_o, measures the randomness of the biosorption system. A positive value of ΔS_o indicates that the heavy metal in the biosorption system possesses high affinity to the biosorption active site in the biosorbent. The higher the value of this thermodynamics property, the higher the randomness of the biosorption system (Farooq et al., 2011; Reddy et al., 2011; Singha and Das, 2011; Subbaiah et al., 2011; Al-Othman et al., 2012; Suksabye and Thiravetyan, 2012; Torab-Moestadi et al., 2013; Manasi et al., 2014). The complexity of biosorbent is the effect of the possession of multiple active groups; different size of the pore, chemical composition, interaction with the solute, etc., cause the biosorption process to become a disorder. However, some studies obtained negative values of ΔS_o (Padmavathy, 2008; Sari and Tuzen, 2008; Gutha et al., 2011; Senthilkumar et al., 2012; Flores-Garnica et al., 2013; Manasi et al., 2014); the negative value of ΔS_o indicates that the biosorption system is more orderliness.

4.7 CONCLUSION

The search for alternative adsorbents from biomass materials for purifying water body from metals is continuing till present. Due to the complexity of the biosorption system, the biosorption mechanisms are still not completely understood, and it makes its application for the real industrial system still far from reality. Most of the biosorption studies focused on single metal, conducted in the static mode using the aqueous solution as the wastewater model. Those conditions are completely different from real wastewater treatment processes.

Different adsorption isotherm expressions are exploited to verify the data fitting; Langmuir equation is still superior for most biosorption systems than other available models. For the kinetics of adsorption, the pseudo-second-order model is still eminent than other equations in representing many biosorption. The complexity of the biosorption process causes some inconsistency of the thermodynamic properties for many biosorption systems.

ACKNOWLEDGMENT

Financial assistant by Indonesia Ministry of Research Technology and Higher Education through PDUPT Grant 2019 is declared.

REFERENCES

Akar, T., and S. Tunali. 2006. Biosorption Characteristics of Aspergillus Flavus Biomass for Removal of Pb(II) and Cu(II) Ions from an Aqueous Solution. Bioresource Technology 97 (15): 1780–1787.

Alomá, I., M. A. Martín-Lara, I. L. Rodríguez, G. Blázquez, and M. Calero. 2012. Removal of Nickel (II) Ions from Aqueous Solutions by Biosorption on Sugarcane Bagasse. Journal of the Taiwan Institute of Chemical Engineers 43 (2): 275–281.

Al-Othman, Z. A., R. Ali, and M. Naushad. 2012. Hexavalent Chromium Removal from Aqueous Medium by Activated Carbon Prepared from Peanut Shell: Adsorption Kinetics, Equilibrium and Thermodynamic Studies. Chemical Engineering Journal 184: 238–247.

Anoop K. K., K. G. Sreejalekshmi, and R. S. Baiju. 2011. Nickel(II) Adsorption onto Biomass-Based Activated Carbon Obtained from Sugarcane Bagasse Pith. Bioresource Technology 102 (22): 10239–10247.

Awwad, A. M., and N. M. Salem. 2014. Kinetics and Thermodynamics of Cd(II) Biosorption onto Loquat (Eriobotrya japonica) Leaves. Journal of Saudi Chemical Society 18: 486–493.

Basha, S., and Z. V. P. Murthy. 2007. Kinetic and Equilibrium Models for Biosorption of Cr(VI) on Chemically Modified Seaweed Cystoseira Indica. Process Biochemistry 42: 1521–1529.

Basu, M., A. K. Guha, and L. Ray. 2017. Adsorption of Lead on Cucumber Peel. Journal of Cleaner Production 151: 603–615.

Bermúdez, Y. G., I. L. R. Rico, O. G. Bermúdez, and E. Guibal. 2011. Nickel Biosorption Using Gracilaria Caudata and Sargassum Muticum. Chemical Engineering Journal 166 (1): 122–131.

Blanchard, G., M. Maunaye, and G. Martin. 1984. Removal of Heavy Metals from Waters by Means of Natural Zeolites. Water Research 18: 1501–1507.

Blazquez, G., M. Calero, A. Ronda, G. Tenorio, and M. A. Martin-Lara. 2014. Study of Kinetics in the Biosorption of Lead onto Native and Chemically Treated Olive Stone. Journal of Industrial and Engineering Chemistry 20: 2754–2760.

Calero, M., A. Perez, G. Blazquez, A. Ronda, and M. A. Martin-Lara. 2013. Characterization of Chemically Modified Biosorbents from Olive Tree Pruning for the Biosorption of Lead. Ecological Engineering 58: 344–354.

Chand, P., A. K. Shil, M. Sharma, and Y. B. Pakade. 2014. Improved Adsorption of Cadmium Ions from Aqueous Solution Using Chemically Modified Apple Pomace: Mechanism, Kinetics, and Thermodynamics. International Biodeterioration and Biodegradation 90: 8–16.

Chen, X. C., Y. P. Wang, Q. Lin, J. Y. Shi, W. X. Wu, Y. X. Chen. 2005. Biosorption of Copper(II) and Zinc(II) from Aqueous Solution by Pseudomonas Putida CZ1. Colloids and Surfaces B: Biointerfaces 46 (2): 101–107.

Chong, H. L. H., P. S. Chia, and M. N. Ahmad. 2013. The Adsorption of Heavy Metal by Bornean Oil Palm Shell and Its Potential Application as Constructed Wetland Media. Bioresource Technology 130: 181–186.

Coelho, G. F., A. C. Goncalves Jr, C. R. T. Tarley, J. Casarin, H. Nacke, and M. A. Francziskowski. 2014. Removal of Metal Ions Cd (II), Pb (II), and Cr (III)

from Water by the Cashew Nut Shell Anacardium Occidentale L. Ecological Engineering 73: 514–525.
Cojocaru, C., M. Diaconu, I. Cretescu, J, Savić, and V. Vasić. 2009. Biosorption of copper(II) Ions from Aqua Solutions Using Dried Yeast Biomass. Colloids and Surfaces A: Physicochemical and Engineering Aspects 335 (1–3): 181–188.
Cui, L., G. Wu, and T. S. Jeong. 2010. Adsorption Performance of Nickel and Cadmium Ions onto Brewer's Yeast. Canadian Journal of Chemical Engineering 88 (1): 109–115.
Ding, Y., D. Jing, H. Gong, L. Zhou, and X. Yang. 2012. Biosorption of Aquatic cadmium(II) by Unmodified Rice Straw. Bioresource Technology 114: 20–25.
Do, D. D. 1998. Adsorption Analysis: Equilibria and Kinetics, Imperial College Press, London SW7 2BT, United Kingdom.
Dubey, A., A. Mishra, and S. Singhal. 2014. Application of Dried Plant Biomass as Novel Low-Cost Adsorbent for Removal of Cadmium from Aqueous Solution. International Journal of Environmental Science and Technology 11 (4): 1043–1050.
Farooq U., M. A. Khan, M. Athar, and J. A. Kozinski. 2011. Effect of Modification of Environmentally Friendly Biosorbent Wheat (Triticum Aestivum) on the Biosorptive Removal of cadmium(II) Ions from Aqueous Solution. Chemical Engineering Journal 171 (2): 400–410.
Ferraz, A. I., C. Amorim, T. Tavares, and J. A. Teixeira. 2015. Chromium(III) Biosorption onto Spent Grains Residual from Brewing Industry: Equilibrium. Kinetics and Column Studies. International Journal of Environmental Science and Technology 12 (5): 1591–1602.
Flores-Garnica, J. G., L. Morales-Barrera, G. Pineda-Camacho, and E. Cristiani-Urbina. 2013. Biosorption of Ni(II) from Aqueous Solutions by Litchi Chinensis Seeds. Bioresource Technology 136: 635–643.
Fomina, M., and G. M. Gadd. 2014. Biosorption: Current Perspectives on Concept, Definition, and Application. Bioresource Technology 160: 3–14.
Ghasemi, M., N. Ghasemi, G. Zahedi, S. R. W. Alwi, M. Goodarzi, and H. Javadian. 2014. Kinetic and Equilibrium Study of Ni(II) Sorption from Aqueous Solutions onto Peganum Harmala-L. International Journal of Environmental Science and Technology 11 (7): 1835–1844.
Giri, A. K., R. Patel, and S. Mandal. 2012. Removal of Cr (VI) from Aqueous Solution by Eichhornia Crassipes Root Biomass-Derived Activated Carbon. Chemical Engineering Journal 185–186: 71–81.
Gogoi, S., S. Chakraborty, and M. D. Saikia. 2018. Surface Modified Pineapple Crown Leaf for Adsorption of Cr(VI) and Cr(III) Ions from Aqueous Solution. Journal of Environmental Chemical Engineering 6 (2): 2492–2501.
González Bermúdez, Y., I. R. L. Rico, O. G. Bermúdez, and E. Guibal. 2011. Nickel Biosorption Using Gracilaria Caudata and Sargassum Muticum. Chemical Engineering Journal 166: 122–131.
Guo, H., L. Yan, D. Song, and K. Li. 2016. Citric Acid Modified Camellia Oleifera Shell for Removal of Crystal Violet and Pb(II): Parameters Study and Kinetic and Thermodynamic Profile. Desalination and Water Treatment 57 (33): 15373–15383.
Guo, H., S. Zhang, Z. Kou, S. Zhai, W. Ma, and Y. Yang. 2015. Removal of cadmium(II) from Aqueous Solutions by Chemically Modified Maize Straw. Carbohydrate Polymers 115: 177–185.
Gupta, V. K., D. Pathania, S. Agarwal, and S. Sharma. 2013. Removal of Cr(VI) onto Ficus Carica Biosorbent from Water. Environmental Science and Pollution Research 20 (4): 2632–2644.
Gupta, V. K., A. Rastogi, and A. Nayak. 2010. Biosorption of Nickel onto Treated Alga (Oedogonium Hatei): Application of Isotherm and Kinetic Models. Journal of Colloid and Interface Science 342 (2): 533–539.

Gutha, Y., V. S. Munagapati, S. R. Alla, and K. Abburi. 2011. Biosorptive Removal of Ni(II) from Aqueous Solution by Caesalpinia Bonducella Seed Powder. Separation Science and Technology 46 (14): 2291–2297.

Gutha, Y., V. S. Munagapati, M. Naushad, and K. Abburi. 2015. Removal of Ni(II) from Aqueous Solution by Lycopersicum Esculentum (Tomato) Leaf Powder as a Low-Cost Biosorbent. Desalination and Water Treatment 54 (1): 200–208.

Hadjittofi, K., M. Prodromou, and I. Pashalidis. 2014. Activated Biochar Derived from Cactus Fibres – Preparation, Characterization and Application on Cu(II) Removal from Aqueous Solutions. Bioresource Technology 159: 460–464.

Haq, F., M. Butt, H. Ali, and H. J. Chaudhary. 2016. Biosorption of Cadmium and Chromium from Water by Endophytic Kocuria Rhizophila: Equilibrium and Kinetic Studies. Desalination and Water Treatment 57 (42): 19946–19958.

Hartono, C. D., K. J. Marlie, J. N. Putro et al. 2016. Levulinic acid from corncob by subcritical water process. International Journal of Industrial Chemistry 7: 401–409.

Hossain, M. A., H. H. Ngo, W. S. Guo, and T. V. Nguyen. 2012. Palm Oil Fruit Shells as Biosorbent for Copper Removal from Water and Wastewater: Experiments and Sorption Models. Bioresource Technology 113: 97–101.

Hou, Y., K. Cheng, Z. Li et al. 2015. Biosorption of Cadmium and Manganese Using Free Cells of Klebsiella Sp. Isolated from Waste Water. PLoS One 10 (10): 1–23.

Hu, X., M. Zhao, G. Song, and H. Huang. 2011. Modification of Pineapple Peel Fibre with Succinic Anhydride for Cu 2+, Cd2+ and Pb2+ Removal from Aqueous Solutions. Environmental Technology 32 (7): 739–746.

Ibrahim, W. M. 2011. Biosorption of Heavy Metal Ions from Aqueous Solution by Red Macroalgae. Journal of Hazardous Materials 192 (3): 1827–1835.

Kalaivani, S. S., T. Vidhyadevi, A. Murugesan et al. 2015. Equilibrium and Kinetic Studies on the Adsorption of Ni(II) Ion from an Aqueous Solution Using Activated Carbon Prepared from Theobroma Cacao (Cocoa) Shell. Desalination and Water Treatment 54 (6): 1629–1641.

Kaya, K., E. Pehlivan, C. Schmidt, and M. Bahadir. 2014. Use of Modified Wheat Bran for the Removal of chromium(VI) from Aqueous Solutions. Food Chemistry 158: 112–117.

Khan, A. A., and R. P. Singh. 1987. Adsorption thermodynamics of carbofuran on Sn(IV) arsenosillicate in H^+, Na^+, and Ca^{2+} forms. Colloid and Surfaces 24: 33–42.

Kim, N., M. Park, and D. Park. 2015. A New Efficient Forest Biowaste as Biosorbent for Removal of Cationic Heavy Metals. Bioresource Technology 175: 629–632.

Kosasih, A. N., J. Febrianto, J. Sunarso, Y. H. Ju, N. Indraswati, and S. Ismadji. 2010. Sequestering of Cu(II) from Aqueous Solution Using Cassava Peel (Manihot Esculenta). Journal of Hazardous Materials 180: 366–374.

Krishnan, K. A., K. G. Sreejalekshmi, and R. S. Baiju. 2011. Nickel(II) Adsorption onto Biomass Based Activated Carbon Obtained from Sugarcane Bagasse Pith. Bioresource Technology 102: 10239–10247.

Kumar, J. I. N, and C. Oommen. 2012. Removal of Heavy Metals by Biosorption Using Freshwater Alga Spirogyra Hyalina. Journal of Environmental Biology 33 (1): 27–31.

Kurniawan, A., V. O. A. Sisnandy, K. Trilestari, J. Sunarso, N. Indraswati, and S. Ismadji. 2011. Performance of Durian Shell Waste as High Capacity Biosorbent for Cr(VI) Removal from Synthetic Wastewater. Ecological Engineering 37: 940–947.

Lagergren, S. 1898. Zur theorie der sogenannten adsorption gelöster stoffe. Kungliga Svenska Vetenskapsakademiens Handlingar 24: 1–39.

Liu, W. J, F. X. Zeng, H. Jiang, and X. S. Zhang. 2011. Adsorption of Lead (Pb) from Aqueous Solution with Typha Angustifolia Biomass Modified by SOCl2 activated EDTA. Chemical Engineering Journal 170 (1): 21–28.

Long, Y., D. Lei, J. Ni, Z. Ren, C. Chen, and H. Xu. 2014. Packed Bed Column Studies on Lead(II) Removal from Industrial Wastewater by Modified Agaricus Bisporus. Bioresource Technology 152: 457–463.

Ma, X., X. Liu, D. P. Anderson, and P. R. Chang. 2015. Modification of Porous Starch for the Adsorption of Heavy Metal Ions from Aqueous Solution. Food Chemistry 181: 133–139.

Mahmood-ul-Hassan, M., V. Suthor, E. Rafique, and M. Yasin. 2015. Removal of Cd, Cr, and Pb from Aqueous Solution by Unmodified and Modified Agricultural Wastes. Environmental Monitoring and Assessment 187: 19.

Manasi, V. Rajesh, A. S. K. Kumar, and N. Rajesh. 2014. Biosorption of Cadmium Using a Novel Bacterium Isolated from an Electronic Industry Effluent. Chemical Engineering Journal 235: 176–185.

Matouq, M., N. Jildeh, M. Qtaishat, M. Hindiyeh, and M. Q. Al Syouf. 2015. The Adsorption Kinetics and Modeling for Heavy Metals Removal from Wastewater by Moringa Pods. Journal of Environmental Chemical Engineering 3: 775–784.

Mishra, A., A. Dubey, and S. Shinghal. 2015. Biosorption of Chromium(VI) from Aqueous Solutions Using Waste Plant Biomass. International Journal of Environmental Science and Technology 12 (4): 1415–1426.

Morsy, F. M., S. H. A. Hassan, and M. Koutb. 2011. Biosorption of Cd(II) and Zn(II) by Nostoc Commune: Isotherm and Kinetics Studies. Clean – Soil, Air, Water 39 (7): 680–687.

Ngabura, M., S. A. Hussain, W. A.W. A. Ghani, M. S. Jami and Y. P. Tan. 2018. Utilization of renewable durian peels for biosorption of zinc from wastewater. Journal of Environmental Chemical Engineering 6: 2528–2539.

Padmavathy, V. 2008. Biosorption of nickel(II) Ions by Baker's Yeast: Kinetic. Thermodynamic and Desorption Studies. Bioresource Technology 99 (8): 3100–3109.

Pakshirajan, K., A. N. Worku, M. A. Acheampong, H. J. Lubberding, and P. N. L. Lens. 2013. Cr(III) and Cr(VI) Removal from Aqueous Solutions by Cheaply Available Fruit Waste and Algal Biomass. Applied Biochemistry and Biotechnology 170 (3): 498–513.

Pandey, G. 2016. Removal of Cd(II) and Cu(II) from Aqueous Solution Using Bengal Gram Husk as a Biosorbent. Desalination and Water Treatment 57 (16): 7270–7279.

Pehlivan, E., T. Altun, and S. Parlayici. 2012. Modified Barley Straw as a Potential Biosorbent for Removal of Copper Ions from Aqueous Solution. Food Chemistry 135 (4): 2229–2234.

Petrović, M., T. Sostaric, M. Stojanovic et al. 2016. Removal of Pb^{2+} Ions by Raw Corn Silk (Zea Mays L.) as a Novel Biosorbent. Journal of the Taiwan Institute of Chemical Engineers 58: 407–416.

Plaza-Cazón, J., M. Viera, E. Donati, and E. Guibal. 2013. Zinc and Cadmium Removal by Biosorption on Undaria Pinnatifida in Batch and Continuous Processes. Journal of Environmental Management 129: 423–434.

Plazinski, W., W. Rudzinski, and A. Plazinska. 2009. Theoretical Models of Sorption Kinetics Including a Surface Reaction Mechanism: A Review. Advance in Colloid and Interface Science 152: 2–13.

Putra, W. P., A. Kamari, S. N. M. Yusofi et al. 2014. Biosorption of Cu(II), Pb(II) and Zn(II) Ions from Aqueous Solutions Using Selected Waste Materials: Adsorption and Characterisation Studies. Journal of Encapsulation and Adsorption Sciences 4 (4): 25–35.

Putro, J. N., F. E. Soetaredjo, S. Y. Lin, Y. H. Ju and S. Ismadji. 2016. Pretreatment and Conversion of Lignocellulose Biomass into Valuable Chemicals. RSC Advances 6: 46834–46852.

Rajesh, Y., M. Pujari, and R. Uppaluri. 2014. Equilibrium and Kinetic Studies of Ni (II) Adsorption Using Pineapple and Bamboo Stem Based Adsorbents. Separation Science and Technology (Philadelphia) 49 (4): 533–544.

Rao, R. A. K., A. Khatoon, and A. Ashfaq. 2016. Application of Terminalia Arjuna as Potential Adsorbent for the Removal of Pb(II) from Aqueous Solution: Thermodynamics, Kinetics and Process Design. Desalination and Water Treatment 57 (38): 17808–17825.

Reddy, D. H. K., D. K. V. Ramana, K. Seshaiah, and A. V. R. Reddy. 2011. Biosorption of Ni(II) from Aqueous Phase by Moringa Oleifera Bark, a Low Cost Biosorbent. Desalination 268: 150–157.

Reddy, D. H. K., K. Seshaiah, A. V. R. Reddy, and S. M. Lee. 2012. Optimization of Cd-(II), Cu(II) and Ni(II) Biosorption by Chemically Modified Moringa Oleifera Leaves Powder. Carbohydrate Polymers 88: 1077–1086.

Romero-Cano, L. A., H. García-Rosero, L. V. Gonzalez-Gutierrez, L. A. Baldenegro-Pérez, and F. Carrasco-Marín. 2017. Functionalized Adsorbents Prepared from Fruit Peels: Equilibrium, Kinetic and Thermodynamic Studies for Copper Adsorption in Aqueous Solution. Journal of Cleaner Production 162: 195–204.

Ronda, A., M. A. Martin-Lara, M. Calero and G. Blazquez. 2013. Analysis of the Kinetics of Lead Biosorption Using Native and Chemically Treated Olive Tree Pruning. Ecological Engineering 58: 278–285.

Sahranavard, M., A. Ahmadpour, and M. R. Doosti. 2011. Biosorption of Hexavalent Chromium Ions from Aqueous Solutions Using Almond Green Hull as a Low-Cost Biosorbent. European Journal of Scientific Research 58 (3): 392–400.

Salem, N. M., and A. M. Awwad. 2014. Biosorption of Ni(II) from Electroplating Wastewater by Modified (Eriobotrya Japonica) Loquat Bark. Journal of Saudi Chemical Society 18 (5): 379–386.

Salihi, I. U., S. R. M. Kutty, M. H. Isa, A. Malakahmad, and U. A. Umar. 2016. Sorption of Zinc Using Microwave Incinerated Sugarcane Bagasse Ash (MISCBA) and Raw Bagasse. Jurnal Teknologi 78: 47–51.

Saqib, A. N. S., A. Waseem, A. F. Khan et al. 2013. Arsenic Bioremediation by Low-Cost Materials Derived from Blue Pine (Pinus Wallichiana) and Walnut (Juglans Regia). Ecological Engineering 51: 88–94.

Sarada, B., M. K. Prasad, K. K. Kumar, and C. V. R. Murthy. 2014. Potential Use of Caulerpa Fastigiata Biomass for Removal of Lead: Kinetics, Isotherms, Thermodynamic, and Characterization Studies. Environmental Science and Pollution Research 21: 1314–1325.

Sari, A., and M. Tuzen. 2008. Biosorption of Pb(II) and Cd(II) from Aqueous Solution Using Green Alga (Ulva Lactuca) Biomass. Journal of Hazardous Materials 152 (1): 302–308.

Sartape, A. S., A. M. Mandhare, P. P. Salvi, D. K. Pawar, and S. S. Kolekar. 2013. Kinetic and Equilibrium Studies of the Adsorption of Cd(II) from Aqueous Solutions by Wood Apple Shell Activated Carbon. Desalination and Water Treatment 51: 4638–4650.

Sathish, T., N. V. Vinithkumar, G. Dharani, and R. Kirubagaran. 2015. Efficacy of Mangrove Leaf Powder for Bioremediation of Chromium (VI) from Aqueous Solutions: Kinetic and Thermodynamic Evaluation. Applied Water Science 5 (2): 153–160.

Saueprasearsit, P. 2011. Adsorption of Chromium (Cr+6) Using Durian Peel. International Conference of Biotechnology and Environment Management IPCBEE 18: 33–38.

Şencan, A., M. Karaboyaci, and M. Kiliç. 2015. Determination of Lead(II) Sorption Capacity of Hazelnut Shell and Activated Carbon Obtained from Hazelnut Shell Activated with ZnCl2. Environmental Science and Pollution Research 22: 3238–3248.

Senthilkumar, P., S. Ramalingam, R. V. Abhinaya, S. D. Kirupha, T. Vidhyadevi, and S. Sivanesan. 2012. Adsorption Equilibrium, Thermodynamics, Kinetics, Mechanism and Process Design of zinc(II) Ions onto Cashew Nut Shell. Canadian Journal of Chemical Engineering 90 (4): 973–982.

SenthilKumar, P., S. Ramalingam, R. V. Abhinaya, K. V. Thiruvengadaravi, P. Baskaralingam, and S. Sivanesan. 2011a. Lead(II) Adsorption onto Sulphuric Acid Treated Cashew Nut Shell. Separation Science and Technology 46 (15): 2436–2449.

Senthilkumar, P., S. Ramalingam, V. Sathyaselvabala, S. D. Kirupha, and S. Sivanesan. 2011b. Removal of copper(II) Ions from Aqueous Solution by Adsorption Using Cashew Nut Shell. Desalination 266 (1–3): 63–71.

Shafique, U., A. Ijaz, M. Salman et al. 2012. Removal of Arsenic from Water Using Pine Leaves. Journal of the Taiwan Institute of Chemical Engineers 43 (2): 256–263.

Shinde, N. R., A. V. Bankar, A. R. Kumar, and S. S. Zinjarde. 2012. Removal of Ni (II) Ions from Aqueous Solutions by Biosorption onto Two Strains of Yarrowia Lipolytica. Journal of Environmental Management 102: 115–124.

Shoaib, A. 2012. Removal of Cr (III) Ions from Tannery Waste Water Through Fungi. The Online Journal of Science and Technology 2 (4): 74–78.

Shroff, K. A. and V. K. Vaidya. 2011. Kinetics and Equilibrium Studies on Biosorption of Nickel from Aqueous Solution by Dead Fungal Biomass of Mucor Hiemalis. Chemical Engineering Journal 171 (3): 1234–1245.

Singha, B., and S. K. Das. 2011. Biosorption of Cr(VI) Ions from Aqueous Solutions: Kinetics. Equilibrium, Thermodynamics and Desorption Studies. Colloids and Surfaces B: Biointerfaces 84 (1): 221–232.

Song, W., B. Gao, T. Zhang et al. 2015. High-Capacity Adsorption of Dissolved Hexavalent Chromium Using Amine-Functionalized Magnetic Corn Stalk Composites. Bioresource Technology 190: 550–557.

Sostaric, T. D., M. S. Petrovic, F. T. Pastor et al. 2018. Study of Heavy Metals Biosorption on Native and Alkali-Treated Apricot Shells and Its Application in Wastewater Treatment. Journal of Molecular Liquids 259: 340–349.

Sreenivas, K. M., M. B. Inarkar, S. V. Gokhale, and S. S. Lele. 2014. Re-Utilization of Ash Gourd (Benincasa Hispida) Peel Waste for Chromium (VI) Biosorption: Equilibrium and Column Studies. Journal of Environmental Chemical Engineering 2 (1): 455–462.

Srinath, T., T. Verma, P. W. Ramteke, and S. K. Garg. 2002. Chromium (VI) Biosorption and Bioaccumulation by Chromate Resistant Bacteria. Chemosphere 48 (4): 427–435.

Subbaiah, M. V., G. Yuvaraja, Y. Vijaya, and A. Krishnaiah. 2011. Equilibrium, Kinetic and Thermodynamic Studies on Biosorption of Pb(II) and Cd(II) from Aqueous Solution by Fungus (Trametes Versicolor) Biomass. Journal of the Taiwan Institute of Chemical Engineers 42 (6): 965–971.

Sugashini, S., and K. M. M. S. Begum. 2015. Preparation of Activated Carbon from Car bonized Rice Husk by Ozone Activation for Cr(VI) Removal. Xinxing Tan Cailiao/ New Carbon Materials 30 (3): 252–261.

Suksabye, P., and P. Thiravetyan. 2012. Cr(VI) Adsorption from Electroplating Plating Wastewater by Chemically Modified Coir Pith. Journal of Environmental Management 102: 1–8.

Tao, H. C., H. R. Zhang, J. B. Li, and W. Y. Ding. 2015. Biomass Based Activated Carbon Obtained from Sludge and Sugarcane Bagasse for Removing Lead Ion from Wastewater. Bioresource Technology 192: 611–617.

Tenorio, G., M. Calero, and G. Bla. 2012. Copper Biosorption by Pine Cone Shell and Thermal Decomposition Study of the Exhausted Biosorbent. Journal of Industrial and Engineering Chemistry 18: 1741–1750.

Torab-Moestadi, M., M. Asadollahzadeh, A, Hemmati, and A. Khosravi. 2013. Equilibrium, Kinetic, and Thermodynamic Studies for Biosorption of Cadmium and Nickel on Grapefruit Peel. Journal of The Taiwan Institute of Chemical Engineers 44: 295–302.

Uslu, G., and M. Tanyol. 2006. Equilibrium and Thermodynamic Parameters of Single and Binary Mixture Biosorption of Lead (II) and Copper (II) Ions onto Pseudomonas Putida: Effect of Temperature. Journal of Hazardous Materials 135 (1–3): 87–93.

Venkata R. D. K., D. H. K. Reddy, J. S. Yu, and K. Seshaiah. 2012. Pigeon Peas Hulls Waste as Potential Adsorbent for Removal of Pb(II) and Ni(II) from Water. Chemical Engineering Journal 197: 24–33.

Wang, C., and H. Wang. 2018. Carboxyl-functionalized Cinnamomum Camphora for Removal of Heavy Metals from Synthetic Wastewater-Contribution to Sustainability in Agroforestry. Journal of Cleaner Production 184: 92–928.

Wang, G., S. Zhang, P. Yao et al. 2018. Removal of Pb(II) from Aqueous Solutions by *Phytolacca americana L.* Biomass as a Low-Cost Biosorbent. Arabian Journal of Chemistry 11: 99–110.

Wang, S., L. Wang, W. Kong et al. 2013. Preparation Characterization of Carboxylated Bamboo Fibers and Their Adsorption for lead(II) Ions in Aqueous Solution. Cellulose 20 (4): 2091–2100.

Wang, X. S., Z. Z. Li, and C. Sun. 2008. Removal of Cr(VI) from Aqueous Solutions by Low-Cost Biosorbents: Marine Macroalgae and Agricultural By-products. Journal of Hazardous Materials 153 (3): 1176–1184.

Wen, X., C. Du, G. Zeng et al. 2018. A Novel Biosorbent Prepared by Immobilized Bacillus Licheniformis for Lead Removal from Wastewater. Chemosphere 200: 173–179.

Witek-Krowiak, A., R. G. Szafran, and S. Modelski. 2011. Biosorption of Heavy Metals from Aqueous Solutions onto Peanut Shell as a Low-Cost Biosorbent. Desalination 265 (1–3): 126–134.

Xu, X., X. Cao, and L. Zhao. 2013. Comparison of Rice Husk- and Dairy Manure-Derived Biochars for Simultaneously Removing Heavy Metals from Aqueous Solutions: Role of Mineral Components in Biochars. Chemosphere 92 (8): 955–961.

Yahya, S. K., Z. A. Zakaria, J. Samin, A. S. S. Raj, and W. A. Ahmad. 2012. Isotherm Kinetics of Cr(III) Removal by Non-Viable Cells of Acinetobacter Haemolyticus. Colloids and Surfaces B: Biointerfaces 94: 362–368.

Yang, Z. H., S. Xiong, B. Wang, Q. Li, and W. C. Yang. 2013. Cr(III) Adsorption by Sugarcane Pulp Residue and Biochar. Journal of Central South University 20 (5): 1319–1325.

Ye, H., Q. Zhu, and D. Du. 2010. Adsorptive Removal of Cd(II) from Aqueous Solution Using Natural and Modified Rice Husk. Bioresource Technology 101 (14): 5175–5179.

Yuvaraja, G., M. V. Subbaiah, and A. Krishnaiah. 2012. Caesalpinia Bonducella Leaf Powder as Biosorbent for Cu(II) Removal from Aqueous Environment: Kinetics and Isotherms. Industrial and Engineering Chemistry Research 51 (34): 11218–11225.

Zafar, M. N., I. Aslam, R. Nadeem, S. Munir, U. A. Rana, and S. U. D. Khan. 2015. Characterization of Chemically Modified Biosorbents from Rice Bran for Biosorption of Ni(II). Journal of the Taiwan Institute of Chemical Engineers 46: 82–88.

Zheng, L., and P. Meng. 2016. Preparation, Characterization of Corn Stalk Xanthates and Its Feasibility for Cd (II) Removal from Aqueous Solution. Journal of the Taiwan Institute of Chemical Engineers 58: 391–400.

Zhou, K., Z. Yang, Y. Liu, and X. Kong. 2015. Kinetics and Equilibrium Studies on Biosorption of Pb(II) from Aqueous Solution by a Novel Biosorbent: Cyclosorus Interruptus. Journal of Environmental Chemical Engineering 3 (3): 2219–2228.

5 Biodegradation of Synthetic Dyes in Wastewaters

S. Ortiz-Monsalve
Laboratory for Leather and Environmental Studies (LACOURO), Department of Chemical Engineering, Universidade Federal do Rio Grande do Sul (UFRGS), Porto Alegre, Brazil
Research group in Mycology (GIM), Faculty of Basic Sciences, Universidad Santiago de Cali (USC), Cali, Colombia
M. Gutterres
Laboratory for Leather and Environmental Studies (LACOURO), Department of Chemical Engineering, Universidade Federal do Rio Grande do Sul (UFRGS), Porto Alegre, Brazil

5.1 INTRODUCTION

Colour is one of the first features that customers or consumers judge in a product. The colour is imparted in the dyeing stage and requires the usage of high concentration of dyes, organic compounds designed to confer different tones and shades on the fibres where they are fixed. Since the synthesis of mauveine in 1856, synthetic dyes have been largely replacing natural dyes (Sabnis 2017). Synthetic dyes are more widely used over natural dyes due to impart greater consistency in shade and strength, produce a wider spectrum of colours and allow higher efficiency in the use of water and energy (Bide 2014). With the ramping industrialisation, the worldwide production of synthetic dyes has been increasing to satisfy the demand of various industrial processes, including paper, pulp and plastic manufacturing and dyeing of textiles and leather.

Dyeing is a process of mass transfer, where the dyes molecules are shifted from an aqueous solution to the fibres into the substrate (Gürses et al. 2016). Dyeing is a complex process occurring in a heterogeneous system and depends on the class of dye, type of fibre, application process and nature of the interactions between dye structure and fibre molecules (Bide 2014; Gürses et al. 2016). In the application of various dyeing techniques, such as batch exhaust or continuous impregnation, an amount of dye is not fixed in the substrate, remaining in the effluent with other chemicals such as salts, surfactants, oils and acids, among others. These hazardous compounds are usually discharged with the wastewaters and must be treated before disposal in water bodies (Katheresan et al. 2018).

Dye-containing wastewaters are characterised for the high biological oxygen demand, chemical oxygen demand, total organic carbon, nitrogen, suspended solids, acidity or alkalinity and colour and odour conditions (Mullai et al. 2017). The layer formed by dyes in the water can also inhibit the aquatic photosynthesis and decrease the amount of dissolved oxygen (Pereira and Alves 2012).

In addition to the toxic nature of dyes, carcinogenic and mutagenic effects have been reported in dye-containing wastewaters. These environmental risks are related to the possible formation of aromatic amines caused by the cleavage of azo chromophores (Vikrant et al. 2018).

Dyes in wastewaters are usually treated by physical, chemical and biological methods. However, dyeing effluents are difficult to treat by traditional methods due to the high structural stability of dyes. Bioremediation appears as an eco-friendly and less expensive alternative for the treatment of industrials wastewaters containing synthetic dyes. Many species of bacteria, fungi (mycoremediation) and algae (phycoremediation) have been used in the bioremediation of synthetic dyes from wastewaters. Different microbial mechanisms can be applied in the treatment of synthetic dyes: biodegradation, bioaccumulation and biotransformation. Biosorption and bioaccumulation are sorption processes that occur in a biomaterial, involving a biosorbent in a solid phase and a sorbate in a liquid phase (Kaushik and Malik 2015). In biodegradation, the chemical structure is partially transformed into less complex fragments (biotransformation) or sometimes can be totally degraded into H_2O, CO_2 and inorganic products (mineralisation). This chapter provides an overview of the microbial decolourisation of dyes from wastewaters through enzymatic biodegradation Recent literature is compiled and reviewed.

5.2 BIODEGRADATION AND BIOREMEDIATION

Biodegradation can be defined as the breakdown of organic compounds by living organisms into smaller components. The breakdown can result in the biotransformation or mineralisation of these compounds. In the mineralisation, organic compounds are complete biodegraded into inorganics, biomass, CO_2 (aerobic) or CH_4 (anaerobic) and H_2O. In biotransformation, chemicals are transformed into less complex metabolites without reaching mineralisation (Crawford 2011).

Biodegradation is performed by many organisms, but from an environmental point of view, the term commonly refers to microbial biodegradation mediated by enzymes, which play an important role in the recycling of the organic matter present in plants and dead animals. Microbial biodegradation is an essential process for the biogeochemical cycles of carbon, nitrogen, sulfur and other elements that participate in the functioning of living systems (Crawford 2011; Das and Dash 2014). Some evolutionary characteristics of the microorganisms such as ubiquity, diversity, metabolic versatility and adaptability are key elements in the maintenance of the ecosystems (Das and Dash 2014).

Microorganisms produce many classes of enzymes, which can be used for the biodegradation of pollutants. The application of microorganisms and their enzymes in the treatment of xenobiotics is called bioremediation. Many bacteria, fungi and algae strains have been reported by the potential in the bioremediation of diverse types of recalcitrant compounds such as natural materials (e.g., lignin), hydrocarbons (e.g., crude oil) and heavy metals (e.g., mercury), and xenobiotics such as polycyclic aromatic hydrocarbons, polychlorinated biphenyls, organochlorides, dyes and other substances (Crawford 2011; Das and Dash 2014; Speight and El-Gendy 2018). Bioremediation is considered an eco-friendly alternative for the treatment of recalcitrant and hazardous pollutants, producing non-toxic by-products.

5.3 BIODEGRADATION OF SYNTHETIC DYES

Dye removal from wastewaters can be reached by different microbial mechanisms: biosorption, bioaccumulation and biodegradation (Kaushik and Malik 2015). Biosorption and bioaccumulation are ways of biodecolourisation by sorption of dyes molecules into the microbial biomass. In this case, the dye structure is not modified. Differently, in the biodegradation mechanism, the structure is enzymatically mineralised or biotransformed into less complex metabolites (Ali 2010).

5.3.1 Mechanism

Overall, the mechanism of biodegradation of synthetic dyes refers to the breakdown of linkages within the dye structure (e.g. azo bond in azo chromophore groups), resulting in the formation of different by-products. It is an energy-dependent process (catabolism) (Kaushik and Malik 2015; Vikrant et al. 2018). Three main groups of enzymes are related to the degradation of dyes: laccases and peroxidases (aerobic mechanism) and azoreductases (anaerobic mechanism) (Mullai et al. 2017; Solís et al. 2012). Biodegradation usually occurs in two steps: (i) cleavage of the chromophore group (e.g. azo), related with the biodecolourisation (Kaushik and Malik 2015) and (ii) biotransformation of the metabolites of biodegradation (e.g. amines and phenolic groups) into smaller and less complex molecules (Ali 2010). Biodegradation may follow different paths, depending on the type of enzyme and dye structure (number, position and identity of functional groups) (Vikrant et al. 2018). In some cases, biodegradation is so efficient that the mineralisation of dyes is achieved (Ali 2010). The efficiency of biodegradation in dye removal usually ranges from 75 to 90% (Katheresan et al. 2018). However, it is important to note that the biodecolourisation related to the cleavage/biotransformation of the chromophore group does not directly imply the total degradation of the dye molecule. To demonstrate that biodegradation occurred is necessary to identify and analyse the breakdown products. Additionally, some metabolites and intermediates produced in the biodegradation (e.g. aromatic amines and phenols) could have higher toxicity and lower biodegradability than the dyes themselves (Brüschweiler and Merlot 2017; Sen et al. 2016).

5.3.2 Microbial Enzymes in Biodegradation of Synthetic Dyes

As previously mentioned, the anaerobic biodegradation of dyes is catalysed by azoreductases. The oxidative degradation is mediated by oxidases (e.g. laccase and phenoloxidases) and peroxidases (e.g. manganese peroxidase, lignin peroxidase, dye-decolourising peroxidase, versatile peroxidase and cellobiose dehydrogenase) (Solís et al. 2012).

5.3.2.1 Azoreductases

Azoreductases (EC 1.7.1.6) are a class flavoproteins involved in the reduction of azo linkages (–N=N–). The reductive cleavage is carried out under anaerobic conditions (Singh et al. 2015). The presence of co-factors such as flavin adenine dinucleotide (FADH), nicotinamide adenine dinucleotide (NADH) or nicotinamide adenine dinucleotide phosphate (NADHP), which act as electron donators, is necessary for the reaction (Singh et al. 2015; Solís et al. 2012). The degradation occurs in two steps: (i) reduction of azo linkages, resulting in the cleavage of the structure producing colourless metabolites (e.g., aromatic amines). The process is mediated by the electron transfer from the enzyme to the mediator compounds (e.g. FADH, NADH, NADHP, among others.) and the dye; (ii) the metabolites of degradation are biotransformed into stable products (Mullai et al. 2017). Dye degrading azoreductases have been reported in bacteria (Guadie et al. 2018; Liu et al. 2016), yeast (Rosu et al. 2017) and algae (Sinha et al. 2016).

5.3.2.2 Laccases

Laccases (EC 1.10.3.2) are a major group of glycoproteins with phenoloxidase activity produced by fungi and bacteria. These enzymes belong to the multi-copper blue oxidases. In nature, phenolics and aromatic compounds are common substrates oxidised by laccases (Cullen 2014; Sen et al. 2016). Most of the known laccases are produced mainly by white-rot fungi (WRF) (Singh et al. 2015). Laccases are part of the lignin-modifying enzyme system of WRF, characterised by their non-specificity and high oxidative potential (Rodríguez-Couto 2015; Solís et al. 2012). Fungal laccases have been extensively reviewed by their potential in bioremediation of several recalcitrant compounds, including dyes (Ali 2010; Kaushik and Malik 2015; Sen et al. 2016). Fungal laccases catalyse the degradation of dyes by accepting an electron from the dye. The final electron acceptor is molecular oxygen (Mullai et al. 2017; Singh et al. 2015). Some redox mediators such as syringaldazine, 1-hydroxy benzotriazole (HBT) and 2,2′-azino-bis (3-ethylbenzothiazoline-6-sulfonic acid) (ABTS) are usually necessary to improve the degradation of dyes by fungal laccases (Ortiz-Monsalve et al. 2017; Solís et al. 2012). Bacterial laccases have also shown efficiency in the biodegradation of dyes (Mendes et al. 2015; Saratale et al. 2011). Bacterial laccases have the advantage of the non-requirement of redox mediators (Mendes et al. 2015). Although algae laccases have been scarcely reported (Otto et al. 2010; Otto and Schlosser 2014), some studies have found laccase activity in biodegradation of dyes mediated by algae (Priya et al. 2011; Xie et al. 2016).

5.3.2.3 Peroxidases

Peroxidases are hemoproteins that catalyse substrates using H_2O_2 or organic hydrogen peroxide as electron-accepting co-substrates (Kües 2015). Manganese peroxidase (MnPs, EC 1.11.1.13) and lignin peroxidase (LiPs, EC 1.11.1.14) are common peroxidases in the ligninolytic enzyme system of white-rot fungi and were first extracted from the fungus *Phanerochaete chrysosporium*. LiPs oxidise non-phenolic aromatic compounds (Cullen 2014; Katheresan et al. 2018). MnPs catalyse phenolic compounds through an intermediary redox reaction: Mn^{2+} is oxidised by MnP and the Mn^{3+} resulting ion acts as an active oxidant of phenolic aromatic compounds (Katheresan et al. 2018; Solís et al. 2012). Fungal LiP and MnP have been reported by the biodegradation of a variety of azo dyes (Bosco et al. 2017; Sodaneath et al. 2017; Zhang et al. 2018). Versatile peroxidases (VPs, EC 1.11.1.16) have also shown ability in the degradation of dyes (Hibi et al. 2012). VPs can be considered a hybrid between LiP and MnP because they oxidise Mn^{2+}, phenolic and non-phenolic substrates in both manganese-dependent and manganese-independent reactions (Cullen 2014; Singh et al. 2015). Dye-decolourising peroxidases (DyPs, EC 1.11.1.19) are enzymes that oxidise phenolic and non-phenolic compounds (Kües 2015). Dye degradation activity of DyPs has been found in fungi (Colpa et al. 2014; Liers et al. 2010) and bacteria (Lončar et al. 2016; Santos et al. 2014).

5.3.2.4 Other Enzymes

Other oxidase enzymes (e.g. tyrosinase) and reductases (e.g. NADH–DCIP reductase) have been involved in dye biodegradation. Tyrosinase (Tyr, EC 1.14.18.1), also known as monophenol monooxygenase, are copper proteins that catalyse the oxidation of phenol (Kües 2015; Saratale et al. 2011). NADH–DCIP reductases (EC 1.6.99.3) are enzymes that reduce the DCIP (2,6-dichloroindophenol) with NADH as an electron donor (Dave et al. 2015). The production of tyrosinases, veratryl alcohol oxidases, NADH–DCIP reductases and riboflavin enzymes can be induced by the presence of dyes, suggesting their role in degradation by yeast (Martorell et al. 2012; Pajot et al. 2011; Song et al. 2017; Waghmode et al. 2012) and bacteria (Chougule et al. 2014; Khandare et al. 2013; Kurade et al. 2013).

5.4 BIODEGRADATION OF SYNTHETIC DYES BY FUNGI

Fungi are a large group of organisms that are evolved to use the energy stored in plant and animal biomass, in order to develop their metabolic activities. Fungi have an important function in nature, mediating the biodegradation of residues and participating in the biogeochemical cycles of nutrients and minerals. They are one of the most suitable agents for the decomposition of organic matter and are an essential component of the flow of energy and carbon in the environment. The use of fungi in the bioremediation of pollutants is called mycoremediation. Several studies proved the efficiency of different fungal species in the biodecolourisation of dye-containing wastewaters by different mechanisms: biosorption, bioaccumulation and biodegradation. A group of wood-decay basidiomycetes, so-called white-rot fungi, has been recognised as the most efficient group of microorganisms in the biodegradation of dyes (Kaushik and Malik 2015). However, the enzymatic biodegradation of dyes has also been demonstrated in other filamentous fungi and yeasts.

5.4.1 Dye Biodegradation by White-rot Fungi and Other Filamentous Fungi

The capacity of white-rot fungi (WRF) in the biodegradation of dyes is related to the extracellular and not-specific lignin-modifying enzyme system (Rodríguez-Couto 2015). Several characteristics make them one of the most suitable organisms for the biodegradation of dyes:

- The enzyme system has a low substrate specificity due to the heterogeneous nature of lignin. The lack of specificity allows the use of WRF in the biodegradation of a number of xenobiotics as polycyclic aromatic hydrocarbons (PAHs) (Taha et al. 2017; Wirasnita and Hadibarata 2016), nitro-aromatic compounds (Anasonye et al. 2015; Levin et al. 2016), polychlorinated biphenyls (Stella et al. 2017), pharmaceutical compounds (Nguyen et al. 2017; Vasiliadou et al. 2016), etc. This characteristic is very convenient in the biodegradation of dyes from wastewaters because, in industrial dyeing, different classes of dyes, or even mixtures between them, are applied. The fungus *Trametes villosa* SCS-10, for instance, showed efficiency in biodegradation of more than six different leather dyes (Ortiz-Monsalve et al. 2017). A strain of *Trametes polyzone* also proved potential in the biodegradation of various dyes, including a mixture of six different textile dyes (Cerrón et al. 2015). Similarly, *Trametes versicolor* CBR43 achieved the biodecolourisation of different classes of dyes (acid, disperse and reactive dyes) (Yang et al. 2017)
- The ligninolytic enzymes are extracellular, allowing the biodegradation of dyes outside the cells, minimising toxic effects related to the high concentration of dyes. Besides, enzymatic extracts can be used in the biodegradation of dyes in vitro. A strain of *Trametes versicolor*, for example, showed tolerance to high initial concentration of dyes (800 mg L^{-1}) during the mycoremediation of Acid Blue 62 and Acid Black 172 (Yang et al. 2017). The fungus *Ganoderma* sp. En3 also showed resistance and efficiency in biodegradation of high concentrations of Remazol Brilliant Blue and Indigo Carmine at 800–1000 m gL^{-1}. The strain reached over 90% of colour removal. Additionally, an enzymatic extract of *Ganoderma* sp. En3 was also efficient in the dye biodegradation (Lu et al. 2016). Similarly, an extracellular enzymatic extract secreted by *Leptosphaerulina* sp. was able to degrade three textile dyes (Novacron Red, Turquoise Blue and Remazol Black) (Plácido et al. 2016).

- The lignin-modifying enzyme system produced by WRF consists mainly of laccases, lignin peroxidases, manganese-dependent peroxidases and versatile peroxidase. However, there is evidence of the participation of other enzymes such as aryl alcohol oxidase and glyoxal oxidase (Watkinson 2016). The enzyme production is not homogeneous, and many fungal strains can produce one or more of these enzymes (Arora and Sharma 2010). This feature has a great advantage in the enzymatic biodegradation of synthetic dyes by WRF, since all of these enzymes have already been reported by the dye-degrading activity. Then, a selected fungus does not necessarily have to be a producer of all the enzymes to biodegrade dyes. Also, not all the enzymes produced should be involved in the dye degradation. *Trametes villosa* SCS-10, for example, showed activity of laccase and manganese-dependent peroxidase. However, only laccase was involved in the biodegradation of leather dyes (Ortiz-Monsalve et al. 2017). Differently, a manganese peroxidase of *Trametes* sp. showed an ability to degrade different types of dyes (Zhang et al. 2016).
- Various low molecular weight compounds, with easy diffusion and high redox potential, are involved as mediators, cofactors and inducers of the enzymatic degradation of lignin. Many of them, such as H_2O_2, Mn^{3+}, oxalate and malic acid, thiols and veratryl alcohol, are secreted by the WRF. These compounds can be added in the fungal treatment of dyes. Peroxidase cofactors as $MnCl_2$ and H_2O_2 and redox mediators as HBT and ABTS have been reported to improve the dye biodegradation activity (Li et al. 2014; Lu et al. 2016; Ma et al. 2014; Plácido et al. 2016).
- Apical growth of fungal hyphae permits the colonisation of heterogeneous microenvironments with dispersed resources. This characteristic allows the use of WRF in the treatment of dyes in different matrices (solid and liquid media). For example, a strain of *Trametes villosa* was used efficiently in the biodegradation of the leather dyes Acid blue 161, Acid Red 357 and Acid Black 210 in solid media. The same strain allowed the biodegradation of the leather dyes in aqueous solution (Ortiz-Monsalve et al. 2017) and real effluents (Ortiz-Monsalve et al. 2019). Similarly, *Trametes versicolor* degraded the dye Red 40. The process was performed in a rotary drum bioreactor under solid-state fermentation conditions (Jaramillo et al. 2017). Another strain of *Trametes versicolor* was used in the treatment of various synthetic dyes (Acid Black 172, Acid Blue 62 and Acid Red 114) under submerged fermentation conditions. The strain reached more than 90% of biodecolourisation (Yang et al. 2017). Table 5.1 summarises recent studies of dye biodegradation by white-rot fungi and other filamentous fungi.

5.4.2 Dye Biodegradation by Yeast

The treatment of dyes by yeast has been mainly focused on the mechanisms of biosorption and bioaccumulation (Sen et al. 2016). Yeasts have a great capacity to accumulate dyes and other xenobiotics (Asfaram et al. 2016; Dil et al.

TABLE 5.1
Biodegradation of synthetic dyes by fungi

Strain	Dye—Concentration (mg L^{-1})	Key enzymes—Maximum enzyme activity (EA)	Colour removal (%)	Conditions (time; pH; temperature, T[°C]; agitation, agt. [rpm])	Reference
Peyronellaea prosopidis	Scarlet RR	Lac (63 U min^{-1} mg) and LiP (120 U min^{-1} mg)	90	5 days; pH = 6.0; T = 35; static	Bankole et al. 2018
Coriolopsis sp.	Cotton Blue—50 Crystal Violet—100	Lac, LiP and NADH-DCIP reductase	>80	14 days; pH = 5.0; static	Munck et al. 2018
Trichoderma harzianum	Mordant Orange-1–50	–	83	30 days; pH = 3.0; agt.= 120	Hadibarata et al. 2018
Ceriporia lacerate	Congo Red—100	MnP (113.82 U min-1 mg)	90	48 h; pH = 8.0; T = 30; agt.= 120	Wang et al. 2017
Trametes villosa	Acid Blue 161, Acid Red 357	Lac (1150–1550 U L^{-1})	>90	168 h; pH = 5.5; T = 30; agt.= 150	Ortiz-Monsalve et al. (2017)
Trametes versicolor	Acid Red 114 and Acid Blue 62	MnP (1.19 U mL^{-1}) and Lac (2.39 U mL^{-1})	>90	6 days; pH = 5.0; T = 28; agt.= 150	Yang et al. (2017)
Leptosphaerulina sp.	Novacron Red and, Remazol Black	Lac (650 U L^{-1}) and MnP (100 U L^{-1})	>90	9 days; pH = 5.0; T = 25; agt.= 150	Plácido et al. 2016
Ganoderma sp.	Remazol Brilliant Blue R—5000	Lac (6700 U L^{-1})	>95	96 h; pH = 5.0; T = 28; agt.= 150	Lu et al. (2016)
Trametes polyzona	Remazol Brilliant—400	Laccase	97	96 h; pH = 6.0; T = 28; agt.= 175	Cerrón et al. (2015)
Trametes sp.	Indigo Carmine—1000	Lac (800 mU g^{-1})	97.53	10 days; pH = 4.5; T = 30; solid state	Li et al. (2014)

2017). Nevertheless, some studies have shown enzymatic biodegradation of dyes by yeast (Martorell et al. 2017; Song et al. 2017). Dye biodegradation has been associated with the primary metabolism of yeast, which can use dyes as a carbon or nitrogen source (Martorell et al. 2017). However, an easily metabolisable carbon source (e.g. glucose) is usually required in many yeast strains (Solís et al. 2012). Yeasts are unicellular organisms characterised by their low metabolic requirements, rapid growth and resistance to environmental stresses. These are advantage characteristics for the treatment

TABLE 5.2
Biodegradation of synthetic dyes by yeast

Strain	Dye—Concentration (mg L^{-1})	Key enzymes Enzyme activity (EA)	Colour removal (%)	Conditions (time; pH; temperature, T[°C]; agitation, agt. [rpm])	Reference
Candida sp.	Reactive Green—100	Azoreductase (1.82 U min^{-1} mg) and Lac	84	96 h; pH = 4.0–6.0; T = 30; agt.= 120	Sinha et al. (2018)
Trichosporon akiyoshidainum	Reactive black 5–200	Phenol oxidase (353 UL^{-1}) and peroxidase (2750 UL^{-1})	100	24 h; pH = 4.0–6.0; T = 25; agt.= 250	Martorell et al. (2017)
Pichia occidentalis	Acid Red B—50	NADH-DCIP reductase (146 to187 μg min^{-1} mg)	98	16 h; pH = 5.0; T = 30; agt. = 160	Song et al. (2017)
Pichia kudriavzevii	Reactive Orange 16–400	NADH-DCIP reductase (110 μg min^{-1} mL); Azoreductase	95	18 h; pH = 6.0; T = 30; agt. = 100	Rosu et al. (2017)
Scheffersomyces spartinae	Acid Scarlet 3R—80	Azoreductase	90	16 h; pH = 5.0–6.0; T = 30; agt. = 160	Tan et al. (2016)
Magnusiomyces ingens	Acid Red B—20	Azoreductase	97.4	8 h; pH = 5.0–6.0; T = 30; agt. = 160	Tan et al. (2014)
Candida tropicalis	Acid Brilliant Scarlet—20	Azoreductase	97.2	10 h; pH = 6.0; T = 35; agt. = 160	Tan et al. (2013)
Candida sp.	Black B-V—200	Tyr (EA = 365 UL^{-1}) and MnP (EA = 447 UL^{-1})	96.38	72 h; pH = 4.0; T = 25; agt. = 250	Martorell et al. (2012)
Galactomyces geotrichum	Remazol Red—50	Tyr (1196–1430 μg min^{-1} mg); NADH-DCIP reductase	96	36 h; pH = 11.0; T = 30; static	Waghmode et al. (2012)
Pichia sp.	Acid Red B—100	Reductase	95	12 h; T = 30; agt. = 150	Qu et al. (2012)
Trichosporon akiyoshidainum	Blue RR-BB—200	MnP (666 to 10,538 U L^{-1}) and Tyr (84 to 786 U L^{-1})	100	12 h; pH = T = 26; agt. = 250	Pajot et al. (2011)

of xenobiotics when compared with filamentous fungi (Ali 2010). Many yeast enzymes such as phenol oxidases, reductases, peroxidases and azoreductases have been related to biodegradation of synthetic dyes. A strain of *Trichosporon akiyoshidainum* reached the oxidative degradation of Reactive Black 5 dye. Degradation was mediated by a phenoloxidase and a peroxidase enzyme. Total biodecolourisation was achieved in 24 h of treatment (Martorell et al. 2017). Similarly, the salt-tolerant yeast strain *Pichia occidentalis* was able to degrade Acid Red B dye. Biodegradation was mediated by an NADH-DCIP reductase enzyme (Song et al. 2017). An azoreductase and an NADH-DCIP reductase of *Pichia kudriavzevii* were related to the degradation of Reactive Orange 16. The metabolites of degradation were analysed (Rosu et al. 2017). Other enzymes such as tyrosinases and manganese peroxidase are induced by the dyes. A strain of *Candida* sp. degraded the synthetic dye Black B-V by the action of tyrosinases and MnPs (Martorell et al. 2012). Recent studies using yeasts to bioremediate dyes are shown in Table 5.2.

5.5 BIODEGRADATION OF SYNTHETIC DYES BY BACTERIA

Beside fungi, bacteria have also shown the ability to biodegrade synthetic dyes. Bacterial oxidoreductive enzymes such as azoreductases, laccases and peroxidases have been associated with dye biodegradation (Chen et al. 2018; Vikrant et al. 2018). Depending on the bacterial species, the biodegradation process can be anaerobic or aerobic (Saratale et al. 2011).

5.5.1 Anaerobic Conditions

Most of study focused on the anaerobic biodegradation of dyes are mediated by azoreductases, enzymes known by the degradation of azo chromophores (Mendes et al. 2015). This mechanism is also known as azo dye reduction. A strain of *Halomonas* sp., for example, was reported as efficient in the biodegradation of Reactive Red 184 dye under anaerobic conditions. Authors associated the biodecolourisation with the activity of azoreductase enzymes (Guadie et al. 2018). Similarly, enzymes azoreductases and NADH-DCIP reductases produced by *Bacillus circulans* biodegraded the dye Methyl Orange under anaerobic conditions (Liu et al. 2017). However, some bacteria strains can degrade azo dyes into colourless aromatic amines those could be hazardous and carcinogenic (Ali 2010; Saratale et al. 2011). The metabolites produced by the dye reduction may be further degraded in aerobic or anaerobic process. These products can also be reduced by other bacterial enzymes such as hydroxylases and oxygenases (Elisangela et al. 2009).

5.5.2 Aerobic Conditions

Other bacterial strains have also shown potential in the aerobic degradation of dyes. Different families of enzymes, such as bacterial laccases and peroxidases, have also been related to the aerobic degradation of dyes (Mendes et al.

TABLE 5.3
Biodegradation of synthetic dyes by bacteria

Strain	Dye—Concentration (mg L^{-1})	Key enzymes Enzyme activity (EA)	Colour removal (%)	Conditions (time; pH; temperature, T[°C]; agitation, agt. [rpm])	Reference
Halomonas sp.	Reactive Red 184–150	Azoreductase (0.43 U mg^{-1}) and Lac (0.369 U mg^{-1})	98	96 h; pH = 10.0; T = 25; static; anaerobic conditions	Guadie et al. (2018)
Bacterial microflora. Genus: *Anoxybacillus, Caloramator* and *Bacillus*	Direct Black G—600	Lac, MnP, LiP and azoreductase	97	8 h; pH = 8.0; T = 55; static; microaerobic conditions	Chen et al. (2018)
Pseudomonas aeruginosa	Remazol Black B—200	Azoreductase (0.4 μg min^{-1} mg)	88.22	32 h; pH = 6.0–9.0; T = 37; static	Hashem et al. (2018)
Bacillus fermus	Acid Blue 9–50	Not assessed. Aerobic degradation with metabolites identified	97	72 h; T = 37; agt. 100	Neetha et al. (2018)
Bacterial consortium. Genus: *Neisseria, Vibrio, Bacillus* and *Aeromonas*	Novacron Brilliant Blue FN-R and Novacron Super Black G—100	Oxidases and reductases	90	6 days; pH = 7.0; T = 37; agt. 160	Karim et al. (2018)
Bacterial consortium: *Pantoea ananatis, Bacillus fortis* and *Alcaligenes faecalis*	Acid Blue 193 and Acid Black 194—(1100–1400 Fed-Batch process)	Laccase, lignin peroxidase, tyrosinase, azoreductase and NADH-DCIP reductase	80	11–1 cycles of 96 h; pH = 7.4	Patel et al. (2017)
Bacillus circulans	Methyl Orange—50	Azoreductase, NADH-DCIP reductase and Lac	>94	12 h; pH = 6.5–8.0; T = 42; agt.= 180	Liu et al. (2017)
Klebsiella oxytoca	Methyl Red—100	Azoreductase (0.846 U mg^{-1})	95.1	3 h; T = 35; agt.= 150	Yu et al. (2015)
Bacillus sp.	Methyl Red—100	Azoreductase, laccase	98	30 min; pH = 8.0; T = 35; agt.= 200	Zhao et al. (2014)

2015). *Bacillus fermus*, for instance, was suitable for the aerobic degradation of Acid Blue 9 at 50 ppm. 97% of colour removal was achieved in 72 h of treatment at 37°C and 100 rpm (Neetha et al. 2018). Similarly, a consortium of thermophilic strains was able to biodegrade aerobically Direct Black G dye. Lac, MnP and Lip were involved in the dye biodegradation (Chen et al. 2018). Furthermore, specialised azoreductases of few bacterial strains have also been related to dye biodegradation under aerobic conditions (Dave et al. 2015). In Table 5.3 are presented recent researches using bacterial degradation of dyes.

5.6 BIODEGRADATION OF SYNTHETIC DYES BY ALGAE

Algae are the organisms responsible for primary production in aquatic ecosystems. Algae have been widely used as a source for the production of biofuel, cosmetics, pharmaceutical compounds and food. Algae also have been reported for the efficiency in bioremediation. The use of algae in the treatment of wastes or wastewaters is defined as phycoremediation (Phang et al. 2015). Many reports have shown efficiency in removal of dyes from wastewaters using macroalgae (seaweed) (Mokhtar et al. 2017), green microalgae (Sinha et al. 2016; Xie et al. 2016) and blue microalgae (prokaryotic cyanobacteria) (Dellamatrice et al. 2017; El-Sheekh et al. 2017).

Phycoremediation of dyes and dye-containing wastewaters may occur by biodegradation/bioconversion, bioaccumulation or biosorption mechanisms. During the biodegradation process, algae degrade the dyes as a carbon source, biotransforming them into low-molecular compounds. In the biosorption process, dyes are adsorbed in the biomass. Both mechanisms can occur simultaneously (Fazal et al. 2018)

Most studies have related the biodecolourisation ability of macroalgae with biosorption and bioaccumulation mechanisms. The large surface area and binding ability of seaweed are advantageous characteristics in biosorption. However, green macroalgae, including *Enteromorpha* sp., *Cladophora* sp. and *Chara* sp. have also been reported by its potential in the biodegradation of dyes (Khataee et al. 2013; Khataee and Dehghan 2011).

Although microalgae have also been used as efficient biosorbents of dyes (da Fontoura et al. 2017; Pathak et al. 2015), certain green microalgae strains have shown ability in the biodegradation of dyes. Algae biodegradation has been mainly related to the activity of algae laccases, polyphenol oxidases and azoreductases (Sinha et al. 2016; Xie et al. 2016). The algae *Chlorella pyrenoidosa*, for instance, showed ability in the phycoremediation of the azo dye Direct Red 31. The colour removal was associated with biosorption coupled with biodegradation by azoreductases enzymes (Sinha et al. 2016). Similarly, a strain of *Chlorella sorokiniana* was able to biodegrade the dye Disperse Blue 2BLN. Biodecolourisation was mediated by laccase and manganese peroxidase enzymes. The dye was broken down into low-molecular metabolites such as ethylbenzene and ethyl acetate. The green microalga achieved 83% colour removal (Xie et al. 2016). *Coelastrella* sp. was used in the biodegradation of the synthetic dye Rhodamine B. The strain achieved dye degradation to non-toxic products. A peroxidase enzyme was involved in dye degradation (Baldev et al. 2013).

Blue microalgae (cyanobacteria) have also been related to biodegradation of dyes. Recently, a strain of *Phormidium autumnal* was evaluated by the ability in the biodegradation of textile dyes (Indigo, RBBR and Sulfur Black). The strain allowed the complete biodegradation of the indigo dye (Dellamatrice et al. 2017). Similarly, a strain of *Microcystis aeruginosa* showed ability in the removal of 65.07% of Disperse Orange 2RL. An

TABLE 5.4

Biodegradation of dyes synthetic by algae

Strain	Dye—Concentration (mg L^{-1})	Mechanism Enzyme activity (EA)	Colour removal (%)	Conditions (time; pH; temperature, T[°C]; agitation, agt. [rpm])	Reference
Scenedesmus sp., *Chlorella* sp. and *Synedra* sp.	Textile wastewater	Biosorption and biodegradation	82.6	20 days; pH = 8.0; T = 30	Aragaw and Asmare (2018)
Chlorella sorokiniana	Disperse Blue 2BLN—60	Lac (3.89 U mL^{-1}) and MnP(4.86 U mL^{1})	83	6 days; pH = 4.0; T = 25	Xie et al. (2016)
Chlorella pyrenoidosa	Direct Red 31–40	Azoreductase	96	30 min; pH = 3.0; T = 28, agt. = 150 rpm	Sinha et al. (2016)
Haematococcus sp.	Congo Red—10	Biodegradation	98	15 days; T = 24	Mahalakshmi et al. (2015)
Coelastrella sp.	Rhodamine B—100	Biodegradation —Peroxidase	80	20 days; pH = 8.0; T = 30	Baldev et al. (2013)
Cyanobacterium phormidium	Indigo blue—200	Biodegradation	91.22	14 days; T = 25	Dellamatrice et al. (2017)
Hydrocoleum oligotrichum	Basic fuschin—5	Biodegradation	92.44	7 days; pH = 7.4; T = 25	Abou-El-Souod and El-Sheekh (2016)
Oscillatoria curviceps	Acid Black 1–100	Lac, polyphenol oxidase and azoreductase	84	5 days; T = 27	Priya et al. (2011)
Enteromorpha sp.	Basic Red 46–15	Biodegradation	83.45	5 h; T = 25	Khataee et al. (2013)
Cladophora sp.	Malachite Green—10	Biodegradation	71	75 min; pH = 8.0	Khataee and Dehghan (2011)

azoreductase enzyme was related to the dye biodegradation into amines by the breakdown of azo bonds (El-Sheekh et al. 2017). The cyanobacterium *Oscillatoria curviceps* also demonstrated high potential to degrade the azo dye Acid Black 1. The strain reached 84% of colour removal. Laccase, polyphenol oxidase, azoreductase and nitrogenase enzymes played a relevant role in biodegradation (Priya et al. 2011). Recent studies applying biodegradation of dyes by algae strains are shown in Table 5.4.

5.7 RECENT ADVANCES IN MICROBIAL DEGRADATION OF DYES

Since Glenn and Gold (1983) used the fungus *Phanerochaete chrysosporium* in the biodecolourisation and biodegradation of polymeric dyes, several researches have been developed to find new microbial strains with potential to biodegrade dyes. As it has been reviewed in this chapter, many studies have reached this aim. Reviews in this area are regularly published, showing advances in the microbial biodecolourisation of synthetic dyes and dye-containing wastewaters (Ali 2010; Katheresan et al. 2018; Saratale et al. 2011; Sen et al. 2016; Solís et al. 2012; Vikrant et al. 2018). However, the major drawback of this biological alternative is the few real industrial applications. There are some disadvantages related to operational conditions (e.g. acidic or alkaline conditions, non-sterility, high temperatures, high concentrations of dyes and heavy metals etc.), little growth of microorganisms, inactivation of enzymes and need of nutrient supply (Kaushik and Malik 2015; Vikrant et al. 2018). Hence, it is necessary to find new approaches of biological remediation of dyes, aiming actual application. Recently advances related with *in vitro* biodegradation, pathways of dye degradation, genetic manipulations of microorganisms, bioreactor configurations, cell and enzyme immobilisation, treatment with mixed microbial and biodegradation of real wastewaters are described in this section.

5.7.1 Biodegradation Pathways

As previously described, some metabolites and intermediates of biodegradation could have higher toxicity and lower biodegradability that the dyes themselves (Brüschweiler and Merlot 2017). Due to this, it is often suggested to identify pathways and by-products of biodegradation. Various analytical techniques such as High-Performance Liquid Chromatography (HPLC), Fourier Transform Infra-Red Spectroscopy (FTIR), UV-Vis spectrophotometry, Gas Chromatography-Mass Spectrometry (GCMS) and Nuclear Magnetic Resonance (NMR) are useful for monitoring and identifying metabolites of degradation. Different techniques have advantages and limitations. The high selectivity of HPLC and GCMS, for instance, allows the identification of degradation by-products using an MS detector. However, the analysis can take a long time, and samples should be pre-treated. In contrast, UV-Vis and FTIR are simple and fast techniques but show low selectivity (Kaushik and Malik 2015; Vikrant et al. 2018). These instrumental procedures have been used in identifying routes of dye degradation achieved by different microorganisms. The biodegradation of Scarlet RR by the enzymes (Lac, MnP and LiP) of the fungus *Peyronellaea prosopidis* was monitored using HPLC, FTIR, GC–MS and UV–Vis. Dye was biotransformed in acetamides and amines (Bankole et al. 2018). Similarly, Tan et al. (2016) used UV-Vis and HPLC to identify the pathway of degradation of Acid scarlet 3R by *Scheffersomyces spartinae.* Authors found that reduction, deamination and desulfonation reactions occurred. Dye was degraded in naphthalene derivatives. Other yeast strain, *Pichia occidentalis* was able to degrade Acid Red dye by the action of NADH-DCIP reductase. Using UV–Vis and HPLC-MS, authors identified many metabolites formed as a result of the decolourisation, deamination and desulfonation of the dye structure (Song et al. 2017). A bacterial strain of *Bacillus fermus* degraded aerobically Acid Blue 9 into benzene sulfonic acid, benzenesulfonate and methanetriyltribenzene. Liquid chromatography-mass spectrometry (LC-MS) and UV-Vis were used to analyse the metabolites produced (Neetha et al. 2018). Similarly, in algae, it has been identified some biodegradation pathways of dyes. *Chlorella sorokiniana* allowed the biodegradation of Disperse Blue 2BLN into various compounds of low molecular weight such as phthalate, ethylbenzene and ethyl acetate. Metabolites were detected using FTIR, GC-MS and UV-Vis techniques. Recent researches showing pathways of biodegradation have been tabulated in Table 5.5.

5.7.2 Dye Biodegradation in Bioreactor

Most studies assessing biodegradation of dyes are carried out on a laboratory scale. For real applications, studies on a pilot and industrial scale are needed to assess the efficiency of the treatment. Different bioreactor configurations are employed in the biodegradation of synthetic dyes. The reactor design depends on the mechanism of the microorganism that will be applied in the treatment (e.g. biodegradation, biosorption, etc.). Common bioreactors used in biological degradation of dyes are stirred tank reactors, fixed or fluidised bed reactors, hybrid reactors, bubble reactors and airlift reactors (Espinosa-Ortiz et al. 2016; Kaushik and Malik 2015). Various studies using different configurations of bioreactors are shown in Table 5.6.

- Stirred tank reactors (STRs) have been extensively used due to their operational advantages related to the transfer of nutrients and oxygen, operation control and low cost. STR reactor type is commonly used for aerobic fermentation (Espinosa-Ortiz et al. 2016; Vikrant et al. 2018). An STR bioreactor configuration was used to improve the dye degradation potential of laccase from the fungi *Marasmiellus palmivorus.* The configuration allowed 90% of colour removal of Reactive Blue 220 dye (Cantele et al. 2017). Similarly, fungal pellets of *Pycnoporus sanguineus* were efficient in the biodegradation of crystal violet mediated by laccase. The study was performed and optimised in laboratory scale and replicated in an STR, achieving 81.4% of colour removal in 12 h of operation (Sulaiman et al. 2013).
- The airlift bioreactor configuration is designed to prevent mechanical agitation, reducing power consumption. These reactors have simple

TABLE 5.5
Pathways and metabolites of synthetic dye biodegradation

Strain	Dye	Key enzymes	Mechanisms	Metabolites	Technique	Reference
Trichoderma harzianum	Mordant Orange 1	–	Azo cleavage, demethylation	Benzoic acid and salicylic acid	UV-Vis, GC-MS and FTIR	Hadibarata et al. 2018
Peyronellaea prosopidis	Scarlet RR	Lac, MnP and LiP	Demethylation, denitrification, ring cleavage and reduction	Acetamides and amines	HPLC, FTIR, GC–MS and UV–Vis	Bankole et al. 2018
Ceriporia lacerate	Congo Red	MnP	Broken down of the azo double bond	Naphthylamine and benzidine	UV-Vis and GC-MS	Wang et al. 2017
Pichia occidentalis	Acid Red B	NADH-DCIP reductase	Biodecolourisation, deamination/ desulfonation	4-amino-naphthalene-1-sulfonic acid, 3,4-dihydroxy-naphthalene-1-sulfonic acid and naphthalene-1,2,4-triol	UV–Vis and HPLC-MS	Song et al. (2017)
Scheffersomyces spartinae	Acid Scarlet 3R	NADH-DCIP reductase	Reduction, deamination and desulfonation	Naphthalene derivatives	UV–Vis and HPLC-MS	Tan et al. (2016)
Bacterial microflora.	Direct Black	Lac, MnP, LiP and azoreductase	Cleavage of azo bonds, desulfonation and deamination	2,7,8-triaminonaphthalen-1-ol, phenylenediamine, aniline and phthalic	LC-ESI-MS /MS	Chen et al. (2018)
Bacillus fermus	Acid Blue 9	Aerobic degradation	Cleavage of azo bonds, desulfonation and deamination	Benzene sulfonic acid, benzenesulfonate and methanetriyltribenzene	LC-MS and UV-Vis	Neetha et al. (2018)
Bacillus circulans	Methyl Orange	Azoreductase, NADH-DCIP reductase and Lac	Cleavage of azo bonds	N,N-dimethyl p-phenylenediamine and 4-amino sulfonic acid	FTIR, LC–MS and UV-Vis	Liu et al. (2017)

TABLE 5.6
Biodegradation of synthetic dyes in bioreactors

Strain	Bioreactor	Dye or wastewater	Enzyme	Immobilisation	Colour removal (%)	Reference
Ganoderma sp.	Airlift	Indigo Carmine	Lac	Copper alginate (*in vitro assays*)	100	Teerapatsakul et al. (2017)
Marasmiellus palmivorus	Stirred tank reactor	Reactive Blue 220 (RB) and Acid Green 28 (AG)	Lac	Free (*in vivo and in vitro assays*)	90 (RB) 75 AG	Cantele et al. (2017)
Activated sludge	Airlift	Aniline	–	Self-immobilzation of consortia. (*in vivo*)	99.93	Jiang et al. (2017)
Irpex lacteus	Packet bed reactors	Reactive Orange 16	Lac	Alginate (*in vivo*)	50–60	Šíma et al. (2017)
Aspergillus bombycis	Aerated reactor	Reactive Red 31	Lac, MnP	Free (*in vivo*)	99	
Pleurotus ostreatus + activated sludge	Packed bed reactor	Remazol Brilliant Blue R	Lac	PUF (*in vivo*)	51–95	Svobodová et al. (2016a)
Pleurotus ostreatus	Trickle-bed reactor	Remazol Brilliant Blue R	Lac	PUF (*in vivo*)	40.6	Svobodová et al. (2016b)
Pseudomonas aeruginosa	Packed bed reactor	Acid Maroon V	–	Agar-agar (*in vivo*)	89	Patel and Gupte (2015)
Aeromonas sp.	Packed bed reactor	Methyl Orange	–	Alginate (*in vivo*)	100	Kathiravan et al. 2014
Coriolus versicolor	Membrane reactor	Acid Orange and synthetic wastewater		Free (*in vivo* and *in vitro* assays)	99	Hai et al. (2013)
Pycnoporus sanguineus	Stirred tank reactor	Crystal violet	Lac	Free (*in vivo*)	81.4	Sulaiman et al. (2013)
Bjerkandera sp.	Packed bed reactor	Remazol Red	Lac	Free (*in vivo*)	65–90	Jonstrup et al. (2012)

designs and low cost of installation and operation (Espinosa-Ortiz et al. 2016). A laccase immobilised from *Ganoderma* sp. allowed 100% of decolourisation of Indigo Carmine dye. The treatment was performed in an airlift reactor (Teerapatsakul et al. 2017). Similarly, the aerobic degradation of aniline by a *Proteobacteria* predominated sludge was successfully achieved in an airlift reactor (batch operation) (Jiang et al. 2017).

- Fluidised bed reactors (FBRs) are characterised by the fluidisation of suspensions inside the reactor. In this case, biomass is moved in a fluid stream of liquid or gas. FBRs can be operated in a batch process and allows uniform mixing of particles. An *Irpex lacteus* fungal strain, for example, was efficient in the treatment of Reactive Orange 16 and Naphthol Blue dyes in a fluidised bed reactor (Šíma et al. 2017). Packet bed reactors (PBRs), in contrast, are characterised for the packed of biomass in a static bed. The fixed bed reactor configuration was adequate for treatment with the fungus *Pleurotus ostreatus*, which achieved 51–95% removal of the RBBR dye (Svobodová et al. 2016a). Similarly, a strain of *Aeromonas* sp. immobilised in alginate reached complete degradation of Methyl Orange. The treatment was carried out in a PBR (Kathiravan et al. 2014).

5.7.3 Biocatalysis of Synthetic Dyes

Microbial enzymes are efficient biocatalyst with the ability to degrade several organic compounds. Many microbial enzymes have shown high potential in biotransforming recalcitrant xenobiotics into simpler fragments. Many advantages, such as substrate specificity, efficiency and reusability potential, allow the direct usage of enzymes in the biodegradation of dyes. This approach is a green, alternative for the treatment of synthetic dyes in industrial wastewater (Katheresan et al. 2018; Vikrant et al. 2018). Some disadvantages related to production costs and instability can be reduced with enzyme immobilisation and genetic engineering (Chatha et al. 2017).

A new laccase purified from *Trametes* sp. F1635, for instance, showed high stability under different pH and temperature conditions. The enzyme extract showed high oxidation activity and allowed more than 60% of colour removal of the dyes Eriochrome Black T, Malachite green and Remazol Brilliant Blue R (RBBR) (Wang et al. 2018). A MnP enzyme from *Trametes* sp. 48424 was also extracted and purified. The enzyme was resistant to Ni^{2+}, Li^{+}, Ca^{2+}, K^{+} and Mn^{2+} metal ions and allowed 85–95% of biodegradation of Indigo Carmine, RBBR, Remazol Brilliant Violet 5R and Methyl Green (Zhang et al. 2016). Similarly, MnP extracted and purified from *Cerrena unicolor* BBP6 was efficient in the biodecolourisation of Methyl Orange and RBBR, achieving 77.6 and 81.0% of colour removal, respectively (Zhang et al. 2018).

5.7.4 Mixed Cultures and Hybrid Treatments

The use of mixed cultures and consortia can improve the biodegradation potential of some microorganisms. Combined anaerobic-aerobic treatment is an efficient alternative to improve dye biodegradation. The process can be performed in single batch reactors or integrated bioreactors with an anaerobic and aerobic phase (Kodam and Kolekar 2015).

A thermophilic microflora, composed by nine different strains, was used in the biodegradation of the azo dye Direct Black G. Authors suggested the involvement of various enzymes (Lac, MnP, LiP and azoreductases) in the degradation of dye (Chen et al. 2018). Another anaerobic-aerobic batch system was used for biodegradation of Reactive Red. The activated sludge reached the complete degradation of the dye (Hameed and Ismail 2018). Similarly, a consortium of bacteria allowed 95.8% of degradation of Remazol Black in a microaerophilic–aerobic batch system. The treatment was carried out in a column reactor operating continuously (Kardi et al. 2016).

Another important approach aiming the industrial application of bioremediation is the coupling with conventional biological and physicochemical processes. A recent study, for instance, combined a biological treatment with ultrafiltration. The biological degradation reached 91% and 85% of total organic carbon and colour removal, respectively. The physiochemical process improved the efficiency to 94% and 97%. Many microbial species were identified in the microbiota, including bacteria and fungi (Korenak et al. 2018). In another study, the yeast *Candida tropicalis* improved the efficiency of an activated sludge system during the biodegradation of Acid Red B. The process was operated in a submerged membrane bioreactor (MBR). 90% and 80% of colour removal and TOC were reached (Li et al. 2015). Similarly, a consortium of bacteria and yeast (*Brevibacillus laterosporus* and *Galactomyces geotrichum*) was used in the treatment of real wastewater from the textile industry. The consortium allowed 89% of dye removal, related to the cumulative action of oxidoreductive enzymes (Kurade et al. 2017).

5.8 CONCLUSION AND FUTURE PERSPECTIVES

The handling of dye-containing wastewaters is a matter of concern due to the damage it can cause to the ecosystems. Dyes in wastewaters are usually treated by physical, chemical and biological methods. However, dyeing effluents are difficult to treat by traditional methods due to the high structural stability of dyes. Biodegradation with microbial enzymes appears as an eco-friendly and less expensive alternative for the treatment of industrials wastewaters containing synthetic dyes.

Further researches should be focused in search for new and stronger microorganisms and enzymes, with better ability in biodegradation. Alternatives for integration between physicochemical techniques and bioremediation should also be explored. Further studies in bioreactors should be developed, focusing applications on a pilot and industrial scale. More studies should be developed to identify routes and metabolites of biodegradation. More analysis of the toxicity of the by-products should be developed. On the other hand, it is important to highlight that sustainable dyeing technologies must also be improved, aiming to decrease toxic effects of dyes, improve fixation rates and reduce water and energy consumption.

Is the use of bioremediation technologies and biodegradation of dyes and other recalcitrant organic compounds still very far from the industrial application? From our point of view, as researchers in this area, we believe that this type of green/eco-friendly solution is going to replace or improve the conventional technologies. There is still much to explore.

REFERENCES

Abou-El-Souod, G. W., and M. M. El-Sheekh. 2016. Biodegradation of basic Fuchsin and Methyl Red by the Blue Green algae *Hydrocoleum oligotrichum* and *Oscillatoria limnetica*. *Environmental Engineering and Management Journal* 15 (2): 279–286.

Ali, H. 2010. Biodegradation of synthetic dyes: a review. *Water, Air, & Soil Pollution* 213 (1–4): 251–273.

Anasonye, F. et al. 2015. Bioremediation of TNT contaminated soil with fungi under laboratory and pilot scale conditions. *International Biodeterioration and Biodegradation* 105: 7–12.

Aragaw, T. A, and A. M. Asmare. 2018. Phycoremediation of textile wastewater using indigenous microalgae. *Water Practice and Technology* 13 (2): 274–284.

Arora, D., and R. Sharma. 2010. Ligninolytic fungal laccases and their biotechnological applications. *Applied Biochemistry and Biotechnology* 160 (6): 1760–1788.

Asfaram, A. et al. 2016. Biosorption of Malachite Green by novel biosorbent *Yarrowia lipolytica* Isf7: application of response surface methodology. *Journal of Molecular Liquids* 214: 249–258.

Baldev, E. et al. 2013. Degradation of synthetic dye Rhodamine B to environmentally non-toxic products using microalgae. *Colloids and Surfaces B: Biointerfaces* 105: 207–214.

Bankole, P. et al. 2018. Biodegradation and detoxification of Scarlet RR dye by a newly isolated filamentous fungus, *Peyronellaea prosopidis*. *Sustainable Environment Research* 28 (5): 214–222.

Bide, M. 2014. Sustainable dyeing with synthetic dyes. In *Roadmap to sustainable textiles and clothing: eco-friendly raw Materials, technologies, and processing methods*, ed. S. Senthilkannan, 81–107. Singapore: Springer Singapore.

Bosco, F, C. Mollea, and B. Ruggeri. 2017. Decolorization of Congo Red by *Phanerochaete chrysosporium*: the role of biosorption and biodegradation. *Environmental Technology* 38 (20): 2581–2588.

Brüschweiler, B. J., and C. Merlot. 2017. Azo dyes in clothing textiles can be cleaved into a series of mutagenic aromatic amines which are not regulated tet. *Regulatory Toxicology and Pharmacology* 88: 214–226.

Cantele, C. et al. 2017. Production, characterization and dye decolorization ability of a high level laccase from *Marasmiellus palmivorus*. *Biocatalysis and Agricultural Biotechnology* 12: 15–22.

Cerrón, L. et al. 2015. Decolorization of textile reactive dyes and effluents by biofilms of *Trametes Polyzona* LMB-TM5 and *Ceriporia* sp. LMB-TM1 isolated from the peruvian rainforest. *Water, Air, and Soil Pollution* 226 (8). 10.1007/s11270-015-2505-4.

Chatha, S. et al. 2017. Enzyme-based solutions for textile processing and dye contaminant biodegradation. A review. *Environmental Science and Pollution Research* 24 (16): 14005–14018.

Chen,Y. et al. 2018. Biodegradation and detoxification of Direct Black G textile dye by a newly isolated thermophilic microflora. *Bioresource Technology* 250: 650–657.

Chougule, A. et al. 2014. Microbial degradation and detoxification of synthetic dye mixture by *Pseudomonas* Sp. SUK 1. *Proceedings of the National Academy of Sciences, India Section B: Biological Sciences* 84 (4): 1059–1068.

Colpa, D. et al. 2014. DyP-type peroxidases: a promising and versatile class of enzymes. *Journal of Industrial Microbiology & Biotechnology* 41 (1): 1–7.

Crawford, R. L. 2011. Biodegradation: principles, scope, and technologies. In *Comprehensive Biotechnology*, ed. M. Moo-Young, 3–13. Amsterdam, Netherlands: Academic Press.

Cullen, D. 2014. Wood decay. In *The ecological genomics of fungi*, ed. F. Martin, 43–62. Oxford, UK: John Wiley & Sons, Inc.

Das, S., and H. R. Dash. 2014. Microbial bioremediation: a potential tool for restoration of contaminated areas. In *Microbial biodegradation and bioremediation*, ed. S. Das, 1–21. Amsterdam, Netherlands: Elsevier.

Dave et al. 2015. Bacterial degradation of azo dye containing wastes. In *Microbial degradation of synthetic dyes in wastewaters*, ed. S. N. Singh, 57–83. Cham, Switzerland: Springer International Publishing.

Dellamatrice, P. et al. 2017. Degradation of textile dyes by Cyanobacteria. *Brazilian Journal of Microbiology* 48 (1): 25–31.

Dil, E. et al. 2017. Multi-responses optimization of simultaneous biosorption of cationic dyes by live yeast *Yarrowial lipolytica* 70562 from binary solution: application of first order derivative spectrophotometry. *Ecotoxicology and Environmental Safety* 139: 158–164.

Elisangela, F. et al. 2009. Biodegradation of textile azo dyes by a facultative *Staphylococcus arlettae* strain VN-11 using a sequential microaerophilic/aerobic process. *International Biodeterioration & Biodegradation* 63 (3): 280–288.

El-Sheekh, M. et al. 2017. Biodegradation of some dyes by the cyanobacteria species *Pseudoanabaena* sp. and *Microcystis aeruginosa kützing*. *The Egyptian Journal of Experimental Biology (Botany)* 13 (2): 233–243.

Espinosa-Ortiz, E. et al. 2016. Fungal pelleted reactors in wastewater treatment: applications and perspectives. *Chemical Engineering Journal* 283: 553–571.

Fazal, T. et al. 2018. Bioremediation of textile wastewater and successive biodiesel production using microalgae. *Renewable and Sustainable Energy Reviews* 82: 3107–3126.

Fontoura, J. et al. 2017. Defatted microalgal biomass as biosorbent for the removal of Acid Blue 161 dye from tannery effluent. *Journal of Environmental Chemical Engineering* 5 (5): 5076–5084.

Glenn, J. K., and M. H. Gold. 1983. Decolorization of several polymeric dyes by the lignin-degrading basidiomycete *Phanerochaete chrysosporium*. *Applied and Environmental Microbiology* 45 (6): 1741–1747.

Guadie, A. et al. 2018. *Halomonas* sp. strain A55, a novel dye decolorizing bacterium from dye-uncontaminated rift valley soda lake. *Chemosphere* 206: 59–69.

Gürses, A. et al. 2016. Dyeing and dyeing technology. In *Dyes and pigments*, ed. A. Gürses, 47–67. Cham, Switzerland: Springer International Publishing.

Hadibarata et al. 2018. Biodegradation of Mordant Orange-1 using newly isolated strain *Trichoderma harzianum* RY44 and its metabolite appraisal. *Bioprocess and Biosystems Engineering* 14: 1–12.

Hai, F. et al. 2013. Degradation of azo dye Acid Orange 7 in a membrane bioreactor by pellets and attached growth of *Coriolus versicolour*. *Bioresource Technology* 141: 29–34.

Hameed, B. B. and Z. Ismail. 2018. Decolorization, biodegradation and detoxification of Reactive Red azo dye using non-adapted immobilized mixed cells. *Biochemical Engineering Journal* 137: 71–77.

Hashem, R. et al. 2018. Optimization and enhancement of textile Reactive Remazol Black B decolorization and detoxification by environmentally isolated PH tolerant *Pseudomonas aeruginosa* KY284155. *AMB Express* 8 (1): 83.

Ilibi, M. et al. 2012. Extracellular oxidases of *Cerrena* Sp. complementarily functioning in artificial dye decolorization including laccase, manganese peroxidase, and novel versatile peroxidases. *Biocatalysis and Agricultural Biotechnology* 1 (3): 220–225.

Jaramillo, A. et al. 2017. Degradation of adsorbed azo dye by solid-state fermentation: improvement of culture conditions, a kinetic study, and rotating drum bioreactor performance. *Water, Air, and Soil Pollution* 228 (6). 10.1007/s11270-017-3389-2

Jiang, Y. et al. 2017. Rapid formation of aniline-degrading aerobic granular sludge and investigation of Its microbial community succession. *Journal of Cleaner Production* 166: 1235–1243.

Jonstrup, M. et al. 2012. Decolorization of textile dyes by *Bjerkandera* sp. BOL 13 using waste biomass as carbon source. *Journal of Chemical Technology & Biotechnology* 88 (3): 388–394.

Kardi, S.N., et al. 2016. Biodegradation of Remazol Black B in sequential microaerophilic–aerobic operations by NAR-2 bacterial consortium. *Environmental Earth Sciences* 75: 1172.

Karim, M. et al. 2018. Decolorization of textile reactive dyes by bacterial monoculture and consortium screened from textile dyeing effluent. *Journal of Genetic Engineering and Biotechnology* 16 (2): 375–380.

Katheresan, V. et al. 2018. Efficiency of various recent wastewater dye removal methods: a review. *Journal of Environmental Chemical Engineering* 6 (4): 4676–4697.

Kathiravan, M. et al. 2014. Biodegradation of Methyl Orange by alginate-immobilized *Aeromonas* sp. in a Packed Bed Reactor: external mass transfer modeling. *Bioprocess and Biosystems Engineering* 37 (11): 2149–2162.

Kaushik, P. and A. Malik. 2015. Mycoremediation of synthetic dyes: an insight into the mechanism, process optimization and reactor design. In *Microbial degradation of synthetic dyes in wastewaters*, ed. S. N. Singh, 1–25. Cham, Switzerland: Springer International Publishing.

Khan, R., and M. H. Fulekar. 2017. Mineralization of a sulfonated textile dye Reactive Red 31 from simulated wastewater using pellets of *Aspergillus bombycis*. *Bioresources and Bioprocessing* 4 (1): 23.

Khandare, R. et al. 2013. Synergistic degradation of diazo dye Direct Red 5B by *Portulaca grandiflora* and *Pseudomonas putida*. *International Journal of Environmental Science and Technology* 10 (5): 1039–1050.

Khataee, A. R. et al. 2013. Degradation of an azo dye using the green macroalgae *Enteromorpha* sp. *Chemistry and Ecology* 29 (3): 221–233.

Khataee, A. R., and G. Dehghan. 2011. Optimization of biological treatment of a dye solution by macroalgae *Cladophora* sp. using response surface methodology. *Journal of the Taiwan Institute of Chemical Engineers* 42 (1): 26–33.

Kodam, K. M. and Y. Kolekar. 2015. Bacterial degradation of textile dyes. In: *Microbial degradation of synthetic dyes in wastewaters*, ed. S. N. Singh, 243–266. Cham: Springer International Publishing.

Korenak, J. et al. 2018. Decolourisations and biodegradations of model azo dye solutions using a sequence batch reactor, followed by ultrafiltration. *International Journal of Environmental Science and Technology* 15 (3): 483–492.

Kües, U. 2015. Fungal enzymes for environmental management. *Current Opinion in Biotechnology* 33: 268–278.

Kurade, M. et al. 2013. Degradation of a xenobiotic textile dye, Disperse Brown 118, by *Brevibacillus laterosporus*. *Biotechnology Letters* 35 (10): 1593–1598.

Kurade, M. et al. 2017. Monitoring the gradual biodegradation of dyes in a simulated textile effluent and development of a novel triple layered fixed bed reactor using a bacterium-yeast consortium. *Chemical Engineering Journal* 307: 1026–1036.

Levin, L. et al. 2016. Degradation of 4-Nitrophenol by the white-rot polypore *Trametes Versicolor*. *International Biodeterioration & Biodegradation* 107: 174–179.

Li, H. et al. 2014. In vivo and in vitro decolorization of synthetic dyes by laccase from solid state fermentation with *Trametes* sp. SYBC-L4. *Bioprocess and Biosystems Engineering* 37 (12): 2597–2605.

Li, H. et al. 2015. Reactor performance and microbial community dynamics during aerobic degradation and detoxification of Acid Red B with activated sludge

bioaugmented by a yeast *Candida tropicalis* TL-FL in MBR. *International Biodeterioration & Biodegradation* 104: 149–156.
Liers, C. et al. 2010. DyP-like peroxidases of the jelly fungus *Auricularia auricula-Judae* oxidize nonphenolic lignin model compounds and high-redox potential dyes. *Applied Microbiology and Biotechnology* 85 (6): 1869–1879.
Liu, W. et al. 2016. Methylene Blue enhances the anaerobic decolorization and detoxication of azo dye by *Shewanella Onediensis* MR-1. *Biochemical Engineering Journal* 110: 115–124.
Liu, W. et al. 2017. Simultaneous decolorization of sulfonated azo dyes and reduction of hexavalent chromium under high salt condition by a newly isolated salt-tolerant strain *Bacillus circulans* BWL1061. *Ecotoxicology and Environmental Safety* 141: 9–16.
Lončar, N. et al. 2016. Exploring the biocatalytic potential of a DyP-type peroxidase by profiling the substrate acceptance of *Thermobifida fusca* DyP Peroxidase. *Tetrahedron* 72 (46): 7276–7281.
Lu, R. et al. 2016. White-rot fungus *Ganoderma* Sp. En3 had a strong ability to decolorize and tolerate the anthraquinone, indigo and triphenylmethane dye with high concentrations. *Bioprocess and Biosystems Engineering* 39 (3): 381–390.
Ma, L. et al. 2014. Efficient decolorization and detoxification of the sulfonated azo dye Reactive Orange 16 and simulated textile wastewater containing Reactive Orange 16 by the white-rot fungus *Ganoderma* sp. En3 isolated from the forest of Tzu-Chin Mountain in China. *Biochemical Engineering Journal* 82: 1–9.
Mahalakshmi, S., D. Lakshmi, and U. Menaga. 2015. Biodegradation of different concentration of dye (Congo Red dye) by using green and blue green algae. *International Journal of Environmental Research* 9 (2): 735–744.
Martorell, M. et al. 2012. Dye-decolourizing yeasts isolated from Las Yungas rainforest. Dye assimilation and removal used as selection criteria. *International Biodeterioration & Biodegradation* 66 (1): 25–32.
Martorell, M. et al. 2017. Biological degradation of Reactive Black 5 dye by yeast *Trichosporon akiyoshidainum*. *Journal of Environmental Chemical Engineering* 5 (6): 5987–5993.
Mendes, S. et al. 2015. Bacterial enzymes and multi-enzymatic systems for cleaning-up dyes from the environment. In *Microbial degradation of synthetic dyes in wastewaters*, ed. S. N. Singh, 27–55. Cham, Switzerland: Springer International Publishing.
Mokhtar et al. 2017. Biosorption of azo-dye using marine macro-alga of *Euchema spinosum*. *Journal of Environmental Chemical Engineering* 5 (6): 5721–5731.
Mullai et al. 2017. Aerobic treatment of effluents from textile industry. In *Biological treatment of industrial effluents*, eds. D. Lee, V. Jegatheesan, H. H. Ngo, P. C. Hallenbeck, and B. T. Ashok, 3–34. Amsterdam, Netherlands: Elsevier.
Munck, C. et al. 2018. Biofilm formation of filamentous fungi *Coriolopsis* sp. on simple muslin cloth to enhance removal of triphenylmethane dyes. *Journal of Environmental Management* 214: 261–266.
Neetha, J. et al. 2018. Aerobic biodegradation of Acid Blue-9 dye by *Bacillus fermus* isolated from *Annona reticulate*. *Environmental Technology & Innovation* 11: 253–261.
Nguyen, T. et al. 2017. Tiamulin removal by wood-rot fungi isolated from swine farms and role of ligninolytic enzymes. *International Biodeterioration & Biodegradation* 116: 147–154.
Ortiz-Monsalve, S. et al. 2017. Biodecolourisation and biodegradation of leather dyes by a native isolate of *Trametes Villosa*. *Process Safety and Environmental Protection* 109: 437–451.
Ortiz-Monsalve, S. et al. 2019. biodecolourisation and biodetoxification of dye-containing wastewaters from leather dyeing by the native fungal strain *Trametes villosa* SCS-10. *Biochemical Engineering Journal* 141: 19–28.

Otto, B. et al. 2010. First description of a laccase-like enzyme in soil algae. *Archives of Microbiology* 192 (9): 759–768.
Otto, B., and D. Schlosser. 2014. First laccase in green algae: purification and characterization of an extracellular phenol oxidase from *Tetracystis Aeria*. *Planta* 240 (6): 1225–1236.
Pajot, J. et al. 2011. Evidence on manganese peroxidase and tyrosinase expression during decolorization of textile industry dyes by *Trichosporon akiyoshidainum*. *International Biodeterioration & Biodegradation* 65 (8): 1199–1207.
Patel, D. et al. 2017. Enzyme mediated bacterial biotransformation and reduction in toxicity of 1:2 chromium complex AB193 and AB194 dyes. *Journal of the Taiwan Institute of Chemical Engineers* 77: 1–9.
Patel, Y, and A Gupte. 2015. Biological treatment of textile dyes by agar-agar immobilized consortium in a packed bed reactor. *Water Environment Research* 87 (3): 242–251.
Pathak, V. et al. 2015. Experimental and kinetic studies for phycoremediation and dye removal by *Chlorella pyrenoidosa* from textile wastewater. *Journal of Environmental Management* 163: 270–277.
Pereira, L., and M. Alves. 2012. Dyes: environmental impact and remediation. In *Environmental protection strategies for sustainable development*, eds. A. Malik and E. Grohmann, 111–162. Dordrecht, Netherlands: Springer Netherlands.
Phang, S.-M., W.-L. Chu, and R. Rabiei. 2015. Phycoremediation. In *The Algae world*, eds. D. Sahoo and J. Seckbach, 357–389. Dordrecht, Netherlands: Springer Netherlands.
Plácido, J., S. Ortiz-Monsalve, S. et al. 2016. Degradation and detoxification of synthetic dyes and textile industry effluents by newly isolated *Leptosphaerulina* sp. from Colombia. *Bioresources and Bioprocessing* 3 (1): 1–14.
Priya, B. et al. 2011. Ability to use the diazo dye, C.I. Acid Black 1 as a nitrogen source by the marine cyanobacterium *Oscillatoria curviceps* BDU92191. *Bioresource Technology* 102 (14): 7218–7223.
Qu, Y. et al. 2012. Aerobic decolorization and degradation of Acid Red B by a newly isolated *Pichia* sp. TCL. *Journal of Hazardous Materials* 223–224: 31–38.
Rodríguez-Couto, S. 2015. Degradation of azo dyes by white-rot fungi. In *Microbial Degradation of Synthetic Dyes in Wastewaters*, ed. S.N. Singh, 315–331. Cham, Switzerland: Springer International Publishing.
Rosu, C. et al. 2017. Biodegradation and detoxification efficiency of azo-dye Reactive Orange 16 by *Pichia kudriavzevii* CR-Y103. *Water, Air, & Soil Pollution* 229 (1): 15.
Sabnis, R. W. 2017. Manufacture of dye intermediates, dyes, and their industrial applications. In *Handbook of industrial chemistry and biotechnology*, eds. J. A. Kent, T. Bommaraju and S. Barnicki, 581–676. Cham, Switzerland: Springer International Publishing.
Santos, A. et al. 2014. New dye-decolorizing peroxidases from *Bacillus subtilis* and *Pseudomonas putida* MET94: towards biotechnological applications. *Applied Microbiology and Biotechnology* 98 (5): 2053–2065.
Saratale, R. et al. 2011. Bacterial decolorization and degradation of azo dyes: a review. *Journal of the Taiwan Institute of Chemical Engineers* 42 (1): 138–157.
Sen, S. K. et al. 2016. Fungal decolouration and degradation of azo dyes: a review. *Fungal Biology Reviews* 30 (3): 112–133.
Šíma, J., R. Milne, Č. Novotný, and P. Hasal. 2017. Immobilization of *Irpex lacteus* to liquid-core alginate beads and their application to degradation of pollutants. *Folia Microbiologica* 62 (4): 335–342.
Singh, R. L. et al. 2015. Enzymatic decolorization and degradation of azo dyes – a review. International *Biodeterioration & Biodegradation* 104: 21–31.
Sinha, S. et al. 2016. Self-sustainable *Chlorella pyrenoidosa* strain NCIM 2738 based photobioreactor for removal of Direct Red-31 dye along with other industrial

pollutants to improve the water-quality. *Journal of Hazardous Materials* 306: 386–394.

Sinha, A. et al. 2018. Degradation of Reactive Green dye and textile effluent by *Candida* sp. VITJASS isolated from wetland paddy rhizosphere soil. *Journal of Environmental Chemical Engineering* 6 (4): 5150–5159.

Sodaneath, H. et al. 2017. Decolorization of textile dyes in an air-lift bioreactor inoculated with *Bjerkandera adusta* OBR105. *Journal of Environmental Science and Health – Part A Toxic/Hazardous Substances and Environmental Engineering* 52 (11): 1099–1111.

Solís, M. et al. 2012. Microbial decolouration of azo dyes: a review. *Process Biochemistry* 47 (12): 1723–1748.

Song, L. et al. 2017. Performance of a newly isolated salt-tolerant yeast strain *Pichia Occidentalis* G1 for degrading and detoxifying azo dyes. *Bioresource Technology* 233: 21–29.

Speight, J. G, and N. S. El-Gendy. 2018. Chemistry of biotransformation. In *Introduction to petroleum biotechnology*, eds. J. G. Speight and N. S. El-Gendy, 287–359. Boston, MA: Gulf Professional Publishing.

Stella, T. et al. 2017. Bioremediation of long-term PCB-contaminated soil by white-rot fungi. *Journal of Hazardous Materials* 324: 701–710.

Sulaiman, S. et al. 2013. Triarylmethane dye decolorization by pellets of *Pycnoporus sanguineus*: statistical optimization and effects of novel impeller geometry. *Bioremediation Journal* 17 (4): 305–315.

Svobodová, K. et al. 2016a. Mutual interactions of *Pleurotus ostreatus* with bacteria of activated sludge in solid-bed bioreactors. *World Journal of Microbiology and Biotechnology* 32 (6). 10.1007/s11274-016-2050-3.

Svobodová, K. et al. 2016b. Selective adhesion of wastewater bacteria to *Pleurotus ostreatus* mycelium in a trickle-bed bioreactor. *AIMS Environmental Science 3* (3): 395–407.

Taha, M. et al. 2017. Bioremediation of biosolids with *Phanerochaete chrysosporium* culture filtrates enhances the degradation of polycyclic aromatic hydrocarbons (PAHs). *Applied Soil Ecology* 124: 163–170.

Tan, L. et al. 2013. Aerobic decolorization and degradation of azo dyes by growing cells of a newly isolated yeast *Candida tropicalis* TL-F1. *Bioresource Technology* 138: 307–313.

Tan, L. et al. 2014. Aerobic decolorization and degradation of azo dyes by suspended growing cells and immobilized cells of a newly isolated yeast *Magnusiomyces ingens* LH-F1. *Bioresource Technology* 158: 321–328.

Tan, L. et al. 2016. Aerobic decolorization, degradation and detoxification of azo dyes by a newly Isolated salt-tolerant yeast *Scheffersomyces spartinae* TLHS-SF1. *Bioresource Technology* 203: 287–294.

Teerapatsakul, C. et al. 2017. Repeated batch for dye degradation in an airlift bioreactor by laccase entrapped in copper alginate. *International Biodeterioration & Biodegradation* 120: 52–57.

Vasiliadou, I. A. et al. 2016. Biological removal of pharmaceutical compounds using white-rot fungi with concomitant FAME production of the residual biomass. *Journal of Environmental Management* 180: 228–237.

Vikrant, K. et al. 2018. Recent advancements in bioremediation of dye: current status and challenges. *Bioresource Technology* 253: 355–367.

Waghmode, T. et al. 2012. Degradation of Remazol Red dye by *Galactomyces geotrichum* MTCC 1360 Leading to increased iron uptake in *Sorghum vulgare* and *Phaseolus mungo* from soil. *Biotechnology and Bioprocess Engineering* 17 (1): 117–126.

Wang, N. et al. 2017. Decolorization and degradation of Congo Red by a newly isolated white rot fungus, *Ceriporia facerata*, from decayed mulberry branches. *International Biodeterioration & Biodegradation* 117: 236–244.

Wang, S. N. et al. 2018. An extracellular yellow Laccase from White Rot Fungus Trametes Sp. {F1635} and its mediator systems for dye decolorization. *Biochimie* 148: 46–54.

Watkinson, S. C. 2016. Physiology and adaptation. In *The Fungi*, eds. S. C. Watkinson, L. Boddy, and N. P. Money, 141–187. Amsterdam, Netherlands: Elsevier Academic Press.

Wirasnita, R., and T. Hadibarata. 2016. Potential of the white-rot fungus *Pleurotus pulmonarius* F043 for degradation and transformation of fluoranthene. *Pedosphere* 26 (1): 49–54.

Xie, L. et al. 2016. Degradation of disperse blue 2BLN by *Oleaginous sorokiniana* XJK. *RSC Advances* 6 (108): 106935–106944.

Yang, S. et al. 2017. Decolorization of acid, disperse and reactive dyes by *Trametes versicolor* CBR43. *Journal of Environmental Science and Health – Part A Toxic/Hazardous Substances and Environmental Engineering* 52 (9): 862–872.

Yu, L. et al. 2015. Intracellular azo decolorization is coupled with aerobic respiration by a *Klebsiella oxytoca* Strain. *Applied Microbiology and Biotechnology* 99 (5): 2431–2439.

Zhang, H. et al. 2016. Characterization of a manganese peroxidase from white-rot fungus *Trametes* Sp.48424 with strong ability of degrading different types of dyes and polycyclic aromatic hydrocarbons. *Journal of Hazardous Materials* 320: 265–277.

Zhang, H. et al. 2018. Purification and characterization of a novel manganese peroxidase from white-rot fungus *Cerrena unicolor* BBP6 and its application in dye decolorization and denim bleaching. *Process Biochemistry* 66: 222–229.

Zhao et al. 2014. Biodegradation of Methyl Red by *Bacillus* sp. strain UN2: decolorization capacity, metabolites characterization, and enzyme analysis. *Environmental Science and Pollution Research* 21 (9): 6136–6145.

6 The Cross-Talk between Bioremediation and Valuation of Residues of the Olive-Oil Production Chain

Ana Filipa Domingues
Department of Biology, University of Aveiro, Aveiro, Portugal

Inês Correia Rosa
CESAM, University of Aveiro, Aveiro, Portugal

Ruth Pereira
GreenUPorto – Sustainable Agrifood Research Centre, Department of Biology, Faculty of Sciences of the University of Porto, Rua do Campo Alegre, Portugal

Joana Luísa Pereira
Department of Biology, University of Aveiro, Aveiro, Portugal
CESAM, University of Aveiro, Aveiro, Portugal

6.1 INTRODUCTION

Earth has been experiencing profound changes throughout the last few centuries. Since the Industrial Revolution in the XVIII century, humanity continues to witness great and fast technological development, in parallel with a substantial improvement in their lifespan and living standards. Consequently, the world population has been increasing exponentially in the last years, and the subsequent environmental impacts are becoming each day more evident. Agricultural and industrial activities represent major routes of pollution that are threatening the normal functioning and structure of ecosystems. Since some biological and chemical processes required to maintain and preserve life have already overcome their resilience capacity (Rockström et al. 2009), it is urgent to develop solutions and more sustainable practices to mitigate these problems and to restore the multifunctionality of ecosystems.

The Mediterranean region is known for a high production of olive oil mainly because of the climatic conditions of this geographical area. Summers are warm and dry, with average temperatures of 30°C, and winters are characterized as moderate due to intense and frequent rainfalls, mainly between October and March/April (Zampounis and Minoans 2006). Olive oil is considered the main vegetable fat in the regional gastronomy, not only because of its therapeutic properties (Obied et al. 2005) but also because of the flavor and aromas it adds to the recipes, thus having a high market value in all the Mediterranean and European countries (Zampounis and Minoans 2006). Moreover, olive oil (and also its associated wastewaters) contains flavonoids and carotenoids. These two relevant groups of phytocompounds are known for their huge benefits in human health, having high market value.

The importance of this industry to the economy of Portugal Spain, Italy and Greece is notorious. In fact, the International Olive Oil Council states a contribution from these countries of about 94% of the worldwide production in 2017; besides these European countries, Tunisia, Morocco, Turkey, Algeria, Argentina, Jordan and Egypt are estimated to contribute with 29% of the total world production (IOOC 2017).

Notwithstanding, olive oil bears a huge water footprint and extremely toxic wastewaters that remain in the end of the production process (Dermeche et al. 2013; El-Abbassi et al. 2012; Gebreyohannes et al. 2016; Roig et al. 2006). The widespread use of the traditional oil extraction system is the main contributor to the elevated water consumption rates registered in south European countries, along with the production of larger volumes of wastewaters when compared with more recent processes (IOOC 2015). The discharge of 1 m^3 (a tonne) of olive oil mill wastewaters (OOMW) in receiving systems has the same environmental impact of about 200 m^3 of domestic wastewater (El-Abbassi et al. 2012). If one considers that 1 tonne of olives generates between 1 and 2 tonnes of OOMW and that Mediterranean countries discharge 30 million m^3 of this effluent each year, the magnitude of the consequences of this practice becomes clearly anticipated (Gebreyohannes et al. 2016; Jalilnejad et al. 2011). Facing these numbers, it is urgent to design solutions that do not only reduce the volume of OOMW but also adequately treat this externality. The present chapter will focus on the current strategies to improve OOMW treatment and management, as well as it aims to propose new avenues that can be further explored in this context.

6.1.1 Regulation for OOMW Discharges in South European Countries

Once recognized the environmental impacts of OOMW, there have been increasing efforts to regulate OOMW discharges in South European countries. Thus, particularly focused in the safety and protection of natural resources (namely water and soils), several regulatory frameworks have been developed and enforced. These can directly or indirectly contribute to mitigate the ecological impacts of OOMW, in terms of ecosystems structure, function and services.

In Europe, the Water Framework Directive (WFD) is focused in the monitoring and maintenance of the ecological status of aquatic systems (surface, transition, coastal and groundwaters) with the aim of promoting all water bodies to a good ecological status by 2027 (European Commission 2000). To achieve this, several measures are encouraged, including sustainable water consumption, reduction in discharges, emissions and/or losses of priority substances to the aquatic environment. Albeit the enforcement of this legislation triggered new environmental responsibilities to the EU member states regarding the continuous monitoring of their waters by the competent entities, there are still gaps in the regulations that allow mismanagement. For example, the noneffectiveness of the “polluter-paying principle”, since the fines of inappropriate wastewater disposal are more cost-efficient than the correct implementation of a water treatment facility, at least for small/medium industries.

Focusing on the disposal of olive oil effluents in particular, different countries have distinct legal frameworks. Whereas no legislation regarding OOMW discharge into the natural environment exists in Greece (Azbar et al. 2004; Inglezakis et al. 2012; Kapellakis et al. 2008), more firm legislation was implemented in Spain, where it is strictly forbidden to discharge untreated OOMW in natural systems (Azbar et al. 2004; Inglezakis et al. 2012; Kapellakis et al. 2008). Also, Spain promoted the construction of about 1000 evaporation ponds to deal with OOMW, along with the implementation of two-phase systems to reduce OOMW volume (Azbar et al. 2004; Inglezakis et al. 2012; Kapellakis et al. 2008). Both Italy and Portugal have intermediate regulatory demands. Italy allows the spread of treated OOMW in agricultural lands under controlled conditions after notification of the competent authorities (Azbar et al. 2004; Inglezakis et al. 2012; Kapellakis et al. 2008). In Portugal, olive oil mills must establish contracts with water and sanitation authorities to deal with produced OOMW, along with the implementation of preventive measures for accidental discharges (Normative Order no. 118/2000; Ordinance 1030/98). Wastewater discharges in waters and soils are allowed, as long as olive oil producers require a license and fulfil regulatory criteria (Normative Order no 626/2000, and more generally Law-by-Decree 236/98).

Despite the development of dedicated environmental policies, it is evident that its enforcement has not been successful, since impacts of OOMW in aquatic and terrestrial ecosystems continue to be reported worldwide (Aharonov-Nadborny et al. 2018; Elhag et al. 2017; Karaouzas et al. 2011a; Pavlidou et al. 2014; Sierra et al. 2001). A consensual framework should be developed with the intent of unifying and disseminating the best environmental practices supported by the evidences generated by the scientific community (Koutsos et al. 2018)—for instance, protective European frameworks directed to both soils and wastewaters are urgent and crucial in the case of OOMW (Komnitsas and Zaharaki 2012). On the other hand, through the development of sustainable solutions promoting the reuse of OOMW in activities such as agriculture, circular economy principles can apply by transforming a waste into a valuable product. Finally, funding to support the implementation of infrastructures for treatment, storage, distribution and disposal of OOMW would certainly boost the commitment of the producers. Thus, a common, integrated regulation addressing treatment, management practices and practical applications, along with the definition of a maximum allowed benchmark value for discharge to be adopted by all EU member states, is critical to assist policy makers in the framing of olive oil mills optimization, ultimately satisfying the social demands toward more sustainable production processes (Koutsos et al. 2018).

6.2 OLIVE OIL MILL WASTEWATERS

6.2.1 Olive Oil Production Processes

Olea europaea L., commonly known as olive, is the species that support the olive oil value chain. Olive oil extraction is a well-tuned process regarding the adequate treatment and possible valuation strategies of solid wastes (olive husk or pomace) and effluents (Morillo et al. 2009; Roig et al. 2006). The production of olive oil is commonly conducted between November and February, and it is achieved by either a continuous centrifugation (modern process) or the application of discontinuous pressure in the olives (oldest and most traditional process).

In general, the traditional process involves the following sequential steps: (1) crushing, (2) pressing and (3) decantation. After crushing the olives, the resulting paste is mixed and homogenized to aggregate oil droplets. The paste is then placed in the hydraulic press for squeezing—a small amount of water is added to help the separation of the oil/water phases. The remaining product is a mixture of olive pulp, peel, stone and water, generally defined as olive pomace. The water mixed with olive oil is later separated by decantation. At the end of this process, three components remain: olive oil, pomace and OOMW. Although this is a simple method involving inexpensive equipment and producing low amount of OOMW, it is a discontinuous process entailing high-energy costs. In addition, the OOMW obtained represents high rates of (bio)chemical oxygen demand (BOD) in receiving environmental compartments (Dermeche et al. 2013), thus a relevant input of organic contaminants.

The continuous process comprises either a three-phase (3P) or a two-phase system (2P), and the main difference between them is the final product. The 3P system yields three final products: olive pomace (solid phase), OOMW and olive oil (liquid phases). The main problem with this system is the enormous amount of water necessary for the centrifugation step that is directly related to the final amount of OOMW produced. The 2P system was developed to minimize the volume of OOMW, thus there is a single addition of water in the crushing phase. In the 2P system, there are only two remaining products: olive oil and wet pomace, commonly designated by TPOMW ("two-phase olive-mill waste"), which is a mixture of OOMW and olive husk that can be reprocessed to enhance the extraction yield (Aggoun et al. 2016; Dermeche et al. 2013). Fig. 6.1 shows the steps of each production processes of olive oil.

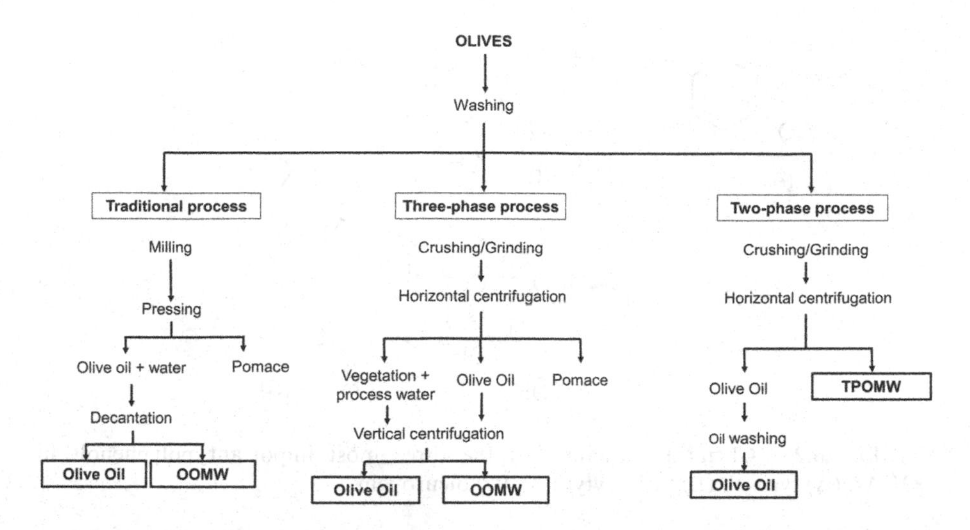

FIGURE 6.1 Production processes for olive oil (adapted from Morillo et al., 2009).

6.2.2 Characterization, Physical and Chemical Properties of OOMW

OOMW properties depend on olive orchards management, olives maturation and extraction processes, as well as on the storage time and regional climatic conditions (Aggoun et al. 2016; Justino et al. 2010; McNamara et al. 2008).

OOMW is known to have a dark color and an intense smell (Justino et al. 2010; McNamara et al. 2008). The main properties of this wastewater are an acidic pH, high turbidity, conductivity, content of suspended solids and high biochemical and chemical oxygen demand (COD and BOD, respectively). BOD typically ranges from 35 to 110 g/L, whereas COD ranges between 40 and 220 g/L (Cassano et al. 2013; Dermeche et al. 2013). These properties can be explained by the high organic content of OOMW, such as lipids, sugars (fructose, mannose, glucose, sucrose, pentose and sucrose), tannins, pectins, polyalcohols and polyphenols, along with some inorganic salts (namely potassium, calcium, sodium, magnesium and iron).

Polyphenols are the major components of OOMW in concentrations of about tens g/L, particularly hydroxytyrosol, tyrosol and oleuropein (Fig. 6.2; Cassano et al. 2013; Justino et al. 2010, 2012; McNamara et al. 2008).

Although when heavily concentrated as in OOMW, polyphenols drive severe toxic effects in different biological receptors (see Section 6.2.3), when used in appropriate doses, these compounds have potential benefits to human health (Gebreyohannes et al. 2016). The benefits of polyphenolic compounds present in OOMW, such as flavonoids and carotenoids, relate to their high activity as antioxidant, anti-inflammatorie, antiallergic, antiviral and antitumor compounds (Aggoun et al. 2016; El-Abbassi et al. 2012; Obied et al. 2005). Among them, hydroxytyrosol is the one with greater antioxidant activity given its great ability

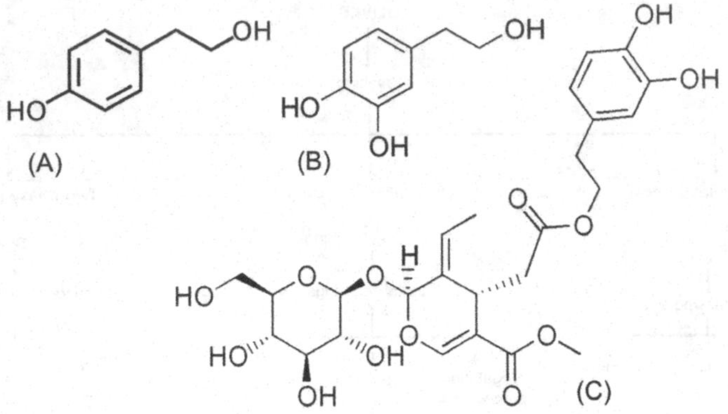

FIGURE 6.2 Chemical structures of the three most important polyphenols in OOMW: (A) tyrosol, (B) hydroxytyrosol, (C) oleuropein.

to protect cells against oxidative stress (Aggoun et al. 2016; El-Abbassi et al. 2012; Kalogerakis et al. 2013). Antioxidant compounds act through the chelation or scavenging of free radicals, such as the superoxide anion, hydrogen peroxide or the hydroxyl radical preventing them to impair biomolecules such as proteins, lipids and DNA.

Regarding oil extraction by-products other than OOMW itself, TPOMW contains essentially olive husk, OOMW and inorganic salts; in the OOMW fraction tyrosol, hydroxytyrosol, p-coumaric and vanillic acids can naturally be found, apart from water. The olive pomace includes cellulose, hemicellulose, lignin, fats and proteins. The main polyphenols existing in the pomace are salidroside, nuezhenide and nuezhenide-oleoside, all of them with more complex chemical structures compared with those found in OOMW (Dermeche et al. 2013).

Olive oil wastes hold relevant microbial communities. The bacterial communities in OOMW and TPOMW are mostly composed by *Alphaproteobacteria, Betaproteobacteria, Gammaproteobacteria, Firmicutes* and *Actinobacteria.* Relatively to fungi, *Candida* spp., *Pichia* and *Sacharomyces* are the most common yeast taxa found in OOMW; while in TPOMW the predominant biota is composed by *Pichia* and *Sacharomyces.* Acidic conditions and high concentrations of salts and sugars favor the growth of osmotolerant yeasts and bacteria due to the higher ability of fungi to decompose organic matter, or given a better adaptation of yeast to the conditions of OOMW as a growth medium (Ben Sassi et al. 2006; Ntougias et al. 2013).

Finally, it is worth noting that the storage of OOMW induces some chemical modifications. Olive oil undergoes lipids oxidation and hydrolysis, with the subsequent degradation of its quality caused by the by-products of these reactions. The quality of the olive oil directly relates to its physical–chemical properties, so the nonfiltered olive oil is more unstable due to the presence of suspended solids. The characteristic dark color of OOMW is also a result of the previous degradation routes that adversely affect the antioxidant activity of the polyphenols, thus promoting the recalcitrant capacity of this wastewater (Kalogerakis et al. 2013).

Volatile compounds are mainly hydroperoxide anions, their production augmenting in the presence of light, metals, high temperature, pigments and unsaturated fatty acids, but also depending on the quantity of natural sterols and antioxidants (Angerosa et al. 2004). These parameters, along with the presence of microorganisms, originate the typical strong smell of stored OOMW (Dermeche et al. 2013; Justino et al. 2009). In a study about the volatilization of OOMW compounds after its spreading in soils (Rana et al. 2003), it was concluded that OOMW emissions are mainly composed by phenol compounds and sulfur dioxide and that these emissions are more likely to occur with higher temperatures.

6.2.3 Environmental Hazardous Potential of OOMW

The environmental hazards of chemical compounds are typically evaluated according to their bioaccumulation, persistence and toxicity. A surrogate commonly used for bioaccumulation is the oil/water partition coefficient (K_{ow}), which evaluates the ease of a chemical to associate with an organic or aqueous phase—when this ratio is higher than 5, the focused compound is lipophilic. Persistence is defined by the ability of a chemical to withstand environmental degradation, thus to be transported over long distances with respect to its original disposal site. The persistence of a given compound can be ranged according to its half-life in water (hydrolysis), sediment or soil, and can be mediated by certain factors such as light (photolysis) or microbial communities (biodegradation). Finally, toxicity evaluates the ability of a given dose of a chemical (parent compound or degradation product) in inducing any type of adverse effects in the biota (Muir and Howard 2006; Wania 2003), clarifying on cause–effect relationships.

Polyphenols are the most concerning compounds in OOMW, as they bioaccumulate and induce several toxic effects. Polyphenols tend to be dominant in the wastewater, according to their K_{ow} established between 6×10^{-4} and 4 (Aggoun et al. 2016; Obied et al. 2005). In addition, K_{ow} is influenced by temperature and by the water load through the olive oil extraction process: under higher temperature, the partition of polyphenols into the oil phase is promoted; if the water volume is increased, they are preferably transferred into the aqueous phase, i.e., OOMW (Obied et al. 2005). Studies focusing on the persistency of OOMW are scarce and the half-life of its phenolic compounds in either soils or water has not been determined yet. Phenolic compounds are the main responsible for the toxicity attributed to the olive oil wastewaters. The hazard potential associated with phenolics has been related to a high ability in inducing the production of reactive oxygen species (ROS), narcosis and respiratory disturbances (Justino et al. 2012).

It has been proved that the most toxic OOMW is the one resulting from the traditional extraction process when compared with 3P and 2P processes, which is likely related to its lowest water content hence highest concentration of phenolic compounds (Ben Sassi et al. 2006). Regardless of the extraction process, OOMW exhibits notorious toxic effects in the biota of water and soil compartments provided their common discharge directly into rivers or in agricultural soils draining into watercourses (Andreozzi et al. 2008; Tafesh et al. 2011; Dhouib et al. 2006; Fiorentino et al. 2003; Isidori et al. 2004; Karaouzas et al. 2011b; Mekki et al. 2008; Priac et al. 2017; Tsioulpas et al. 2002; Venieri et al. 2010). Moreover, following deposition in soils, these wastes may leach to groundwaters, triggering a contamination cycle that may contribute to additionally impact aquatic ecosystems.

Fiorentino et al. (2003) exposed four freshwater organisms (*Pseudokirchneriella subcapitata, Daphnia magna, Brachionus calcyflorus* and *Thamnocephalus platyurus*) to 15 phenolic compounds highly abundant in OOMW (including tyrosol and hydroxytyrosol), and all compounds exhibited toxic potential to one or more organisms. Growth inhibition and mortality were the main toxic effects reported in several aquatic species after exposure to OOMW including *P. subcapitata* (Andreozzi et al. 2008), *D. magna* (Isidori et al. 2004), *D. longispina* (Justino et al. 2009), *Gammarus pulex* (Karaouzas et al. 2011b), *Hydropsyche peristerica* (Karaouzas et al. 2011b) and *Danio rerio* (Venieri et al. 2010).

Beyond the impacts recorded directly in organisms, the discharge of OOMW into the environment may transform clear and transparent waters into an opaque and highly contaminated system. This scenario may translate into severe impacts on aquatic and terrestrial ecosystems. Another common downstream environmental problem resulting from OOMW discharge is eutrophication in rivers and lakes, primarily because these effluents contain high concentrations of nutrients, promoting the overgrowth of aquatic microalgae. OOMW has a high amount of sugars in its composition which induce microbial metabolism, as well as stimulate the growth of aquatic flora (McNamara et al. 2008). Lipids and polyphenols composing the OOMW can block the entrance of sunlight and oxygen in the aquatic systems due to the formation of a biofilm in the water surface; this constrains the growth of pelagic and benthic producers and compromises oxygen levels available for the subsurface biota (Dermeche et al. 2013; McNamara et al. 2008). Eutrophication does not only deteriorate the ecological and chemical status of natural ecosystems but may also negatively impact recreational and touristic activities, fishing and sailing, raising economic and cultural concerns.

Regarding the specific case of discharge of OOMW in soils, it leads to the impairment of the soil structure by changing the soil resistance and increasing salinity levels. In addition, phenolic compounds trigger toxic effects and have the potential of inhibiting the germination and growth of plants (Barbera et al. 2013; Chatzistathis and Koutsos 2017; Dermeche et al. 2013; Karpouzas et al. 2010), as well as the reproduction of some invertebrates (Hentati et al. 2016). Regarding adverse effects in soil species, plants growth impairment and seeds germination inhibition were the most commonly observed effects in terrestrial species with agronomic importance as *Latuca sativa* (Priac et al. 2017),

Hordeum vulgare (Ben Sassi et al. 2006), *Lepidium sativum* (Tsioulpas et al. 2002), *Raphanus sativus* and *Cucumis sativus* (Andreozzi et al. 2008). Moreover, microorganisms such as *Bacillus subtilis, B. megaterium, Escherichia coli, Pseudomonas aeruginosa, P. fluorescens, Staphylococcus aureus, Streptococcus pyogenes, Klebsiella pneumoniae* and *Vibrio fischeri* were all found to be negatively affected by exposure to OOMW through the inhibition of growth rates and luminescence (Tafesh et al. 2011; Dhouib et al. 2006; Mekki et al. 2008; Obied et al. 2007). Shifts in the physiological profile of microbial communities were also retrieved after the exposure of soils to high concentrations of polyphenols (Hentati et al. 2016), leading to modifications in local microflora diversity.

6.3 OOMW MANAGEMENT

6.3.1 OOMW VALUATION STRATEGIES

Valuation can be defined as a strategy that aims to attain a sustainable development through the management of industrial by-products, wastes and wastewaters until their ultimate reuse for industrial processes or agriculture, or through the recovery of its fine chemicals (El-Abbassi et al. 2017). The Mediterranean region has the most significant olive oil production worldwide, thus the major volumes of OOMW to deal with. This region is characterized by arid soils, water scarcity and low-energy resources. Therefore, it is important to the region that proper management strategies based on the reuse of OOMW are developed and applied to better achieve economic and environmental sustainability. The valuation of OOMW can be done by using it for different applications, or by taking profit of some of its components.

Application of olive oil by-products in soils as substitutes of chemical fertilizers has been a hot topic within the scientific community (Chatzistathis and Koutsos 2017; El-Abbassi et al. 2017). Research in this field is challenging, and several avenues are worth exploring. For example, application of OOMW in olive orchards might be a management solution for this effluent, and this was the basis of the work by Mechri et al. (2009). Their focus was on the putative consequences of the application of this waste as a fertilizer in the olive trees development, the olive and the olive oil quality. Their findings suggest that arbuscular mycorrhizal fungi were negatively affected by OOMW, reducing the colonization of the olive tree roots. This process gradually promoted other side effects contrasting to the increase of olive oil phenol load, namely the reduction of the photosynthetic rates, as well as the decrease in nitrogen, phosphorus and oil contents of the olive fruit. Overall, this study indicates that the use of OOMW as a fertilizer should be reasoned considering its effects on the final characteristics of the olive fruit, thus in the overall olive oil quality.

Nevertheless, the application of OOMW in field crops should always be limited to low volumes to prevent profound changes in soil structure and properties (Barbera et al. 2013). This control should not be decided based only on the polyphenol content of OOMW but must also take into account a series of other variables such as soil type, pH, electrical conductivity, total organic matter, total nitrogen and nitrates, available phosphorus, exchangeable potassium and soil moisture. Indeed, the toxic effects of OOMW are a result of a complex interaction between all these variables and not the direct consequence of the increase in polyphenols (Hentati et al. 2016; Komnitsas and Zaharaki 2012). A long- and short-term assessment of the impacts of spreading OOMW in soils was presented by Di Bene et al. (2013). They concluded that soils tend to reach their initial chemical state after being in contact with OOMW, although recognizing the risk of applying these effluents in terrestrial ecosystems. Nevertheless, OOMW may bear assets to enriched soils, namely improvement in porosity and aggregate's stability; minimization of runoffs (and a consequent increase of water retention capacity) and erosion; enhancement of soil organic matter; temporary carbon enrichment as well as that of other nutrients, which enhances soil fertility and may reduce the need of synthetic fertilizers; stimulation of microflora development; inhibition of microorganisms; and action as a control agent against phytopathogens (Barbera et al. 2013; Chatzistathis and Koutsos 2017; Karpouzas et al. 2010).

Composting is another solution for OOMW that relies essentially in the value of biomass as a natural resource with several environmental benefits. It is a seasonably abundant, renewable feedstock that preserves carbon dioxide atmospheric balance, contains small amounts of nitrogen and sulfur, and it is an important energetic alternative in the future because of the low footprint and human health impacts of its reuse (Hernández et al. 2014). Composting is an aerobic process that takes advantage on the high metabolic efficiency of the microbial communities in OOMW. It has three key stages: (1) activation phase, which presumes the stabilization of humidity and aeration conditions; (2) thermophilic phase, where there is an increase in the reaction temperature that promotes the activity of bacteria and fungi; and (3) mesophilic phase, where the biomass reaches the surrounding temperature. At the end of the process, carbon dioxide and humus remain, the latter also called compost, with large potential as either a soil amender or a fertilizer (Chowdhury et al. 2013). The compost is mostly composed by organic matter, nitrogen, phosphorus and trace elements. Humic substances result from the degradation of large organic compound chains, such as cellulose, hemicellulose and proteins, which are also main components of the olive oil pomace (Tortosa et al. 2014). All these compounds are crucial to the structure and fertility of soils by affecting, for example water retention capacity, cationic exchange capacity, microbiologic activity and degradation of pesticide residues; and also for plants physiology, nutrient uptake and root development (Chowdhury et al. 2013; Tortosa et al. 2014). The soil microbiota may greatly benefit with both OOMW and olive oil compost input, through the increase in total population diversity (bacteria and yeasts), along with enhancements in extracellular enzymatic activities and in functional and catabolic activities (Chowdhury et al. 2013). In particular, compost stimulates the proliferation of specific bacterial communities and microbial consortia specialized in the degradation of organic compounds (polyphenols, tannins and lipids), which further enhances OOMW importance in soils bioremediation (Ntougias et al. 2013). OOMW can be composted alone or mixed with other agroresidues, which is an advantage for small olive oil mill owners (Aquilanti et al. 2014; Chowdhury et al. 2014). The benefits of this compost in agriculture, along with the production of large volumes of olive oil wastes and the reduced organic matter contents of Mediterranean soils, may render composting a powerful sustainable solution for OOMW.

Biosorption can play a major role in the olive oil production chain, namely by valuing solid olive residues in the treatment of contaminated solutions (Table 6.1) for their strong adsorption properties, low cost and local availability in large amounts. The main solid residues that result from the olive oil production are (1) olive stones, which refer to the olive pits, cores or kernels; (2) pruning material, which includes the woody wastes of the olive trees such as the branches and twigs; (3) olive leaves from the trees resulting from the deleafing of olives before washing; (4) pomaces, already described in Section 1.2.1; and (5) exhausted pomaces, also called exhausted cakes, that result from pomaces that have been further processed to extract the remaining oil that may exist (Anastopoulos et al. 2015). Each of these residues has its own physicochemical characteristics conferring unique biosorption properties. Moreover, the kind of activation (physical or chemical) used and the processing conditions are key determinants of the sorbent capacity of olive solid wastes (Bhatnagar et al. 2014). There are already a couple of reviews on the sorption characteristics of these residues, as well as on the different treatment methods and other conditions that may be applied to improve yields (Anastopoulos et al. 2015; Bhatnagar et al. 2014). Therefore, here we will mostly summarize the suitability of each type of waste to treat aqueous solutions contaminated with specific sorbates and update on the most recent advances that were not covered in the previous reviews.

Olive stones (i.e., olive pits, cores or kernels) seem to be the most widely explored by-products as biosorbents (Anastopoulos et al. 2015; references within Table 6.1). This is probably due to the fact that these wastes are highly lignified, rendering them an advantageous source to produce activated carbon (Anastopoulos et al. 2015). Olive stones were proven to successfully adsorb some metals (Al, Cd, Cr, Cu, Fe, Ni, Pb, U and Zn) and dyes (Alizarin Red, Methylene Blue, Safranine and Remazol Red B), as well as the pesticide aldrin, the pharmaceutical amoxicillin and phenols (Table 6.1).

On the other hand, olive pruning and leaves have been suggested as sorbents of metals (Cd, Cu, Ni and Pb; Table 6.1), but they are the least studied wastes in this context despite being among the major wastes of the olive oil production chain. This incongruence is likely to be due to difficulties of collection, handling and shredding prior to use (Anastopoulos et al. 2015). Pomace has been suggested as a successful biosorbent for several metals and related elements (Cd, Cu, Cr, Fe, Hg, Ni, Pb, U, TH and Zn), dyes (Safranine, Remazol Red B, Reactive Blue 19 and Reactive Red 198), pesticides (aldrin, endrin and dieldrin), radioisotopes (Ga-67, TI-201 and Cs-137) and phenols (Table

TABLE 6.1
Contaminants reported to be biosorbed by the solid residues of the olive oil production chain. Information on studies published before 2014 was retrieved from the review by Anastopoulos et al. (2015). The residues considered in this summary include derivatives such as ashes

Residue	Contaminant				References
	Metals	Dyes	Pesticides	Others	
Olive stones	Al Cd Cr Cu Fe Ni Pb U Zn	Alizarin Red Methylene blue Safranine Remazol Red B	Aldrin	Pharmaceuticals (amoxicillin) Phenols	Anastopoulos et al. (2015) Albadarin and Mangwandi (2015) Calero et al. (2016) Calero et al. (2018) Dardouri and Sghaier (2017) Hodaifa et al. (2014) Limousy et al. (2017) Moubarik and Grimi (2015) Ronda et al. (2015) Trujillo et al. (2016)
Pruning material	Cd Cu Ni Pb	—	Dimethoate Imidacloprid Diuron Tebuconazole Oxyfluorfen	—	Anastopoulos et al. (2015) Almendros Molina et al. (2016) Delgado-Moreno et al. (2017b) Delgado-Moreno et al. (2017a)
Olive leaves	Cd Cu Pb	—	—	—	Anastopoulos et al. (2015)
Pomace	Cd Cu Cr Fe	Safranine Remazol Red B	Aldrin Endrin Dieldrin	Radioisotopes (Ga-67, TI-201, Cs-137) Phenols Spilled crude oil	Anastopoulos et al. (2015) Martinez-Garcia et al. (2006)

(Continued)

TABLE 6.1 (Cont.)

Residue	Contaminant				References
	Metals	Dyes	Pesticides	Others	
	Hg Ni Pb U TH Zn	Reactive Blue 19 Reactive Red 198			Venegas et al. (2015) Dashti et al. (2015) Petrella et al. (2018) Kučić and Simonič (2017)
Exhausted cake	Cd Cr CrO_4^{2-} Fe Ni Pb Zn	Lanaset Grey G Black Dycem TTO	Dimethoate Imidacloprid Diuron Tebuconazole Oxyfluorfen	—	Anastopoulos et al. (2015) El-Kady et al. (2016) Delgado-Moreno et al. (2017b) Delgado-Moreno et al. (2017a)

6.1). Still, the potential of olive pomace can be extended further since it can be used as an inoculum of hydrocarbonoclastic bacteria, which are useful for hydrocarbon-bioremediation, e.g., in cases of olive oil spill (Dashti et al. 2015). The exhausted cake has been suggested as a sorbent of different metals (Cd, Cr, CrO_4^{2-}, Fe, Ni, Pb and Zn), dyes (Lanaset Grey G and Black Dycem TTO) and pesticides (dimethoate, imidacloprid, diuron, tebuconazole and oxyfluorfen). More recently, some studies arose combining different debris to increase the potential of the by-products to treat contaminated solutions. For example, Delgado-Moreno et al. (2017b) proposed the use of wet olive cake and vermicompost, together with olive tree pruning, as substitutes of some of the components of biomixtures used in biobed systems to adsorb pesticides.

The valuation of residues of the olive oil production chain should be a priority given the innumerous problems associated with their disposal. Moreover, considering the increasingly serious environmental contamination scenarios worldwide, the possibility of using these wastes to biosorb some problematic sorbates is worth to explore.

Further applications that significantly increase the commercial value of OOMW are the use of these wastes as biofuels—biodiesel, biohydrogen, biomethane and bioethanol (Christoforou and Fokaides 2016, Dermeche et al. 2013; Hernández et al. 2014)—and as alternative energy resources (Chouchene et al. 2012; Jeguirim et al. 2012). OOMW could also be the source of many added value products. Single cell oils, mannitol and citric acid are examples of carbon sources that can be produced biotechnologically through the use of OOMW as a substrate for the microorganisms (Dourou et al. 2016). Moreover, several bioactive compounds can be recovered from OOMW that are highly valuable to human health (Aggoun et al. 2016). Polyphenols have been gaining interest as bioactive compounds due to their high antioxidant potential, which makes them adequate to be applied in food, cosmetics and pharmaceuticals development, either as additives or nutrients (Aggoun et al. 2016; Obied et al. 2005). Fractionation of the existing fatty acids in the OOMW allows the recovery of oleic and linoleic acids, which may be very useful in pharmaceutical applications requiring products of high purity, such as pure glycerol and biodegradable soap (Elkacmi et al. 2017).

Overall, all valuation possibilities using olive oil by-products discussed above are the most explored ones, and agricultural applications are the most relevant. In fact, through the reuse of olive oil residues to replace chemical fertilizers, directly or following composting, crop nutrition and development can be enhanced. Still, such applications should be made under controlled conditions, appraising for example appropriate dosage and application timing and after testing for their environmental safety. The development of sustainable biosorbents, as well as the identification of new sources for added value bioproducts, has been increasingly receiving attention, with olive oil by-products being relevant in this context. Considering the marked presence of the olive oil industry in south European countries, an opportunity is set to stimulate solutions to the related wastes, reflecting a circular economy approach toward a greener industry.

6.3.2 OOMW Treatment Processes

It is imperative to improve methods to treat OOMW to mitigate their nuisance effects, such as odor and the contamination of soils, groundwaters and superficial waters (Ioannou-Ttofa et al. 2017; Kavvadias et al. 2010; Paraskeva and Diamadopoulos 2006). OOMW is commonly treated under a set of general premises: (1) a good treatment process must take into account the seasonality of olive oil production, the distribution and local operations; (2) the efficiency of the method; and (3) the simplicity of the method (McNamara et al. 2008). Technological processes have been developed to mitigate the nuisance potential of OOMW (mainly focused on polyphenol removal) that include physicochemical techniques, biological techniques, or a combination between them (Figure 6.3). Nevertheless, the implementation of available treatment solutions remains a challenge to small producers because of economic constraints (Ioannou-Ttofa et al. 2017).

6.3.2.1 Physicochemical Processes

Physicochemical processes are based on solvent extraction, oxidation, coagulation, flocculation, adsorption onto resins, membrane filtration or an integrative approach comprising several of these techniques (Ochando-Pulido et al. 2017).

Solvent extraction can be applied as liquid–liquid or solid-phase extraction. Regarding liquid–liquid extraction, the most efficient solvents for polyphenols are in decreasing order, ethyl acetate, a mixture of chloroform with isopropyl

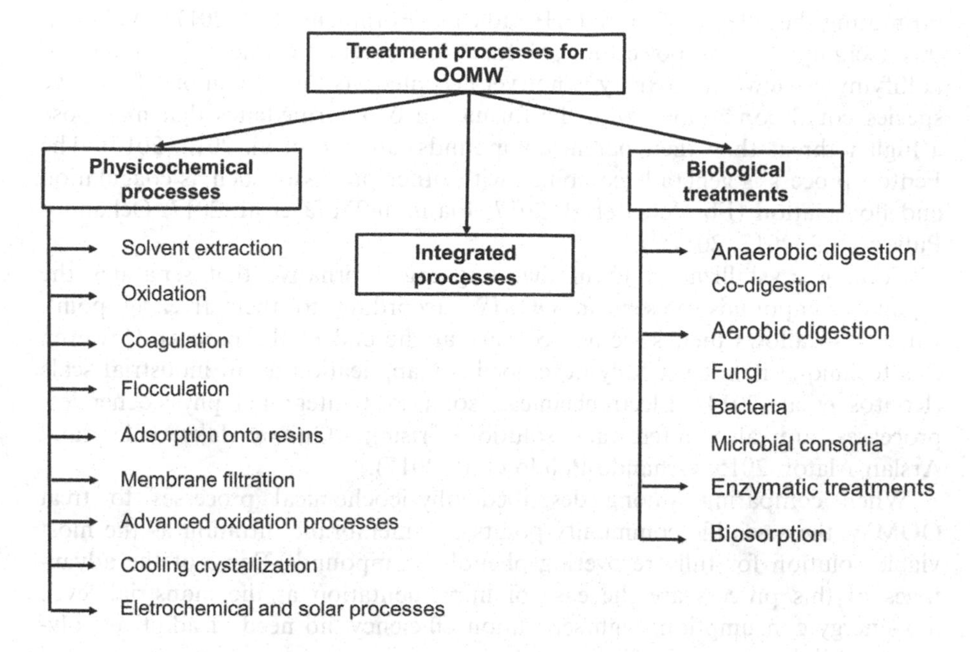

FIGURE 6.3 Summary of treatment processes currently described for OOMW.

alcohol and diethyl ether. Ethyl acetate is the most used because it combines a potent polyphenol recovery with a low environmental footprint. This methodology is applied in many small olive oil mills mostly because of the low economic costs and the easiness of the technique, but the whole process viability is reduced when compared with other solutions (Kalogerakis et al. 2013).

Oxidation, coagulation and flocculation can be applied alone or combined, within a huge variety of treatment possibilities. Advanced oxidation processes are based on technologies that use ozonation, photolysis, photocatalysis, ultrasounds, wet air oxidation or the Fenton reaction (Nogueira et al. 2016), basically for purifying waters and wastewaters by removing the organic and inorganic particles. Ozonation is widely used in wastewater treatment plants because of its high removal efficacy, but it is not well established that ozone is capable of reducing effluents toxicity (Justino et al. 2012). Photocatalysis is based on semiconductors coupled with oxidants, light or both, with the intent of releasing reactive oxygen species capable of degrading both organic and inorganic contaminants (Nogueira et al. 2016). The most used oxidation process is the Fenton process because of its wider advantages: it does not generate residues or need special equipment; it is easy to be scaled up; it is highly efficient and concomitantly unspecific, which means that it can be applied to a wide range of matrices (Justino et al. 2010). The photo-Fenton reaction consists in the irradiation of the Fenton reagent with UV light (like solar energy, abundant in Mediterranean countries) that enhances the whole process by promoting the release of more OH• radicals (Ebrahiem et al. 2017). Although this technique seems powerful tool for wastewaters treatment, its ability of nullifying wastewaters toxicity is not yet accepted, as the formation of reactive species could come along with the formation of intermediates that may pose a higher threat than their parent compounds (Justino et al. 2012, 2010). The Fenton process is generally combined with other processes such as coagulation and flocculation (Ebrahiem et al. 2017; Ioannou-Ttofa et al. 2017; Ochando-Pulido et al. 2013, 2017).

Cooling crystallization is another ongoing alternative that separates the organic compounds present in OOMW according to their freezing point, which also allows their selective recovery at the end of the process. However, this technique is not yet fully developed for application at an industrial scale (Kontos et al. 2014). Electrochemical, solar and integrated physicochemical processes are also interesting solutions rising (Gursoy-Haksevenler and Arslan-Alaton 2015; Ochando-Pulido et al. 2017).

When comparing among described physicochemical processes to treat OOMW, the scientific community points out membrane filtration as the most viable solution for fully recovering phenolic compounds. Some of the advantages of this process are the ease of implementation at the industrial level, low-energy consumption, high separation efficiency, no need of additives/solvents, possibility to operate at room temperature, small area requirements and low maintenance costs (Bazzarelli et al. 2016; Cassano et al. 2013; La Scalia et al. 2017).

6.3.2.2 Biological Treatments for OOMW

Biological processes that may be used to treat OOMW include anaerobic and aerobic digestion, and biosorption. Generally, biotreatments are able to remove organic and inorganic particles and have a more sustainable, environmentally friendlier and reliable nature, as well as better cost-efficiency ratios compared with physicochemical approaches (Daâssi et al. 2014; Paraskeva and Diamadopoulos 2006). A review covering the advantages and problems of aerobic and anaerobic bioreactors was presented by McNamara et al. (2008).

Anaerobic digestion is an eco-friendly treatment process that offers many advantages, including the low production of nutrients, wastes, greenhouse gases and the production of both fertilizers and biogas (Sampaio et al. 2011). OOMW consists in a highly recalcitrant solution that cannot be directly subjected to anaerobiosis because of its high toxicity to methanogenic consortia (Hamdi 1992)—most of the times this treatment demands dilution, nutrient input or pH adjustment to prevent microorganisms' growth inhibition. On the one hand, the application of these pretreatments may pose some alterations in the organic composition of the wastewater, which decreases its energetic potential and increases the operation costs; on the other hand, it may dilute the OOMW to reduce phenolics' concentration at the expenses of an increment of wastewater volumes for further treatment (McNamara et al. 2008; Paraskeva and Diamadopoulos 2006). Alternatives based on complementary effluents have been exploited, namely co-digestion. This approach basically consists in the mixture of two different effluents, allowing the reduction of COD through the overall treatment process, rendering a better cost-efficiency ratio and enabling optimized conditions (e.g., pH and nutrients) for microorganisms. Moreover, the combination of OOMW with annually produced effluents enhances the viability of this process. There are several effluents that may be applied to co-digest OOMW, such as piggery, household waste/sewage sludge, abattoir, municipal wastewater and manure (Paraskeva and Diamadopoulos 2006; Sampaio et al. 2011).

Aerobic treatments may apply to the degradation of phenolics in OOMW. Studies using fungi (*Geotrichum candidum, Candida boidinii, Penicillium* sp., *Aspergillus niger* and *Trametes versicolor*) claim that the most efficient removal yields regarding color, COD and phenolic concentrations were obtained when these strains were gradually exposed to the wastewater, as this stimulates the microorganisms' metabolism toward increasing levels of detoxifying enzymes (Aissam et al. 2007; Assas et al. 2002; Ergül et al. 2009; Fadil et al. 2003). Nonetheless, studies that do not include this adaptation step also reached good treatment results regarding OOMW (Daâssi et al. 2014; García García et al. 2000; Karakaya et al. 2012).

Laccase belongs to a group of enzymes capable of degrading aromatic compounds known as ligninolytic enzymes—these are produced extracellularly and include manganese and lignin peroxidases, among others. The oxidation ability of this enzyme toward high- and low-molecular-weight polyphenols makes them biotechnologically valuable, and the microorganisms that produce them an efficient environmental solution, not only for OOMW but also for other effluents. As the aromatic compounds are quite similar to lignin monomers in terms of structure, and fungi are the main degraders of ligninolytic compounds, it is straightforward that fungi are considered suitable and powerful microorganisms to treat matrices contaminated with these types of chemicals (Duarte et al. 2014). In addition, the residue that remains after enzymatic degradation could still be processed to further obtain added value products (Zerva et al. 2017).

Pleurotus spp. is a widely studied genus regarding OOMW treatment (Aggelis et al. 2003). Experiments with *P. ostreatus* showed this fungus' ability to grow on sterilized OOMW and to reduce its toxicity toward *Artemia* sp. and toward seeds of *L. sativum* (Aggelis et al. 2003). Despite that the detoxifying action promoted by *P. ostreatus* laccase enzymes generates reaction products with higher toxicity than the initial ones. Tsioulpas et al. (2002) added evidences on the ability of *Pleurotus* spp. to remove high amounts of phenolic compounds; moreover, they showed that treated OOMW bare reduced phytotoxicity, although it contained more toxic phenols when compared with the untreated OOMW. OOMW decoloration and detoxification were also achieved by *Pleurotus* and *Ganoderma* spp. because of their intense enzymatic activities. In this context, despite *Pleurotus* spp. took more time to produce laccase (which means a longer acclimation period when compared with other fungus genera), at the end it translated in better yields compared with *Ganoderma* spp. (Ntougias et al. 2012).

One aspect deserving detailed attention is the fact that by applying fungi in the treatment of environmental matrices, their metabolic action can be inhibited due to the high concentration of contaminants. In addition, there are concerns with the possible introduction of allochthonous species in the receptor media that could induce a local imbalance. This was the motto of the work by Duarte et al. (2014), who addressed this drawback by developing a silica-alginate-fungi biocomposite, using *P. sajor caju*, that prevents fungi release, protecting both the organism and the surrounding environmental matrix. Besides the capacity of this biocomposite to reduce COD, concentration of phenolic compounds, sterols and fatty acids, it can be reused in other treatments. However, the biocomposite was unable to remove high-molecular-weight compounds, which is a disadvantage in what regards OOMW wastewaters.

Specific bacteria have also been considered for OOMW biotreatment approaches. For example, a study using *Ralstonia eutropha* toward the degradation of both phenolic compounds and COD in OOMW was conducted, on the basis of previous evidences on the ability of these bacteria to grow in a 20% dilution of OOMW (Jalilnejad et al. 2011). The authors detailed that previous acclimation to phenol and aeration enabled the growth of the microorganism, promoting a better biodegradation performance, as the inhibitory effect of OOMW was reduced. In another study, strains of *Ralstonia* sp. and *Pseudomonas* sp., both individually and combined in a co-culture, have also been analyzed toward a sample of natural OOMW (which have been treated by reverse osmosis) and a sample of OOMW provided by an anaerobic treatment facility. The results showed that the bacteria were able to mutually complement each other in completely removing the monocyclic aromatic compounds from the first sample and some low-molecular-weight aromatic compounds found in the second sample, respectively (Di Gioia et al. 2001). *Paenibacillus jamilae* was tested toward the production of exopolysaccharides (microbial polymers with great biotechnological potential at the cellular level) in OOMW as growth medium (Aguilera et al. 2008). The results showed that *P. jamilae* both produced the polymer and detoxified OOMW. Furthermore, strains of *Raoultella terrigena* and *Pantoea agglomerans* were isolated, selected and used for the biotreatment of OOMW. The success of this approach was demonstrated as the remaining residue was in agreement with the EU legislation, the authors then arguing on its potential to irrigate agricultural lands as a fertilizer (Maza-Márquez et al. 2013).

An integrated approach based on aerobic and anaerobic treatments was proposed by González-González and Cuadros (2015). First, the aerobic pretreatment was performed by using already existing microorganisms in OOMW, thus profiting from adaptation. Then, fractions of this treated effluent were exposed to the anaerobic treatment. Polyphenolic removal reached 90% after inoculated OOMW aeration for 7 days, and it was possible to recover ~30% of water following anaerobic digestion. Another promising microbial–bacterial consortium was proposed as an OOMW biotreatment combining the bacteria *R. terrigena* and *P. agglomerans* with the microalgae *Chlorella vulgaris* and *Scenedesmus obliquus.* Considering that microalgae consume carbon dioxide and that bacteria needs oxygen to further biodegrade pollutants, the reduction of the phenolic concentration in OOMW was promoted under these conditions (Maza-Márquez et al. 2014). In summary, treatments with microbial consortia rather than single microorganisms seem to be more advantageous: first, a higher abundance and diversity of organisms normally render the process more efficient since different organisms target a wider group of organic compounds; second, their enhanced stability and metabolic activity can improve the final outcome of the process (Maza-Márquez et al. 2014; Sivasubramanian and Namasivayam 2015).

Enzymatic treatments have also been suggested as an interesting solution to remediate OOMW provided the release of polyphenolic compounds and fermented carbohydrates, the latter susceptible to go under anaerobiosis and produce biogas. In addition, this treatment does not require chemical additives nor a controlled operating equipment to monitor basic parameters such as pH and temperature (Dammak et al. 2016).

Biosorption is a physicochemical process by which a solid surface of a biological matrix interacts with a sorbate resulting in the decrease of its concentration (Fomina and Gadd 2014). This process is a common approach used in the treatment of contaminants (i.e., the sorbates) including compounds such as metals and metalloids, dyes, pesticides, phenols, radioisotopes and pharmaceuticals (Fomina and Gadd 2014). The literature on biosorption to treat OOMW is scarcely available and suggests its low potential compared to other methods described above. The few studies that exist are based in the use of bioreactors with microorganisms that are able to adsorb phenolic compounds which can then be valued by being reused in different industrial activities, namely in the pharmaceutical and cosmetic sector (Roig et al. 2006). For example, Chiavola et al. (2014) proposed a sequencing batch reactor based on active sludge. This bioreactor could achieve high performances, including complete removal of biodegradable organic contents of the tested OOMW but residual polyphenol levels were still unsatisfactory even when the bioreactor was coupled with a membrane separation stage.

6.3.2.3 Integrated Treatments for OOMW

Physicochemical treatments are more efficient toward discolored OOMW but their incapacity of detoxifying waters and wastewaters is an important shortcoming. To overcome it, integrated systems combining chemical and biological treatments have been investigated. There are several studies covering this thematic, with a diversity of combinations being suggested: photo-Fenton and fungi (Justino et al. 2009); electrochemical and aerobic treatment (Hanafi et al. 2011); ozonation and ultraviolet irradiance, ozonation and aerobic treatment, and ozonation/ultraviolet followed by aerobic degradation (Lafi et al. 2009); anaerobic digestion with Fenton and electro-Fenton (El-Gohary et al. 2009; Khoufi et al. 2006). Overall, these studies concluded that an integrated approach enhances the removal of phenolic compounds, COD, and allows a reduction in the toxicity of the wastewater. Justino et al. (2009) suggested that the photo-Fenton process could be applied as a tertiary treatment in a wastewater treatment plant due to its large efficiency in decolorizing OOMW, and also a lower consumption of water. However, when photo-Fenton was applied before the fungi treatment, *P. sajor caju* was not capable of achieve the same COD, phenols and toxicity removal; this was attributed to the increasing wastewater toxicity promoted by the Fenton process. Lafi et al. (2009) referred that ultraviolet/ozonation treatment is more effective than the ozonation followed by aerobic degradation in removing phenols. This experiment also recognized ultraviolet/ozonation followed by aerobic degradation as the best-known approach regarding COD decrease, as they achieved a removal rate of 90.7%.

6.4 CONCLUSIONS

This chapter presents a detailed analysis regarding management and treatment processes that can be applied to OOMW. Taking into consideration, the technological costs associated with some treatments, the lack of legislation, the lack of environmental literacy and the financial limitations of small producers, it is obvious that there is a long way to go regarding the olive oil industry and the management of its externalities. In fact, major reasons behind the lack of efficient treatments implementation in olive oil mills are the existence of small-scale industries combined with the seasonality of the production process, mill's geographical dispersion and the reduced economic power of their managers, who in most of the cases run a familiar business and cannot afford significant equipment investments (Ioannou-Ttofa et al. 2017). The fact that a commonly accepted cost-efficient treatment fulfilling environmental quality standards is still lacking renders the environmental hazardous potential of OOMW an ongoing problem, especially in the Mediterranean region (Paraskeva and Diamadopoulos 2006).

Still, the use of treated OOMW in enhancing agricultural activities is a relevant opportunity, thus reducing dilution water needs and its costs to the olive mill (Ioannou-Ttofa et al. 2017). Other opportunities are into place, such as the use of olive oil by-products to produce effective biosorbents; this can both contribute to the reduction of the amount of these industrial residues and to improve parallel processes to the removal of several emerging contaminants, such as metals and pesticides. Furthermore, the possibility of recovering phenolic compounds from OOMW opens another valuation route, as these compounds can be part of a plethora of applications by industries focusing on nutraceutics, cosmetics, biofuels and agrochemicals.

Regarding OOMW treatment technologies, intense research efforts dedicating to different types of strategies have been made. The overall conclusion that can be taken is that integrated processes represent the most efficient strategies to achieve satisfactory performances on the removal of phenolic compounds, COD and toxicity. Notwithstanding, investment in a given treatment process should not be made at the expenses of an underdevelopment of valuation strategies because the latter will always represent an added value in increasing the life cycle of the olive oil value chain products. The optimization and implementation of these ideas can lead the olive oil industry to reach zero waste and, ultimately, a sustainable solution for both economy and environment.

REFERENCES

Aggelis, G. et al., 2003. Phenolic removal in a model olive oil mill wastewater using *Pleurotus ostreatus* in bioreactor cultures and biological evaluation of the process. Water Research, 37(16), pp. 3897–3904.

Aggoun, M. et al., 2016. Olive mill wastewater microconstituents composition according to olive variety and extraction process. Food Chemistry, 209, pp. 72–80.

Aguilera, M. et al., 2008. Characterisation of *Paenibacillus jamilae* strains that produce exopolysaccharide during growth on and detoxification of olive mill wastewaters. Bioresource Technology, 99(13), pp. 5640–5644.

Aharonov-Nadborny, R. et al., 2018. Mechanisms governing the leaching of soil metals as a result of disposal of olive mill wastewater on agricultural soils. Science of the Total Environment, 630, pp. 1115–1123.

Aissam, H., Penninckx, M.J. & Benlemlih, M., 2007. Reduction of phenolics content and COD in olive oil mill wastewaters by indigenous yeasts and fungi. World Journal of Microbiology and Biotechnology, 23(9), pp. 1203–1208.

Albadarin, A.B. & Mangwandi, C., 2015. Mechanisms of Alizarin Red S and Methylene blue biosorption onto olive stone by-product: Isotherm study in single and binary systems. Journal of Environmental Management, 164, pp. 86–93.

Almendros Molina, A.I. et al., 2016. Study of Ni(II) removal by olive tree pruning and pine cone shell by experimental design methodology. Desalination and Water Treatment, 57(32), pp. 15057–15072.

Anastopoulos, I., Massas, I. & Ehaliotis, C., 2015. Use of residues and by-products of the olive-oil production chain for the removal of pollutants from environmental media: A review of batch biosorption approaches. Journal of Environmental Science and Health – Part A Toxic/Hazardous Substances and Environmental Engineering, 50(7), pp. 677–718.

Andreozzi, R. et al., 2008. Effect of combined physico-chemical processes on the phytotoxicity of olive mill wastewaters. Water Research, 42(6–7), pp. 1684–1692.

Angerosa, F. et al., 2004. Volatile compounds in virgin olive oil: Occurrence and their relationship with the quality. Journal of Chromatography A, 1054(1–2), pp. 17–31.

Aquilanti, L. et al., 2014. Integrated biological approaches for olive mill wastewater treatment and agricultural exploitation. International Biodeterioration and Biodegradation, 88, pp. 162–168.

Assas, N. et al., 2002. Decolorization of fresh and stored-black olive mill wastewaters by *Geotrichum candidum*. Process Biochemistry, 38(3), pp. 361–365.

Azbar, N. et al., 2004. A review of waste management options in olive oil production. Critical Reviews in Environmental Science and Technology, 34(3), pp. 209–247.

Barbera, A.C. et al., 2013. Effects of spreading olive mill wastewater on soil properties and crops, a review. Agricultural Water Management, 119, pp. 43–53.

Bazzarelli, F. et al., 2016. Advances in membrane operations for water purification and biophenols recovery/valorization from OMWWs. Journal of Membrane Science, 497, pp. 402–409.

Ben Sassi, A. et al., 2006. A comparison of olive oil mill wastewaters (OMW) from three different processes in Morocco. Process Biochemistry, 41(1), pp. 74–78.

Bhatnagar, A. et al., 2014. Valorization of solid waste products from olive oil industry as potential adsorbents for water pollution control: —A review. Environmental Science and Pollution Research, 21(1), pp. 268–298.

Calero, M. et al., 2016. The scale-up of Cr^{3+} biosorption onto olive stone in a fixed bed column. Desalination and Water Treatment, 57(52), pp. 25140–25152.

Calero, M. et al., 2018. Neural fuzzy modelization of copper removal from water by biosorption in fixed-bed columns using olive stone and pinion shell. Bioresource Technology, 252, pp. 100–109.

Cassano, A. et al., 2013. Fractionation of olive mill wastewaters by membrane separation techniques. Journal of Hazardous Materials, 248–249(1), pp. 185–193.

Chatzistathis, T. & Koutsos, T., 2017. Olive mill wastewater as a source of organic matter, water and nutrients for restoration of degraded soils and for crops managed with sustainable systems. Agricultural Water Management, 190, pp. 55–64.

Chiavola, A., Farabegoli, G. & Antonetti, F., 2014. Biological treatment of olive mill wastewater in a sequencing batch reactor. Biochemical Engineering Journal, 85, pp. 71–78.

Chouchene, A. et al., 2012. Energetic valorisation of olive mill wastewater impregnated on low cost absorbent: Sawdust versus olive solid waste. Energy, 39(1), pp. 74–81.

Chowdhury, A.K.M.M.B. et al., 2013. Olive mill waste composting: A review. International Biodeterioration and Biodegradation, 85, pp. 108–119.

Chowdhury, A.K.M.M.B. et al., 2014. Composting of three phase olive mill solid waste using different bulking agents. International Biodeterioration and Biodegradation, 91, pp. 66–73.

Christoforou, E. & Fokaides, P.A., 2016. A review of olive mill solid wastes to energy utilization techniques. Waste Management, 49, pp. 346–363.

Daâssi, D. et al., 2014. Enhanced reduction of phenol content and toxicity in olive mill wastewaters by a newly isolated strain of *Coriolopsis gallica*. Environmental Science and Pollution Research, 21(3), pp. 1746–1758.

Dammak, I., Khoufi, S. & Sayadi, S., 2016. A performance comparison of olive oil mill wastewater enzymatic treatments. Food and Bioproducts Processing, 100, pp. 61–71.

Dardouri, S. & Sghaier, J., 2017. Adsorptive removal of methylene blue from aqueous solution using different agricultural wastes as adsorbents. Korean Journal of Chemical Engineering, 34(4), pp. 1037–1043.

Dashti, N. et al., 2015. Olive-pomace harbors bacteria with the potential for hydrocarbon-biodegradation, nitrogen-fixation and mercury-resistance: Promising material for waste-oil-bioremediation. Journal of Environmental Management, 155, pp. 49–57.

Delgado-Moreno, L., Nogales, R. & Romero, E., 2017a. Biodegradation of high doses of commercial pesticide products in pilot-scale biobeds using olive-oil agroindustry wastes. Journal of Environmental Management, 204, pp. 160–169.

Delgado-Moreno, L., Nogales, R. & Romero, E., 2017b. Wastes from the olive oil production in sustainable bioremediation systems to prevent pesticides water contamination. International Journal of Environmental Science and Technology, 14(11), pp. 2471–2484.

Dermeche, S. et al., 2013. Olive mill wastes: Biochemical characterizations and valorization strategies. Process Biochemistry, 48(10), pp. 1532–1552.

Dhouib, A. et al., 2006. Pilot-plant treatment of olive mill wastewaters by *Phanerochaete chrysosporium* coupled to anaerobic digestion and ultrafiltration. Process Biochemistry, 41(1), pp. 159–167.

Di Bene, C. et al., 2013. Short- and long-term effects of olive mill wastewater land spreading on soil chemical and biological properties. Soil Biology and Biochemistry, 56, pp. 21–30.

Di Gioia, D. et al., 2001. Biodegradation of synthetic and naturally occurring mixtures of mono-cyclic aromatic compounds present in olive mill wastewaters by two aerobic bacteria. Applied Microbiology and Biotechnology, 55(5), pp. 619–626.

Dourou, M. et al., 2016. Bioconversion of olive mill wastewater into high-added value products. Journal of Cleaner Production, 139, pp. 957–969.

Duarte, K.R. et al., 2014. Removal of phenolic compounds in olive mill wastewater by silica-alginate-fungi biocomposites. International Journal of Environmental Science and Technology, 11(3), pp. 589–596.

Ebrahiem, E.E., Al-Maghrabi, M.N. & Mobarki, A.R., 2017. Removal of organic pollutants from industrial wastewater by applying photo-Fenton oxidation technology. Arabian Journal of Chemistry, 10, pp. S1674–S1679.

El-Abbassi, A. et al., 2017. Potential applications of olive mill wastewater as biopesticide for crops protection. Science of the Total Environment, 576, pp. 10–21.

El-Abbassi, A., Kiai, H. & Hafidi, A., 2012. Phenolic profile and antioxidant activities of olive mill wastewater. Food Chemistry, 132(1), pp. 406–412.

El-Gohary, F.A. et al., 2009. Integrated treatment of olive mill wastewater (OMW) by the combination of Fenton's reaction and anaerobic treatment. Journal of Hazardous Materials, 162(2–3), pp. 1536–1541.

Elhag, M. et al., 2017. Stream network pollution by olive oil wastewater risk assessment in Crete, Greece. Environmental Earth Sciences, 76(7), pp. 1–12.

Elkacmi, R. et al., 2017. Techno-economical evaluation of a new technique for olive mill wastewater treatment. Sustainable Production and Consumption, 10(November 2016), pp. 38–49.

El-Kady, A.A. et al., 2016. Kinetic and adsorption study of Pb (II) toward different treated activated carbons derived from olive cake wastes. Desalination and Water Treatment, 57(18), pp. 8561–8574.

Ergül, F.E. et al., 2009. Dephenolisation of olive mill wastewater using adapted *Trametes versicolor*. International Biodeterioration and Biodegradation, 63(1), pp. 1–6.

European Commission, 2000. Directive 2000/60/EC of the European Parliament and of the Council of 23 October 2000 establishing a framework for community action in the field of water policy. Official Journal of the European Parliament, L 327/1 (October 23), pp. 1–82.

Fadil, K. et al., 2003. Aerobic biodegradation and detoxification of wastewaters from the olive oil industry. International Biodeterioration and Biodegradation, 51(1), pp. 37–41.

Fiorentino, A. et al., 2003. Environmental effects caused by olive mill wastewaters: Toxicity comparison of low-molecular-weight phenol components. Journal of Agricultural and Food Chemistry, 51(4), pp. 1005–1009.

Fomina, M. & Gadd, G.M., 2014. Biosorption: Current perspectives on concept, definition and application. Bioresource Technology, 160, pp. 3–14.

García García, I. et al., 2000. Removal of phenol compounds from olive mill wastewater using *Phanerochaete chrysosporium, Aspergillus niger, Aspergillus terreus* and *Geotrichum candidum*. Process Biochemistry, 35(8), pp. 751–758.

Gebreyohannes, A.Y., Mazzei, R. & Giorno, L., 2016. Trends and current practices of olive mill wastewater treatment: Application of integrated membrane process and its future perspective. Separation and Purification Technology, 162, pp. 45–60.

González-González, A. & Cuadros, F., 2015. Effect of aerobic pretreatment on anaerobic digestion of olive mill wastewater (OMWW): An ecoefficient treatment. Food and Bioproducts Processing, 95(September 2012), pp. 339–345.

Gursoy-Haksevenler, B.H. & Arslan-Alaton, I., 2015. Evidence of inert fractions in olive mill wastewater by size and structural fractionation before and after thermal acid cracking treatment. Separation and Purification Technology, 154, pp. 176–185.

Hamdi, M., 1992. Toxicity and biodegradability of olive mill wastewaters in batch anaerobic digestion. Applied Biochemistry and Biotechnology, 37(2), pp. 155–163.

Hanafi, F. et al., 2011. Augmentation of biodegradability of olive mill wastewater by electrochemical pre-treatment: Effect on phytotoxicity and operating cost. Journal of Hazardous Materials, 190(1–3), pp. 94–99.

Hentati, O. et al., 2016. Soil contamination with olive mill wastes negatively affects microbial communities, invertebrates and plants. Ecotoxicology, 25(8), pp. 1500–1513.

Hernández, D. et al., 2014. Biodiesel production from an industrial residue: Alperujo. Industrial Crops and Products, 52, pp. 495–498.

Hodaifa, G. et al., 2014. Iron removal from liquid effluents by olive stones on adsorption column: Breakthrough curves. Ecological Engineering, 73, pp. 270–275.

Inglezakis, V.J., Moreno, J.L. & Doula, M., 2012. Olive oil waste management EU legislation: Current situation and policy recommendations. International Journal of Chemical Enviromental Engineering Systems, 3(2), pp. 65–77.

Ioannou-Ttofa, L. et al., 2017. Treatment efficiency and economic feasibility of biological oxidation, membrane filtration and separation processes, and advanced oxidation for the purification and valorization of olive mill wastewater. Water Research, 114, pp. 1–13.

IOOC, 2015. International olive oil production – Costs study. International Olive Oil Council, pp. 1–11. http://www.internationaloliveoil.org/documents/index/339-economy/1815-international-olive-oil-production-costs-study/

IOOC, 2017. International olive oil council – Market newsletter. International Olive Oil Council, 18(119), pp. 1–6.

Isidori, M. et al., 2004. Chemical and toxic evaluation of a biological treatment for olive-oil mill wastewater using commercial microbial formulations. Applied Microbiology and Biotechnology, 64(5), pp. 735–739.

Jalilnejad, E., Mogharei, A. & Vahabzadeh, F., 2011. Aerobic pretreatment of olive oil mill wastewater using *Ralstonia eutropha*. Environmental Technology, 32(10), pp. 1085–1093.

Jeguirim, M. et al., 2012. A new valorisation strategy of olive mill wastewater: Impregnation on sawdust and combustion. . Conservation and Recycling, 59, pp. 4–8.

Justino, C. et al., 2010. Degradation of phenols in olive oil mill wastewater by biological, enzymatic, and photo-Fenton oxidation. Environmental Science and Pollution Research, 17(3), pp. 650–656.

Justino, C.I. et al., 2009. Toxicity and organic content characterization of olive oil mill wastewater undergoing a sequential treatment with fungi and photo-Fenton oxidation. Journal of Hazardous Materials, 172(2–3), pp. 1560–1572.

Justino, C.I.L. et al., 2012. Olive oil mill wastewaters before and after treatment: A critical review from the ecotoxicological point of view. Ecotoxicology, 21(2), pp. 615–629.

Kalogerakis, N. et al., 2013. Recovery of antioxidants from olive mill wastewaters: A viable solution that promotes their overall sustainable management. Journal of Environmental Management, 128, pp. 749–758.

Kapellakis, I.E., Tsagarakis, K.P. & Crowther, J.C., 2008. Olive oil history, production and by-product management. Reviews in Environmental Science and Biotechnology, 7(1), pp. 1–26.

Karakaya, A., Laleli, Y. & Takaç, S., 2012. Development of process conditions for biodegradation of raw olive mill wastewater by *Rhodotorula glutinis*. International Biodeterioration and Biodegradation, 75, pp. 75–82.

Karaouzas, I. et al., 2011a. Spatial and temporal effects of olive mill wastewaters to stream macroinvertebrates and aquatic ecosystems status. Water Research, 45(19), pp. 6334–6346.

Karaouzas, I. et al., 2011b. Bioassays and biochemical biomarkers for assessing olive mill and citrus processing wastewater toxicity. Environmental Toxicology, 26, pp. 669–676.

Karpouzas, D.G. et al., 2010. Olive mill wastewater affects the structure of soil bacterial communities. Applied Soil Ecology, 45(2), pp. 101–111.

Kavvadias, V. et al., 2010. Disposal of olive oil mill wastes in evaporation ponds: Effects on soil properties. Journal of Hazardous Materials, 182(1–3), pp. 144–155.
Khoufi, S., Aloui, F. & Sayadi, S., 2006. Treatment of olive oil mill wastewater by combined process electro-Fenton reaction and anaerobic digestion. Water Research, 40(10), pp. 2007–2016.
Komnitsas, K. & Zaharaki, D., 2012. Pre-treatment of olive mill wastewaters at laboratory and mill scale and subsequent use in agriculture: Legislative framework and proposed soil quality indicators. Resources Conservation and Recycling, 69, pp. 82–89.
Kontos, S.S., Koutsoukos, P.G. & Paraskeva, C.A., 2014. Removal and recovery of phenolic compounds from olive mill wastewater by cooling crystallization. Chemical Engineering Journal, 251, pp. 319–328.
Koutsos, T.M., Chatzistathis, T. & Balampekou, E.I., 2018. A new framework proposal, towards a common EU agricultural policy, with the best sustainable practices for the re-use of olive mill wastewater. Science of the Total Environment, 622–623, pp. 942–953.
Kučić, D. & Simonič, M., 2017. Batch adsorption of Cr(VI) ions on zeolite and agroindustrial waste. Chemical and Biochemical Engineering Quarterly, 31(4), pp. 497–507.
La Scalia, G. et al., 2017. A sustainable phenolic compound extraction system from olive oil mill wastewater. Journal of Cleaner Production, 142, pp. 3782–3788.
Lafi, W.K. et al., 2009. Treatment of olive mill wastewater by combined advanced oxidation and biodegradation. Separation and Purification Technology, 70(2), pp. 141–146.
Limousy, L. et al., 2017. Amoxicillin removal from aqueous solution using activated carbon prepared by chemical activation of olive stone. Environmental Science and Pollution Research, 24(11), pp. 9993–10004.
Martinez-Garcia, G. et al., 2006. Olive oil waste as a biosorbent for heavy metals. International Biodeterioration and Biodegradation, 58(3–4), pp. 231–238.
Maza-Márquez, P. et al., 2013. Biodegradation of olive washing wastewater pollutants by highly efficient phenol-degrading strains selected from adapted bacterial community. International Biodeterioration and Biodegradation, 82, pp. 192–198.
Maza-Márquez, P. et al., 2014. Biotreatment of olive washing wastewater by a selected microalgal-bacterial consortium. International Biodeterioration and Biodegradation, 88, pp. 69–76.
McNamara, C.J. et al., 2008. Bioremediation of olive mill wastewater. International Biodeterioration and Biodegradation, 61(2), pp. 127–134.
Mechri, B. et al., 2009. Olive orchard amended with olive mill wastewater: Effects on olive fruit and olive oil quality. Journal of Hazardous Materials, 172(2–3), pp. 1544–1550.
Mekki, A. et al., 2008. Assessment of toxicity of the untreated and treated olive mill wastewaters and soil irrigated by using microbiotests. Ecotoxicology and Environmental Safety, 69(3), pp. 488–495.
Morillo, J.A. et al., 2009. Bioremediation and biovalorisation of olive-mill wastes. Applied Microbiology and Biotechnology, 82(1), pp. 25–39.
Moubarik, A. & Grimi, N., 2015. Valorization of olive stone and sugar cane bagasse by-products as biosorbents for the removal of cadmium from aqueous solution. Food Research International, 73, pp. 169–175
Muir, D.C.G. & Howard, P.H., 2006. Are there other persistent organic pollutants? A challenge for environmental chemists. Environmental Science and Technology, 40(23), pp. 7157–7166.
Nogueira, V. et al., 2016. Photocatalytic treatment of olive oil mill wastewater using TiO_2 and Fe_2O_3 Nanomaterials. Water, Air, and Soil Pollution, 227(3), p. 227:88.

Ntougias, S. et al., 2012. Biodegradation and detoxification of olive mill wastewater by selected strains of the mushroom genera *Ganoderma* and *Pleurotus*. Chemosphere, 88(5), pp. 620–626.
Ntougias, S., Bourtzis, K. & Tsiamis, G., 2013. The microbiology of olive mill wastes. BioMed Research International, 2013, pp. 1–16.
Obied, H.K. et al., 2005. Bioactivity and analysis of biophenols recovered from olive mill waste. Journal of Agricultural and Food Chemistry, 53(4), pp. 823–837.
Obied, H.K. et al., 2007. Bioscreening of Australian olive mill waste extracts: Biophenol content, antioxidant, antimicrobial and molluscicidal activities. Food and Chemical Toxicology, 45(7), pp. 1238–1248.
Ochando-Pulido, J.M. et al., 2013. Reuse of olive mill effluents from two-phase extraction process by integrated advanced oxidation and reverse osmosis treatment. Journal of Hazardous Materials, 263, pp. 158–167.
Ochando-Pulido, J.M. et al., 2017. A focus on advanced physico-chemical processes for olive mill wastewater treatment. Separation and Purification Technology, 179, pp. 161–174.
Paraskeva, P. & Diamadopoulos, E., 2006. Technologies for olive mill wastewater (OMW) treatment: A review. Journal of Chemical Technology & Biotechnology, 81 (9), pp. 1475–1485.
Pavlidou, A. et al., 2014. Effects of olive oil wastes on river basins and an oligotrophic coastal marine ecosystem: A case study in Greece. Science of the Total Environment, 497–498, pp. 38–49.
Petrella, A. et al., 2018. Heavy metals retention (Pb(II), Cd(II), Ni(II)) from single and multimetal solutions by natural biosorbents from the olive oil milling operations. Process Safety and Environmental Protection, 114(Ii), pp. 79–90.
Priac, A., Badot, P.-M. & Crini, G., 2017. Treated wastewater phytotoxicity assessment using *Lactuca sativa*: Focus on germination and root elongation test parameters. Comptes Rendus Biologies, 340(3), pp. 188–194.
Rana, G., Rinaldi, M. & Introna, M., 2003. Volatilisation of substances after spreading olive oil waste water on the soil in a Mediterranean environment. Agriculture Ecosystems & Environment, 96(1–3), pp. 49–58.
Rockström, J. et al., 2009. A safe operating space for humanity. Nature, 461(7263), pp. 472–475.
Roig, A., Cayuela, M.L. & Sánchez-Monedero, M.A., 2006. An overview on olive mill wastes and their valorisation methods. Waste Management, 26(9), pp. 960–969.
Ronda, A. et al., 2015. Complete use of an agricultural waste: Application of untreated and chemically treated olive stone as biosorbent of lead ions and reuse as fuel. Chemical Engineering Research and Design, 104, pp. 740–751.
Sampaio, M.A., Gonçalves, M.R. & Marques, I.P., 2011. Anaerobic digestion challenge of raw olive mill wastewater. Bioresource Technology, 102(23), pp. 10810–10818.
Sierra, J. et al., 2001. Characterisation and evolution of a soil affected by olive oil mill wastewater disposal. Science of the Total Environment, 279(1–3), pp. 207–214.
Sivasubramanian, S. & Namasivayam, S.K.R., 2015. Phenol degradation studies using microbial consortium isolated from environmental sources. Journal of Environmental Chemical Engineering, 3(1), pp. 243–252.
Tafesh, A. et al., 2011. Synergistic antibacterial effects of polyphenolic compounds from olive mill wastewater. Evidence-based Complementary and Alternative Medicine, 2011 Article ID 431021, 9 p.
Tortosa, G. et al., 2014. Strategies to produce commercial liquid organic fertilisers from 'alperujo' composts. Journal of Cleaner Production, 82, pp. 37–44.
Trujillo, M.C. et al., 2016. Simultaneous biosorption of methylene blue and trivalent chromium onto olive stone. Desalination and Water Treatment, 57(37), pp. 17400–17410.

Tsioulpas, A. et al., 2002. Phenolic removal in olive oil mill wastewater by strains of *Pleurotus spp.* in respect to their phenol oxidase (laccase) activity. Bioresource Technology, 84(3), pp. 251–257.

Venegas, A., Rigol, A. & Vidal, M., 2015. Viability of organic wastes and biochars as amendments for the remediation of heavy metal-contaminated soils. Chemosphere, 119, pp. 190–198.

Venieri, D., Rouvalis, A. & Iliopoulou-Georgudaki, J., 2010. Microbial and toxic evaluation of raw and treated olive oil mill wastewaters. Journal of Chemical Technology and Biotechnology, 85(10), pp. 1380–1388.

Wania, F., 2003. Assessing the potential of persistent organic chemicals for long-range transport and accumulation in polar regions. Environmental Science and Technology, 37(7), pp. 1344–1351.

Zampounis, V. & Minoans, T., 2006. Olive oil in the world market. In *Olive Oil Chemistry and Technology [e-book]*, American Oil Chemists' Society Press. Available at: https://books.google.pt/books?id=Nc9VCgAAQBAJ&printsec=frontcover&hl=ptPT#v=onepage&q&f=false [Acessed March 31, 2018].

Zerva, A. et al., 2017. Degradation of olive mill wastewater by the induced extracellular ligninolytic enzymes of two wood-rot fungi. Journal of Environmental Management, 203, pp. 791–798.

7 Applications of Biosorption in Dyes Removal

Jaqueline Benvenuti, Santiago Ortiz-Monsalve, Bianca Mella, and Mariliz Gutterres
Federal University of Rio Grande do Sul (UFRGS), Chemical Engineering Department, Laboratory for Leather and Environmental Studies (LACOURO).
Porto Alegre, Brazil

7.1 INTRODUCTION

Dyes are organic compounds with functional groups able to interact with different materials to provide color. They are employed across a wide variety of industry segments. Since most of them are water soluble, their applications in liquid medium generate colored wastewaters. Dyes are organic pollutants usually non-biodegradable due to their complex chemical structure, synthetic source, and some may be mutagens and carcinogens, such as the widely used azo-dyes (Fuck and Gutterres 2008). They are stable under different conditions and consequently tend not to undergo degradation (Ngulube et al. 2017). Therefore, conventional treatment techniques do not always achieve satisfactory results of color removal, one of the biggest challenges in wastewater management and treatment.

Biosorption is a technique that has been widely studied for dye removal. Biosorption is, in general, the removal of a substance from aqueous solutions by biological materials and may include several mechanisms, such as ion exchange, adsorption, complexation and precipitation (Abdolali et al. 2014). The low cost of the biosorbent materials, their renewable origin and high availability are the main advantages of biosorption. The sources that have been studied as alternative biosorbents are mainly related to living or dead microorganisms and agro-industrial wastes.

Bacteria and fungi have been extensively studied for dye biosorption (Kaushik and Malik 2015) and presented good prospects for applications in wastewater treatment. Solid waste from industries as tanneries holds out great promise for potential application in wastewater treatment systems (Gomes et al. 2016). Moreover, leather industry is an example of industry that can take advantage of these biosorbent materials, by using them to treat their own wastewater.

7.2 DYES AND ENVIRONMENTAL IMPACT

Dyes are colored compounds that are used to color a product generally to increase its visibility in the market, improving its appearance and attractiveness to the consumers. According to a theory proposed in 1876, the color of the dyes comes from two groups present in the dye molecule. The *chromophore*, color-bearing group, and the *auxochromes*, color helpers, that can influence dye solubility and shift the color of a colorant (Christie 2001; Witt 1876).

There are two common classifications for dyes, the chemical classification, where dyes are arranged in accordance with their chemical structure; and the classification by usage, the principal system adopted by the *Colour Index*, a publication of the Society of Dyers and Colourists, where dyes are classified by the method of application. The most important chemical classes are the azo, carbonyl, phthalocyanine, arylcarbonium ion, sulfur, polymethine and nitro dyes (Gürses et al. 2016; Hunger 2003).

The application classes may be separate in dyes for protein fibers—acid dyes, mordant dyes and premetallized dyes; dyes for cellulosic fibers—direct dyes, reactive dyes and vat dyes; dyes for acrylic fibers—basic (cationic) dyes; and dyes for polyester—disperse dyes. Alternatively, colorants can be classified in terms of the electronic excitation process mechanism (theoretical classification) and specified in terms of color space coordinates (CIELAB system) (Christie 2001; Ejder-Korucu et al. 2015).

Industries are responsible for about 20% of global water consumption (WWAP 2014). The real volume of wastewater generated by industrial sector is unknown because it is still a limited and sporadically collected data. Globally, it is estimated that 80% of the wastewater is released into the environment without any treatment leading to ecosystem degradation and consequently harmful impacts on human health (WWAP 2017).

The disposal of large concentrations of potentially toxic compounds into the environment is the major consequence of global industrialization. Among these compounds, synthetic dyes represent a matter of growing concern regarding their impact into the environment and the health of consumers. Many industries use dyes in their production chains. The textile, leather, paper, paint and dye manufacturing industries are known for the discharge of effluents with dyes (Katheresan et al. 2018). The colored effluent generated if not treated properly and discharged into the water bodies does not represent only an esthetic issue but can also cause problems to aquatic life and human health.

Furthermore, the environmental problems caused by the effluents of these industries are not restricted to the dyes themselves. In the various industrial unit operations, other different chemical compounds such as chemical auxiliaries, acids, salts, reducing and oxidizing agents, oils, fats and enzymes are also added to the dyeing process to provide some properties to the material to be dyed and to guarantee the good absorption and fixation of dye. The presence of these compounds in wastewaters is related with high biological oxygen demand, chemical oxygen demand, total organic carbon, nitrogen, suspended solids, dissolved solids, alkalinity, pH and odorous conditions in the wastewater (Mullai et al. 2017).

A major environmental concern of the industries mentioned above is the incomplete fixation of the dyes, related with problems of diffusion and affinity between dyes and cellulosic, protein, polyamide or polyester fibers. In textile industries, for instance, it is estimated that 10–60% of the dyes remain in the effluents, producing highly colored wastewater (Hessel et al. 2007; Rosa et al. 2015). Color is usually the first characteristic provided by small concentrations of dyes (1 mg L^{-1}) in wastewaters (Chequer et al. 2013). In addition to the aesthetic effect, the presence of color in the water may also cause the inhibition of aquatic photosynthesis by absorbing and reflecting the sunlight. Further issues such as reduction of water reoxygenation capacity and depletion of dissolved oxygen have been also associated with the presence of dyes in wastewaters (Katheresan et al. 2018; Pereira and Alves 2012).

In addition to the incomplete fixation of dyes, the great deal of water and energy used in dyeing is other environmental concern of these industries. Most of the dyeing processes are performed in aqueous solution, so are consequently produced large volumes of dye-containing wastewaters. Moreover, conventional dyeing processes consume high amount of energy in the form of electricity and fossil fuels (Chequer et al. 2013).

In the last decades a considerable concern over the carcinogenic and mutagenic effects related with the toxic nature of certain dyes, mainly azo dyes, has been seen (Katheresan et al. 2018; Vikrant et al. 2018). The acute and chronic toxicity of dyes has been extensively studied (Brüschweiler and Merlot 2017; Neumann 2010; Rawat et al. 2016). The use of azo dyes has been increasingly regulated by legislation. A number of aromatic amines formed by the reductive cleavage of the azo group have been recognized carcinogens and have been banned of the industry (Brüschweiler and Merlot 2017; Pereira and Alves 2012).

The mutagenic, carcinogenic and allergenic effects of dyes, previously mentioned, have made the legislation of several countries more restrictive with respect to the limits of dyes in the effluents. Currently, physicochemical and biological technologies are used to handle dye containing wastewaters. However, the color removal is one of the most difficult challenges of wastewater treatment plants. The conventional or traditional treatment is based in: (i) a pre-treatment for the equalization and neutralization of dye-containing wastewater, (ii) a primary treatment for the removal of suspended solids. In this step, physicochemical technologies, such as coagulation, flocculation and filtration are usually applied, (iii) a secondary treatment, for the biological stabilization of waste components using activated sludge and (iv) a tertiary method, such as adsorption using activated carbon materials or reverse osmosis, for the total removal of dye particles (Katheresan et al. 2018). However, the main disadvantages of these technologies are related with high cost, limited applicability, low efficiency and handling of sludge remaining (Kaushik and Malik 2009; Pereira and Alves 2012). Then, the treatment of dye-containing wastewaters remains a challenge that encourages the development of new advanced technologies, based on chemical, physical or biological methods. Chemical methods are photocatalysis (Wang et al. 2018), chemocatalysis (Du and Chen 2018), ozonation (Ghuge and Saroha 2018; Mella et al. 2017a) and Fenton reaction (Shi et al. 2018). Physical methods include adsorption with alternative materials (Gomes et al. 2016; Mella et al. 2017b; Piccin et al. 2016a), coagulation or flocculation (Mella et al. 2017a), micro and nanofiltration (Foorginezhad and Zerafat 2017) and reverse osmosis (Liu et al. 2019). Biological methods comprise biodegradation, bioaccumulation and biosorption. Biodegradation of dyes can occur by means of bacteria (Guadie et al. 2018), fungi (Ortiz-Monsalve et al. 2017) and algae (Dellamatrice et al. 2017) strains. Bioaccumulation occurs in living biomass as bacteria (Chandra et al. 2011) and biosorption takes place mostly in non-living biomass as agricultural waste (Albadarin and Mangwandi 2015) among others.

7.3 SORPTION MECHANISMS AND SORBENT MATERIALS

Sorption is a generic term used to designate the absorption and adsorption phenomena in separation and purification processes. The interaction between the sorbent (solid) and certain solutes (gas or liquid) may cause selective retention of the solute, resulting in its separation from the fluid phase. After sorption, the solute (chemical species to be sorbed) is called "sorbate" (Hiester et al. 1963). It is usually called sorption when the exact interaction mechanism of the solute species on the surface of the sorbent cannot be distinguished or when absorption and adsorption occur simultaneously (Borda and Sparks 2008).

Absorption phenomenon occurs when the substance to be sorbed is incorporated into another of different physical state. While in adsorption, a molecule or ion of the sorbate is physically adhered (physical adsorption) or chemically bound (chemisorption) to another molecule on the surface of the sorbent material (Gadd 2009). Physical adsorption occurs due to weak van der Waals forces and electrostatic interactions. It is a fast and reversible process that presents low adsorption energy, since it does not dissociate the adsorbed species (Dąbrowski 2001). Chemisorption occurs by the strong chemical bonds between functional groups of the adsorbate molecule and the adsorbent. This phenomenon involves high adsorption energy and may be an irreversible process (Ruthven 1984).

Classified as a subcategory of adsorption, biosorption is broadly described as the removal of target substances from aqueous solutions by means of biological matrix sorbents. It is a fast and generally reversible process (Michalak et al. 2013). In fact, biosorption includes several mechanisms besides adsorption, such as ion exchange, complexation and precipitation (Abdolali et al. 2014).

If a surface functional group of the sorbent reacts with an ion or molecule of the solute to form a stable molecular unit, this can be described as surface complexation (Sposito 2008). Surface precipitate is formed when the adsorption occurs and continues through the creation of three-dimensional surface species (Borda and Sparks 2008; Fomina and Gadd 2014). Ion exchange is a mechanism in which ions present in a liquid phase are replaced by ions that are electrically connected to the functional groups of a solid (organic or inorganic) matrix insoluble in the liquid phase. The redistribution of ions between the two phases occurs by diffusion, and chemical factors are less significant or even absent (Cobzaru and Inglezakis 2015). Non-reversible events such as precipitation reactions, chelating effects, phase changes and surface sorption can occur in the ion exchange process although, theoretically, it is a process totally reversible (Dyer 2000).

Biosorption mechanisms can be investigated by understanding the surface structure of biosorbent material and functional groups, applying different techniques to characterize the biosorbents and through kinetic and thermodynamic studies (Abdolali et al. 2014). Sorption phenomena and the mechanisms involved in the different sorbent materials are deeply present in the literature (Aksu 2005; Crini et al. 2019; Hassan and Carr 2018; Speight 2018; Tran et al. 2015). A schematic representation of main mechanisms involved in sorption phenomena is shown in Scheme 7.1 that focuses on biosorption mechanisms, the aim of this chapter.

Four types of sorbent materials are widely used: activated carbon, zeolites, silica gel and activated alumina (Yang 2003). Commercial activated carbon is characterized by high specific surface area, physical-chemical stability and mechanical strength, high porosity and adsorption capacity, as well as the possibility of regeneration (Yahya et al. 2015). Zeolites are crystalline aluminosilicates with an orderly distribution of micropores with uniform size (pore diameter <2 nm) (Li and Yu 2014). Silica gel is an amorphous form of silicon dioxide that has high thermal and mechanical stability and several functional groups on the surface of its micropores (Wang et al. 2014). Alumina presents high specific surface area, mechanical and thermal resistance, good stability over a wide pH range and does not disintegrate when immersed in water

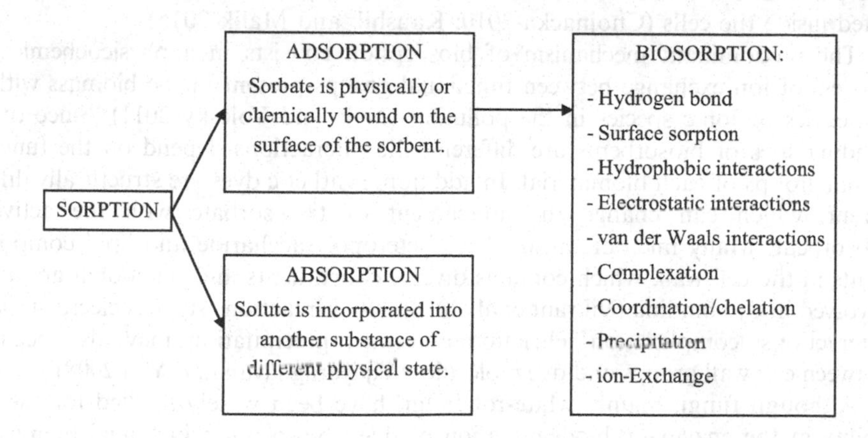

SCHEME 7.1 Principal biosorption mechanisms involved in pollutant removal from aqueous medium.

(Abou-Ziyan et al. 2017). Activated carbon is one of the oldest adsorbents and, because of its versatility, is widely used in the removal of organic and inorganic pollutants in water and sewage treatment (Gamal et al. 2018). It is effective in removing color, odor, taste and other impurities from the water, recovering solvents and controlling air pollution (Danish and Ahmad 2018). The raw materials most used in the production of activated carbon are coconut shell, mineral coal, peat, wood among others (Lee et al. 2014).

However, the cost of commercial activated carbon is still considered high, and this limits its use in applications such as effluent treatment. Therefore, many studies have been carried out in the search for alternative sorbent materials to conventional activated carbon, such as in the case of biosorbents.

7.4 BIOSORPTION

Biosorption is an adsorption process that occurs in a biomaterial. The process involves a solid adsorbent phase (biosorbent) and a liquid phase containing the chemical (sorbate). Many recalcitrant pollutants, such as heavy metals and organic compounds (e.g. dyes), are common pollutants that can be removed from wastewater by biosorption. Biomaterials used as biosorbents are generally derived from raw biomass of microorganisms or plants (Naja and Volesky 2011).

It is important to note that bioaccumulation and biosorption are different biological methods for removal of xenobiotics. Biosorption is a passive process which does not involve metabolic energy. Although biosorption is usually performed with dead biomass, it may occur in living or dead cells (Kaushik and Malik 2009). Bioaccumulation is a two-step process where a microorganism is cultivated in the presence of a pollutant, aiming the accumulation into the actively growing cells. In the first stage occurs a passive process of biosorption of the pollutant, which is subsequently transported and accumulated into the cytoplasm, in an active process dependent of cell metabolism. In biosorption the pollutants are bound to the cell wall and in bioaccumulation are accumulated inside the cells (Chojnacka 2010; Kaushik and Malik 2015).

The predominant mechanism of biosorption consists in a physicochemical process of ion exchange between functional groups present in the biomass with molecules or ionic species in the pollutant (Naja and Volesky 2011). Since the binding sites of biosorbents are different, the interactions depend on the functional groups of each biomaterial. In addition, synthetic dyes are structurally different, which can change the attachment of the sorbate with the active biosorbent. Many microorganisms have heteropolysaccharide and lipid components in the cell wall, which contains diverse constituents and functional groups involved in dye binding (Vikrant et al. 2018). Other reactions such as electrostatic interactions, complexation, chelation and microprecipitation may also occur between cell wall groups and dye molecules (Vijayaraghavan and Yun 2008).

Although fungi, mainly white-rot fungi, have been widely studied for their ability in the enzymatic biodegradation of dyes, biosorption in fungal biomass is also a mechanism used for the removal of dyes. The mechanism consists of physical-chemical interactions such as adsorption, deposition and ion-exchange, related with the binding of various functional groups present on the fungal cell wall such as amino, carboxyl, thiol and phosphate with the dye molecules. This is a process that can be completed in few hours (Kaushik and Malik 2015; Sen et al. 2016). The fungi cell wall can be attached to both anionic and cationic dyes (Chojnacka 2010).

With respect to dye biosorption by bacteria, the chemical groups of the cell wall also play a vital role in dye biosorption. The cell wall of Gram-positive bacteria is composed of a peptidoglycan layer, connected by amino acid bridges. Gram-negative bacteria have a thinner peptidoglycan layer and an additional outer membrane composed of phospholipids and lipopolysaccharide (Srinivasan and Viraraghavan 2010). Charged groups present on the bacterial cell wall such as carboxyl are related with the binding of dye cations in solution. Amine groups also can bind dyes molecules by electrostatic interaction or hydrogen bonding (Vijayaraghavan and Yun 2008).

Dye biosorption depends not only on the properties of the cell wall of biomass. Other operational parameters such as temperature, pH, biosorbent dose and particle size, initial dye concentration, class nature of dye, ionic strength, contact time and temperature are involved in the efficiency of treatment (Safa and Bhatti 2010). The biosorption capacity of the microbial biomass can be enhanced by optimizing these variables (Kaushik and Malik 2015). Additionally, pre-treatment processes can also improve the adsorption capacity of biomass. These treatments are focused on increasing or activating the binding sites in the biomass. The pre-treatments include physical methods (autoclaving or drying) or chemical methods (organic or inorganic chemicals) (Sen et al. 2016).

Biosorption is mainly applied for the removal of pollutants from wastewaters, especially metals and dyes. The main advantage of biosorption is the low cost of the biosorbents. Furthermore, biosorbents are renewable materials, their adsorptive potential can be enhanced by physicochemical or chemical activation treatments and regenerated by chemical methods and reused (Kaushik and Malik 2009).

Differently of biodegradation and bioaccumulation, the operational process of biosorption is simple and, because of the use of dead biomass, there are advantages in comparison to the use of live microorganisms. Dead biomass can be stored for longer periods, does not suffer from the toxicity of the sorbates, costs related with supply of nutrient and maintenance of the microorganism are not necessary (Chojnacka 2010) and, in general, the sorbed compounds can be easily desorbed from the material. However, the disadvantages of employing dead biomass, in dry and powder form are the difficulty of biomass separation from the biosorption system, beyond loss in mass after regeneration and low mechanical resistance (Michalak et al. 2013). Using immobilization techniques these problems can be avoided (Copello et al. 2013; Huang et al. 2010). Biosorption of dyes are also performed with different reactor configurations, aiming the development of technology that allows the application of biosorption at industrial scale (Kaushik and Malik 2015).

7.5 BIOSORBENT MATERIALS

The biological matrix involved in the biosorption process may be classified according to their origin. It may be from live or dead microbial biomass (bacteria, fungi, yeasts, microalgae) (Javanbakht et al. 2013; Montazer-Rahmati et al. 2011; Puchana-Rosero et al. 2016b), industrial waste (fermentation, food waste, sludge, etc.) (El-Sayed and El-Sayed 2014; Piccin et al. 2016b), agricultural waste (residues from fruit and vegetable crops) (Bharathi and Ramesh 2013), animal waste (Maruyama et al. 2007), natural residues (plant waste, tree bark, weeds, peat) (Kuppusamy et al. 2015) or from ashes of these materials (Buema et al. 2014; Harja et al. 2013), among others (chitosan, cellulose, etc.) (Anirudhan et al. 2016).

Biosorbent materials have been studied in order to determine their application and process conditions. These materials, to be used, should present some important characteristics for adsorption, such as availability, specific high area, sorption capacity, selectivity, low cost, easy regeneration and compatibility with the process (Oliveira et al. 2008). A wide variety of biosorbents are available for application in the wastewater bioremediation, and their general characteristics are deeply present in the literature (Kharat 2015; Park et al. 2010; Saratale et al. 2011; Solís et al. 2012; Srinivasan and Viraraghavan 2010). In the following sections, the studies developed applying the biosorbents from industrial and agricultural wastes and from dead and living biomass for wastewater remediation will be highlighted.

7.5.1 Industrial and Agricultural Wastes, Plant Materials and Natural Residues

Due to high availability and problem of disposal of industrial and agricultural wastes, the research by means of reutilization of these renewable materials has increased in the field of alternative biosorbents in pollution control (Gisi et al. 2016; Gupta and Suhas 2009). The pre-treatment for use of industrial and agricultural wastes, plant materials and natural residues is usually washing, milling and sieving to select the particle size for biosorption. These materials can also be modified by means of well-known modification techniques (Bhatnagar and Sillanpää 2010).

Agricultural solid wastes have been studied to remove many dyes from aqueous solutions. Examples include agave bagasse (Rosas-Castor et al. 2014), almond shell (Deniz 2013), cashew nut shell (Subramaniam and Ponnusamy 2015), corncob (Gardazi et al. 2016), grass waste (Hameed 2009), garlic peel (Hameed and Ahmad 2009), grapefruit peel (Saeed et al. 2010), lentil straw (Çelekli et al. 2012), luffa sponge (Demir et al. 2008; Demir and Deveci 2018), mango seeds (Alencar et al. 2012), melon peel (Djelloul and Hamdaoui 2014), pine sawdust (Cheng et al. 2013), olive stone (Albadarin and Mangwandi 2015), pomegranate peel (Ahmad et al. 2014), potato peel (Guechi and Hamdaoui 2016), rice husk (Tavlieva et al. 2013), rice straw (Sangon et al. 2018), sugarcane bagasse (Adebayo et al. 2014), walnut shell (Cao et al. 2014), banana peel (Amela et al. 2012), coir pith (Etim et al. 2016), orange peel (Mafra et al. 2013) and sunflower piths (Baysal et al. 2018).

Industrial solid wastes can also be used as biosorbent. Several materials have been investigated for the removal of pollutants from wastewaters (Ahmaruzzaman 2011; Bhatnagar and Sillanpää 2010), such as pulp and paper waste and iron sand (Iakovleva et al. 2015), raw hide trimmings from tanneries (Fathima et al. 2009), black liquor lignin from pulp mill (Ge et al. 2014), grape stalk and bagasse from winemaking bagasse (Olivella et al. 2012), fly ash generated in the sugar industry (Freitas and Farinas 2017) and slurry derived from fertilizer plant (Mohan et al. 2001).

7.5.2 Tannery Solid Wastes

The leather and textile industry are examples of industries that can use their solid waste for the treatment of wastewater (Gomes et al. 2016; Mella et al. 2017b; Rangabhashiyam et al. 2013). Tanneries produce large amounts of solid waste such as bovine hair, raw hide trimmings, wet-blue leather shavings and trimmings, and sludge from wastewater treatment plant. When environmentally friendly technologies are employed in the leather process, more of the solid waste constituents can be used as byproducts or as inputs for other industries.

TABLE 7.1
Tannery solid waste used as biosorbent for dyes

Sorbents	Dyes	Maximum dye capacity (mg g^{-1})	References
Activated carbon obtained from chromium-tanned shavings and buffing dust (BD)	Methylene blue	$q = 166.7$ $q_{BD} = 200.0$	(Yılmaz et al. 2007)
	Acid Brown	$q_{BD} = 6.240$	(Sekaran et al. 1998)
Activated carbon obtained from the sludge of tannery-treatment effluent plant	Acid Black 210 Acid Red 357	$q_{black} = 1108$ $q_{red} = 589.5$	(Puchana-Rosero et al. 2016a)
	Acid Black 210 Acid Red 357 Acid Yellow 194	$q_{black} = 156.6$ $q_{yellow} = 550.0$ $q_{red} = 413.9$ $q_{red} = 24.74$	(Piccin et al. 2012) (Gomes et al. 2016)
Natural non-tanned leather and chromium-containing leather waste	Acid Blue 45	$q = 100.0$	(Arthy and Saravanakumar 2013)
	Methylene blue Commercial reactive red	$q_{blue} = 80.00$ $q_{red} = 163.0$	(Oliveira et al. 2008)
	Acid Yellow 11 Direct Red 31	$q_{yellow} = 980.1$ $q_{red} = 1369$	(Zhang and Shi 2004)
Fleshing	Acid Blue 113	$q = 70.00$	(Fathima et al. 2010)
Cattle hair waste	Acid Blue 161 Acid Black 210	$q_{blue} = 104.9$ $q_{black} = 26.29$	(Mella et al. 2017b)

If these practices are implemented, there is a reduction in the potential environmental impact and in the disposal costs of the solid waste in landfills (Andrioli et al. 2015).

Landfilling cannot be considered as an environmentally friendly solution for tanned leather waste. The treatment of tanned leather waste has been the subject of comprehensive studies, mainly including the reutilization of the chromium extracted from the leather waste in the tanning process itself (Cassano et al. 1997; Petruzzelli et al. 1995) and isolation of protein fractions (Cabeza et al. 1997, 1999). Over the years, much attention has been given to the methods of converting these materials into useful products.

According to the literature, different biosorbents from tannery industry have been proposed as alternative to commercial activated carbon for the removal of different types of dyes, as observed in Table 7.1, where q is the amount of solute adsorbed per unit mass of adsorbent.

7.5.3 Winery Solid Waste

Winemaking generates solid residues that represent up to 30% of the total amount of vinified grapes. Almost 20 million tons of these wastes are discharged every year worldwide (Melo et al. 2015). These wastes represent both ecological and economical impasse for the wineries in relation to their storage and disposal (Barcia et al. 2014). Although grape wastes are considered biodegradable, they need a minimum time to be mineralized, being a factor of pollution (Cataneo et al. 2008). The use of these wastes for different purposes represents a significant advance in the search for the process with zero discard and valorization of by-products.

Waste from winemaking and its by-products is usually intended for composting and subsequent application to soils as fertilizer, but its composition is not considered ideal for this practice, since there is the presence of polyphenols that inhibit plantations germination (Northup et al. 1998). They are also incorporated into the animal feed because of their nutritional quality, although the animals may be intolerant to some components of the residue (Brenes et al. 2016; Laufenberg et al. 2003).

Winemaking waste may also be used as biosorbent material in natural form, as a cheap alternative sorbent material, or as activated carbon, since it has many organic components, a desirable characteristic to produce activated carbon (Nayak et al. 2016). According to the literature, winemaking waste in biosorption field is still a matter of research, once most of the studies reports the sorption of metals by grape wastes. Regarding the use of grape waste in dyes removal, there are few studies in this area (Benvenuti et al. 2019; Geçibesler and Toprak 2017; Olivella et al. 2012; Perez-Ameneiro et al. 2014; Mechati et al. 2015).

7.5.4 Living or Dead Microorganisms

Microbial biomass and other biomaterials are produced as waste or by products of other industries, such industrial large-scale fermentations (winery,

TABLE 7.2

Removal of dyes by fungal strains

Strain	Dye—concentration (mg L^{-1})	Mechanism	Color removal (%)	Maximum dye capacity (mg g^{-1})	Sorbent dosage (g L^{-1})	Contact time (min)	Reference
Filamentous fungi							
Penicillium glabrum	Congo Red—50	Biosorption	–	101.01	0.33	180	(Bouras et al. 2017)
Aspergillus niger	Congo Red—50	Biodegradation and biosorption	>98.5	263.2	80	12 h	(Lu et al. 2017)
Mucor circinelloides	Congo Red—150	Biosorption	94	169.49	10	180	(Azin and Moghimi 2018)
Aspergillus lentulus	Reactive Remazol Red—2000 Reactive Blue—2000	Bioaccumulation and biosorption	74.6 62.2	151 189	4–5	150	(Mathur et al. 2018)
Trametes villosa	Acid Blue 161—100	Biosorption	89.47	221.6	1.2	24 h	(Puchana-Rosero et al. 2016b)
Panus tigrinus	Reactive Blue 19—50	Biosorption	83.18	–	15	90	(Mustafa et al. 2017)
Penicillium janthinellum	Congo Red, Naphthol Green B, Chrome Black—150	Biosorption	>99	344.83	20	24 h	(Wang et al. 2015)
Trametes pubescens (alginate-immobilized)	Congo Red—100	Biosorption	97.96	495.24	0.5	60	(Si et al. 2015)
Aspergillus lentulus	Acid Blue 120—100	Biosorption	91	50	2	60	(Kaushik et al. 2014)

(*Continued*)

TABLE 7.2 (Cont.)

Strain	Dye—concentration (mg L^{-1})	Mechanism	Color removal (%)	Maximum dye capacity (mg g^{-1})	Sorbent dosage (g L^{-1})	Contact time (min)	Reference
Aspergillus niger	Procion Red—200	Biodegradation and biosorption	30	–	3	180	(Almeida and Corso 2014)
Thamnidium elegans	Reactive Red 198—200	Biosorption	97.83	234.24	0.8	75	(Akar et al. 2013)
Yeast							
Yarrowia lipolytica	Brilliant Green—10	Biosorption and Bioaccumulation	99.93	65.35	–	16 h	(Dil et al. 2017)
Yeast slurry from brewery	Reactive Black B—100 Direct Blue 85—100	Biosorption	100 96	162.7 139.2	0.63	30 60	(Castro et al. 2017)
Trichosporon akiyoshidainum	Reactive Black 5—300	Biodegradation, bio-accumulation and biosorption	100	–	–	12 h	(Martorell et al. 2017)
Diutina rugosa	Indigo Vat Blue I—10	Biosorption and biodegradation	99.97	–	2	5 days	(Bankole et al. 2017)
Saccharomyces cerevisiae	Ramazol Blue—100	Biosorption	100	100	1	60	(Mahmoud 2016)
Yarrowia lipolytica	Malachite Green—35	Biosorption and Bioaccumulation	100	–	–	24 h	(Asfaram et al. 2016)

TABLE 7.3

Removal of dyes by bacteria strains

Strain	Dye—concentration (mg L^{-1})	Mechanism	Color removal (%)	Maximum dye capacity (mg g^{-1})	Sorbent dosage (g L^{-1})	Contact time (min)	Reference
Pseudomonas aeruginosa (Rice husk ash silica immobilized)	Methylene Blue—45	Biosorption	95	75.7	10	60	(Iqbal et al. 2018)
Aerobic granular sludge	Reactive Yellow 15—10	Biodegradation and biosorption	85	–	25	72 h	(Sarvajith et al. 2018)
Active sludge (PVA immobilized)	Reactive Red 2—10	Biodegradation and biosorption	100	–	10	20 h	(Hameed and Ismail 2018)
Aerobic granular sludge	Methyl Orange—40	Biosorption	–	741.6	4	6 h	(Huang et al. 2018)
Agrobacterium fabrum (alginate-immobilized)	Methylene Blue—200	Biosorption	–	91	1	60	(Sharma et al. 2018)
Acidithiobacillus thiooxidans	Sulfur Blue 15—2000	Biodegradation and biosorption	91.4	769.2	0.38–0.76	20	(Nguyen et al. 2016)
Bacillus catenulatus	Basic Blue 3—2000	Biosorption	58	139.74	10	10	(Kim et al. 2015)
Shewanella onediensis	Congo Red—2000	Biodegradation and biosorption	99.25	–	–	24 h	(Liu et al. 2016)
Aerobic granular sludge	Methylene Blue—100	Biosorption	90.10	381.7	2.5	60	(Wei et al. 2015)

antibiotics, enzyme and brewery). Microorganisms as some bacteria, fungi, yeast and algae can carry out the biodegradation and biosorption of dyes in wastewater.

Fungi have proved to be suitable organisms for the removal of dyes from wastewaters by various mechanisms: biosorption, biodegradation and bioaccumulation. White-rot fungi have been extensively reported by their ability in the biodegradation of dyes (Ali 2010; Sen et al. 2016; Vikrant et al. 2018), however, some species have been employed as effective biosorbents. Other filamentous fungi and yeast have mainly been studied by the potential in the biosorption of dyes (Kaushik and Malik 2015). A summary of recent studies using filamentous fungi and yeast as biosorbents of dyes is shown in Table 7.2

Similar to fungi, bacteria have been mainly reported by the potential in biodegradation of synthetic dyes. The degradation occurs most efficiently under anaerobic conditions and in the presence of light, decolorizing high concentrations of dyes. Photobioreactor can be used to avoid the problems of wastewater treatment with suspended bacterial cultures (Frigaard 2016). The cleavage of dyes is mediated by oxidoreductive enzymes, such as azoreductases, laccases and peroxidases (Chen et al. 2018; Vikrant et al. 2018). The application of bacterial biomass in biosorption of dyes has been less explored (Srinivasan and Viraraghavan 2010). However, the use of biosorption with consortia of bacteria and activated granular sludge is a promising biotechnology for the treatment of dye-containing wastewaters (Wang et al. 2018). Table 7.3 shows recent studies using bacterial biomass as biosorbent.

The bioremediation of dye containing wastewater by microalgae may involve biosorption and bioaccumulation/bioconversion processes, which can occur simultaneously. Microalgae can adsorb the dyes to its surface (biosorption); use the dyes as carbon source converting them into metabolites (bioconversion) and bioaccumulate the dyes by a mechanism that comprise adsorption and/or enzyme degradation (Fazal et al. 2018). Dead and living microalgae can be used in the bioremediation of wastewater. Some examples are studies reported the biomass of *Spirulina platensis* microalgae for the removal of textile dyes (Cardoso et al. 2012), *Chlorella pyrenoidosa* biomass for the removal of rhodamine B dye from dyeing stones effluents (Rosa et al. 2018), *Cosmarium* sp microalgae for the removal of malachite green dye (Daneshvar et al. 2007) and using *Scenedesmus* sp. defatted microalgal biomass (waste from microalgal biofuel extraction) for the removal of leather dyes from wastewater (Fontoura et al. 2017).

7.6 CONCLUDING REMARKS

There is concern from society regarding the discharge of dyes in water bodies and effort by industries which utilize dyes in order to implement clean technologies and treat the effluents generated in their processes. Biosorption is a potential method of wastewater treatment since the biosorbents are generally wastes or by-products of other process, highly available, low-cost and renewable materials. However, the research in this field should be continued to improve biosorption capacity of these materials. The application of these residual materials as biosorbents in industrial wastewater treatment is of interest in reusing waste from the production process and post-consumption, and reducing global raw materials consumption.

REFERENCES

Abdolali, A., W. S. Guo, H. H. Ngo, S. S. Chen, N. C. Nguyen, and K. L. Tung. 2014. "Typical Lignocellulosic Wastes and By-Products for Biosorption Process in Water and Wastewater Treatment: A Critical Review." *Bioresource Technology* 160: 57–66.

Abou-Ziyan, H., D. Abd El-Raheim, O. Mahmoud, and M. Fatouh. 2017. "Performance Characteristics of Thin-Multilayer Activated Alumina Bed." *Applied Energy* 190: 29–42.

Adebayo, M. A., L. D. T. Prola, E. C. Lima et al. 2014. "Adsorption of Procion Blue MX-R Dye from Aqueous Solutions by Lignin Chemically Modified with Aluminium and Manganese." *Journal of Hazardous Materials* 268: 43-50.

Ahmad, M. A., N. A. A. Puad, and O. S. Bello. 2014. "Kinetic, Equilibrium and Thermodynamic Studies of Synthetic Dye Removal Using Pomegranate Peel Activated Carbon Prepared by Microwave-Induced KOH Activation." *Water Resources and Industry* 6: 18–35.

Ahmaruzzaman, M. 2011. "Industrial Wastes as Low-Cost Potential Adsorbents for the Treatment of Wastewater Laden with Heavy Metals." *Advances in Colloid and Interface Science* 166 (1–2): 36–59.

Akar, T., S. Arslan, and S. T. Akar. 2013. "Utilization of Thamnidium Elegans Fungal Culture in Environmental Cleanup: A Reactive Dye Biosorption Study." *Ecological Engineering* 58: 363–370.

Aksu, Z. 2005. "Application of Biosorption for the Removal of Organic Pollutants: A Review." *Process Biochemistry* 40 (3–4): 997–1026.

Albadarin, A. B., and C. Mangwandi. 2015. "Mechanisms of Alizarin Red S and Methylene Blue Biosorption onto Olive Stone By-Product: Isotherm Study in Single and Binary Systems." *Journal of Environmental Management* 164: 86–93.

Alencar, W. S., E. Acayanka, E. C. Lima et al. 2012. "Application of Mangifera Indica (Mango) Seeds as a Biosorbent for Removal of Victazol Orange 3R Dye from Aqueous Solution and Study of the Biosorption Mechanism." *Chemical Engineering Journal* 209: 577–588.

Ali, H. 2010. "Biodegradation of Synthetic Dyes: A Review." *Water, Air, & Soil Pollution* 213 (1–4): 251–273.

Almeida, E. J. R, and C. R. Corso. 2014. "Comparative Study of Toxicity of Azo Dye Procion Red MX-5B Following Biosorption and Biodegradation Treatments with the Fungi Aspergillus Niger and Aspergillus Terreus." *Chemosphere* 112: 317–322.

Amela, K., M. A. Hassen, and D. Kerroum. 2012. "Isotherm and Kinetics Study of Biosorption of Cationic Dye onto Banana Peel." *Energy Procedia* 19: 286–295.

Andrioli, E., L. Petry, and M. Gutterres. 2015. "Environmentally Friendly Hide Unhairing: Enzymatic-Oxidative Unhairing as an Alternative to Use of Lime and Sodium Sulfide." *Process Safety and Environmental Protection* 93: 9–17.

Anirudhan, T. S., J. R. Deepa, and J. Christa. 2016. "Nanocellulose/Nanobentonite Composite Anchored with Multi-Carboxyl Functional Groups as an Adsorbent for the Effective Removal of Cobalt(II) from Nuclear Industry Wastewater Samples." *Journal of Colloid and Interface Science* 467: 307–320.

Arthy, M., and M.P. Saravanakumar. 2013. "Isotherm Modeling, Kinetic Study and Optimization of Batch Parameters for Effective Removal of Acid Blue 45 Using Tannery Waste." *Journal of Molecular Liquids* 187: 189–200.

Asfaram, A., M. Ghaedi, G. R. Ghezelbash et al. 2016. "Biosorption of Malachite Green by Novel Biosorbent Yarrowia Lipolytica Isf7: Application of Response Surface Methodology." *Journal of Molecular Liquids* 214: 249–258.

Azin, E., and H. Moghimi. 2018. "Efficient Mycosorption of Anionic Azo Dyes by Mucor Circinelloides: Surface Functional Groups and Removal Mechanism Study." *Journal of Environmental Chemical Engineering* 6 (4): 4114–4123.

Bankole, P. O., A. A. Adekunle, O. F. Obidi, O. D. Olukanni, and S. P. Govindwar. 2017. "Degradation of Indigo Dye by a Newly Isolated Yeast, Diutina Rugosa from Dye Wastewater Polluted Soil." *Journal of Environmental Chemical Engineering* 5 (5): 4639–4648.

Barcia, M. T., P. B. Pertuzatti, S. Gómez-Alonso, H. T. Godoy, and I. Hermosín-Gutiérrez. 2014. "Phenolic Composition of Grape and Winemaking By-Products of Brazilian Hybrid Cultivars BRS Violeta and BRS Lorena." *Food Chemistry* 159: 95–105.

Baysal, M., K. Bilge, B. Yılmaz, M. Papila, and Y. Yürüm. 2018. "Preparation of High Surface Area Activated Carbon from Waste-Biomass of Sunflower Piths: Kinetics and Equilibrium Studies on the Dye Removal." *Journal of Environmental Chemical Engineering* 6 (2): 1702–1713.

Benvenuti, J., M. Gutterres, and J. H. Z. dos Santos. 2019. "Sol-Gel Hybrid Material Synthesized from Winemaking Waste as Adsorbent for Textile Cationic Dye Removal from Wastewater." In 6th International Conference on Multifunctional, Hybrid and Nanomaterials. Sitges: Elsevier.

Bharathi, K. S., and S. T. Ramesh. 2013. "Removal of Dyes Using Agricultural Waste as Low-Cost Adsorbents: A Review." *Applied Water Science* 3 (4): 773–790.

Bhatnagar, A., and M. Sillanpää. 2010. "Utilization of Agro-Industrial and Municipal Waste Materials as Potential Adsorbents for Water Treatment—A Review." *Chemical Engineering Journal* 157 (2–3): 277–296.

Borda, M. J, and D. L Sparks. 2008. "Kinetics and Mechanisms of Sorption-Desorption in Soils: A Multiscale Assessment." In *Biophysico-Chemical Processes of Heavy Metals and Metalloids in Soil Environments* edited by A. Violante, P. M. Huang, and G. M. Gadd, 97–124. Hoboken, NJ: John Wiley & Sons, Inc.

Bouras, H. D., A. R. Yeddou, N. Bouras et al. 2017. "Biosorption of Congo Red Dye by Aspergillus Carbonarius M333 and Penicillium Glabrum Pg1: Kinetics, Equilibrium and Thermodynamic Studies." *Journal of the Taiwan Institute of Chemical Engineers* 80: 915–923.

Brenes, A., A. Viveros, S. Chamorro, and I. Arija. 2016. "Use of Polyphenol-Rich Grape By-Products in Monogastric Nutrition. A Review." *Animal Feed Science and Technology* 211: 1–17.

Brüschweiler, B. J., and C. Merlot. 2017. "Azo Dyes in Clothing Textiles Can Be Cleaved into a Series of Mutagenic Aromatic Amines Which Are Not Regulated Yet." *Regulatory Toxicology and Pharmacology* 88: 214–226.

Buema, G., F. Noli, P. Misaelides, D. M. Sutiman, I. Cretescu, and M. Harja. 2014. "Uranium Removal from Aqueous Solutions by Raw and Modified Thermal Power Plant Ash." *Journal of Radioanalytical and Nuclear Chemistry* 299 (1): 381–386.

Cabeza, L. F., M. M. Taylor, E. Brown, and W. N. Marmer. 1997. "Influence of Pepsin and Trypsin on Chemical and Physical Properties of Isolated Gelatin from Chrome Shavings." *Journal of the American Leather Chemists Association* 92: 200–207.

Cabeza, L. F., M. M. Taylor, E. Brown, and W. N. Marmer. 1999. "Isolation of Protein Products from Chromium-Containing Leather Waste Using Two Consecutive

Enzymes and Purification of Final Chromium Product: Pilot Plant Studies." *Journal of the Society of Leather Technologies and Chemists* 83: 14–19.
Cao, J. S., J. X. Lin, F. Fang, M. T. Zhang, and Z. R. Hu. 2014. "A New Absorbent by Modifying Walnut Shell for the Removal of Anionic Dye: Kinetic and Thermodynamic Studies." *Bioresource Technology* 163: 199–205.
Cardoso, N. F., E. C. Lima, B. Royer, M. V. Bach, G. L. Dotto, L. A.A. Pinto, and T. Calvete. 2012. "Comparison of Spirulina Platensis Microalgae and Commercial Activated Carbon as Adsorbents for the Removal of Reactive Red 120 Dye from Aqueous Effluents." *Journal of Hazardous Materials* 241–242: 146–153.
Cassano, A., E. Drioli, R. Molinari, and C. Bertolutti. 1997. "Quality Improvement of Recycled Chromium in the Tanning Operation by Membrane Processes." *Desalination* 108 (1–3): 193–203.
Castro, K. C. de, A. S. Cossolin, C. O. Dos Reis, and E. B de Morais. 2017. "Biosorption of Anionic Textile Dyes from Aqueous Solution by Yeast Slurry from Brewery." *Brazilian Archives of Biology and Technology* 60, e17160101. Epub May 11, 2017.
Cataneo, C. B., V. Caliari, L. V. Gonzaga, E. M. Kuskoski, and R. Fett. 2008. "Antioxidant activity and phenolic content of agricultural by-products from wine production." *Semina:CienciasAgrarias* 29 (1): 93–102.
Çelekli, A., B. Tanrıverdi, and H. Bozkurt. 2012. "Lentil Straw: A Novel Adsorbent for Removing of Hazardous Dye – Sorption Behavior Studies." *CLEAN – Soil, Air, Water* 40 (5): 515–522.
Chandra, R, R. N Bharagava, A. Kapley, and H. J. Purohit. 2011. "Bacterial Diversity, Organic Pollutants and Their Metabolites in Two Aeration Lagoons of Common Effluent Treatment Plant (CETP) during the Degradation and Detoxification of Tannery Wastewater." *Bioresource Technology* 102: 2333–2341.
Chen, Y., L. Feng, H. Li, Y. Wang, G. Chen, and Q. Zhang. 2018. "Biodegradation and Detoxification of Direct Black G Textile Dye by a Newly Isolated Thermophilic Microflora." *Bioresource Technology* 250: 650–657.
Cheng, G., L. Sun, L. Jiao et al. 2013. "Adsorption of Methylene Blue by Residue Biochar from Copyrolysis of Dewatered Sewage Sludge and Pine Sawdust." *Desalination and Water Treatment* 51 (37–39): 7081–7087.
Chequer, F. M. D., G. A. R. de Oliveira, E. R. A. Ferraz, J. C. Cardoso, M. V. B. Zanoni, and D. P. de Oliveira. 2013. "Textile Dyes: Dyeing Process and Environmental Impact." In *Eco-Friendly Textile Dyeing and Finishing* edited by M. Günay, 260.
Chojnacka, K. 2010. "Biosorption and Bioaccumulation–the Prospects for Practical Applications." *Environment International* 36 (3): 299–307.
Christie, R. M. 2001. *Colour Chemistry.* Cambridge, UK: Royal Society of Chemistry.
Cobzaru, C., and V. Inglezakis. 2015. "Chapter Ten – Ion Exchange." In *Progress in Filtration and Separation* edited by S. Tarleton, 425–498. Oxford: Academic Press.
Copello, G. J., M. P. Pesenti, M. Raineri et al. 2013. "Polyphenol-SiO2 Hybrid Biosorbent for Heavy Metal Removal. Yerba Mate Waste (Ilex Paraguariensis) as Polyphenol Source: Kinetics and Isotherm Studies." *Colloids and Surfaces. B, Biointerfaces* 102: 218–226.
Crini, G., E. Lichtfouse, L. D Wilson, and N. Morin-Crini. 2019. "Conventional and Non-Conventional Adsorbents for Wastewater Treatment." *Environmental Chemistry Letters* 17: 195.
Dąbrowski, A. 2001. "Adsorption — from Theory to Practice." *Advances in Colloid and Interface Science* 93 (1–3): 135–224.
Daneshvar, N., M. Ayazloo, A. R. Khataee, and M. Pourhassan. 2007. "Biological Decolorization of Dye Solution Containing Malachite Green by Microalgae *Cosmarium* Sp." *Bioresource Technology* 98 (6): 1176–1182.

Danish, M., and T. Ahmad. 2018. "A Review on Utilization of Wood Biomass as a Sustainable Precursor for Activated Carbon Production and Application." *Renewable and Sustainable Energy Reviews* 87: 1–21.

Dellamatrice, P. M., M. E. Silva-Stenico, L. A. B. de Moraes, M. F. Fiore, and R. T. R. Monteiro. 2017. "Degradation of Textile Dyes by Cyanobacteria." *Brazilian Journal of Microbiology* 48 (1): 25–31.

Demir, H., and M. A. Deveci. 2018. "Comparison of Ultrasound and Conventional Technique for Removal of Methyl Orange by Luffa Cylindrica Fibers." *Arabian Journal for Science and Engineering* 43 (11): 5881–5889.

Demir, H., A. Top, D. Balköse, and S. Ülkü. 2008. "Dye Adsorption Behavior of Luffa Cylindrica Fibers." *Journal of Hazardous Materials* 153 (1–2): 389–394.

Deniz, F. 2013. "Dye Removal by Almond Shell Residues: Studies on Biosorption Performance and Process Design." *Materials Science and Engineering: C* 33 (5): 2821–2826.

Dil, E. A., M. Ghaedi, G. R. Ghezelbash, and A. Asfaram. 2017. "Multi-Responses Optimization of Simultaneous Biosorption of Cationic Dyes by Live Yeast Yarrowia Lipolytica 70562 from Binary Solution: Application of First Order Derivative Spectrophotometry." *Ecotoxicology and Environmental Safety* 139: 158–164.

Djelloul, C., and O. Hamdaoui. 2014. "Removal of Cationic Dye from Aqueous Solution Using Melon Peel as Nonconventional Low-Cost Sorbent." *Desalination and Water Treatment* 52 (40–42): 7701–7710.

Du, W. N., and S. T. Chen. 2018. "Photo- and Chemocatalytic Oxidation of Dyes in Water." *Journal of Environmental Management* 206: 507–515.

Dyer, A. 2000. "ION EXCHANGE." In *Encyclopedia of Separation Science* edited by I. D Wilson, 156–173. Oxford: Academic Press.

Ejder-Korucu, M., A. Gürses, Ç. Doğar, S. K. Sharma, and M. Açıkyıldız. 2015. "Removal of Organic Dyes from Industrial Effluents: An Overview of Physical and Biotechnological Applications." In *Green Chemistry for Dyes Removal from Wastewater,* edited by S. K. Sharma, 1–34. Massachusetts: Wiley Online Books.

El-Sayed, H. E. M., and M. M. H. El-Sayed. 2014. "Assessment of Food Processing and Pharmaceutical Industrial Wastes as Potential Biosorbents: A Review." *BioMed Research International* 2014: 1467–1469.

Etim, U. J., S. A. Umoren, and U. M. Eduok. 2016. "Coconut Coir Dust as a Low Cost Adsorbent for the Removal of Cationic Dye from Aqueous Solution." *Journal of Saudi Chemical Society* 20: S67–76.

Fathima, N. N., R. Aravindhan, J. R. Rao, and B. U. Nair. 2009. "Utilization of Organically Stabilized Proteinous Solid Waste for the Treatment of Coloured Waste-Water." *Journal of Chemical Technology & Biotechnology* 84 (9): 1338–1343.

Fathima, N. N., R. Aravindhan, J. R. Rao, and B. U. Nair. 2010. "Stabilized Protein Waste as a Source for Removal of Color from Wastewaters." *Journal of Applied Polymer Science* 120 (3): 1397–1402.

Fazal, T., A. Mushtaq, F. Rehman, A. U. Khan, N. Rashid, W. Farooq, M. S. U. Rehman, and J. Xu. 2018. "Bioremediation of Textile Wastewater and Successive Biodiesel Production Using Microalgae." *Renewable and Sustainable Energy Reviews* 82: 3107–3126.

Fomina, M., and G. M. Gadd. 2014. "Biosorption: Current Perspectives on Concept, Definition and Application." *Bioresource Technology* 160: 3–14.

Fontoura, J. T. da, G. S. Rolim, B. Mella, M. Farenzena, and M. Gutterres. 2017. "Defatted Microalgal Biomass as Biosorbent for the Removal of Acid Blue 161 Dye from Tannery Effluent." *Journal of Environmental Chemical Engineering* 5 (5): 5076–5084.

Foorginezhad, S., and M. M. Zerafat. 2017. "Microfiltration of Cationic Dyes Using Nano-Clay Membranes." *Ceramics International* 43 (17): 15146–15159.

Freitas, J. V., and C. S. Farinas. 2017. "Sugarcane Bagasse Fly Ash as a No-Cost Adsorbent for Removal of Phenolic Inhibitors and Improvement of Biomass Saccharification." *ACS Sustainable Chemistry & Engineering* 5 (12): 11727–11736.

Frigaard, N.-U. 2016. "Biotechnology of Anoxygenic Phototrophic Bacteria." In *Anaerobes in Biotechnology* edited by R. Hatti-Kaul, G. Mamo, and B. Mattiasson, 139–154. Cham: Springer International Publishing.

Fuck, W. F., and M. Gutterres. 2008. "Produtos Químicos Perigosos e de Uso Restrito No Couro." *Tecnicouro* Novembro/Dezembro: 82–89.

Gadd, G. M. 2009. "Biosorption: Critical Review of Scientific Rationale, Environmental Importance and Significance for Pollution Treatment." *Journal of Chemical Technology and Biotechnology* 84: 13–28.

Gamal, M. E., H. A. Mousa, M. H. El-Naas, R. Zacharia, and S. Judd. 2018. "Bio-Regeneration of Activated Carbon: A Comprehensive Review." *Separation and Purification Technology* 197: 345–359.

Gardazi, S. M. H., T. A. Butt, N. Rashid et al. 2016. "Effective Adsorption of Cationic Dye from Aqueous Solution Using Low-Cost Corncob in Batch and Column Studies." *Desalination and Water Treatment* 57 (59): 28981–28998.

Ge, Y., Z. Li, Y. Kong, Q. Song, and K. Wang. 2014. "Heavy Metal Ions Retention by Bi-Functionalized Lignin: Synthesis, Applications, and Adsorption Mechanisms." *Journal of Industrial and Engineering Chemistry* 20 (6): 4429–4436.

Geçibesler, İ. H., and M. Toprak. 2017. "Azure A Removal from Aqueous System Using Natural and Modified (Grape Stalk and Pomegranate Peel) Adsorbents." *Proceedings of the National Academy of Sciences India Section A - Physical Sciences* 87 (2): 171–179.

Ghuge, S. P., and A. K. Saroha. 2018. "Ozonation of Reactive Orange 4 Dye Aqueous Solution Using Mesoporous Cu/SBA-15 Catalytic Material." *Journal of Water Process Engineering* 23: 217–229.

Gisi, S. De, G. Lofrano, M. Grassi, and M. Notarnicola. 2016. "Characteristics and Adsorption Capacities of Low-Cost Sorbents for Wastewater Treatment: A Review." *Sustainable Materials and Technologies* 9: 10–40.

Gomes, C. S., J. S. Piccin, and M. Gutterres. 2016. "Optimizing Adsorption Parameters in Tannery-Dye-Containing Effluent Treatment with Leather Shaving Waste." *Process Safety and Environmental Protection* 99: 98–106.

Guadie, A., A. Gessesse, and S. Xia. 2018. "Halomonas Sp. Strain A55, a Novel Dye Decolorizing Bacterium from Dye-Uncontaminated Rift Valley Soda Lake." *Chemosphere* 206: 59–69.

Guechi, E. K., and O. Hamdaoui. 2016. "Biosorption of Methylene Blue from Aqueous Solution by Potato (Solanum Tuberosum) Peel: Equilibrium Modelling, Kinetic, and Thermodynamic Studies." *Desalination and Water Treatment* 57 (22): 10270–10285.

Gupta, V. K., and J. Suhas. 2009. "Application of Low-Cost Adsorbents for Dye Removal – A Review." *Journal of Environmental Management* 90 (8): 2313–2342.

Gürses, A., M. Açıkyıldız, K. Güneş, and M. S. Gürseş. 2016. "Dyes and Pigments: Their Structure and Properties." In *Dyes and Pigments.* SpringerBriefs in Molecular Science, edited by A. Gürses, M. Açıkyıldız, K. Güneş and M. S. Gürses, 13–29. Cham: Springer International Publishing.

Hameed, B. B., and Z. Z. Ismail. 2018. "Decolorization, Biodegradation and Detoxification of Reactive Red Azo Dye Using Non-Adapted Immobilized Mixed Cells." *Biochemical Engineering Journal* 137: 71–77.

Hameed, B. H. 2009. "Grass Waste: A Novel Sorbent for the Removal of Basic Dye from Aqueous Solution." *Journal of Hazardous Materials* 166 (1): 233–238.

Hameed, B. H., and A. A. Ahmad. 2009. "Batch Adsorption of Methylene Blue from Aqueous Solution by Garlic Peel, an Agricultural Waste Biomass." *Journal of Hazardous Materials* 164 (2–3): 870–875.

Harja, M., G. Buema, D. M. Sutiman, and I. Cretescu. 2013. "Removal of Heavy Metal Ions from Aqueous Solutions Using Low-Cost Sorbents Obtained from Ash." *Chemical Papers* 67 (5): 497–508.

Hassan, M. M., and C. M. Carr. 2018. "A Critical Review on Recent Advancements of the Removal of Reactive Dyes from Dyehouse Effluent by Ion-Exchange Adsorbents." *Chemosphere* 209: 201–219.

Hessel, C., C. Allegre, M. Maisseu, F. Charbit, and P. Moulin. 2007. "Guidelines and Legislation for Dye House Effluents." *Journal of Environmental Management* 83 (2): 171–180.

Hiester, N. K., T. Vermeulen, and G. Klein. 1963. "Adsorption and Ion Exchange." In *Chemical Engineers' Handbook* edited by R. H. Perry, C. H. Chilton, and S. D. Kirkpatrick, 4th ed., 16.1–16.40. Tokyo: McGraw-Hill Book Company, Inc.

Huang, X., Y. Wang, X. Liao, and B. Shi. 2010. "Adsorptive Recovery of Au3+ from Aqueous Solutions Using Bayberry Tannin-Immobilized Mesoporous Silica." *Journal of Hazardous Materials* 183 (1): 793–798.

Huang, X., D. Wei, L. Yan, B. Du, and Q. Wei. 2018. "High-Efficient Biosorption of Dye Wastewater onto Aerobic Granular Sludge and Photocatalytic Regeneration of Biosorbent by Acid TiO_2 Hydrosol." *Environmental Science and Pollution Research* 25 (27): 27606-27613.

Hunger, K. 2003. *Industrial Dyes Chemistry, Properties, Applications.* edited by K. Hunger. Kelkheim: Wiley-VCH Verlag GmbH & Co. KGaA.

Iakovleva, E., E. Mäkilä, J. Salonen, M. Sitarz, and M. Sillanpää. 2015. "Industrial Products and Wastes as Adsorbents for Sulphate and Chloride Removal from Synthetic Alkaline Solution and Mine Process Water." *Chemical Engineering Journal* 259: 364–371.

Iqbal, A., S. Sabar, M. K. Mun-Yee, M. N. N. Asshifa, A. R. M. Yahya, and F. Adam. 2018. "Pseudomonas Aeruginosa USM-AR2/SiO2 Biosorbent for the Adsorption of Methylene Blue." *Journal of Environmental Chemical Engineering* 6 (4): 4908–4916.

Javanbakht, V., S. A. Alavi, and H. Zilouei. 2013. "Mechanisms of Heavy Metal Removal Using Microorganisms as Biosorbent." *Water Science and Technology* 69 (9): 1775–1787.

Katheresan, V., J. Kansedo, and S. Y. Lau. 2018. "Efficiency of Various Recent Wastewater Dye Removal Methods: A Review." *Journal of Environmental Chemical Engineering* 6 (4): 4676–4697.

Kaushik, P., and A. Malik. 2009. "Fungal Dye Decolourization: Recent Advances and Future Potential." *Environment International* 35 (1): 127–141.

Kaushik, P., and A. Malik. 2015. Mycoremediation of Synthetic Dyes: An Insight into the Mechanism, Process Optimization and Reactor Design. In *Microbial Degradation of Synthetic Dyes in Wastewaters SE - 1* edited by S. N. Singh, 1st ed., 1–25. Environmental Science and Engineering. Cham: Springer International Publishing.

Kaushik, P., A. Mishra, A. Malik, and K. K. Pant. 2014. "Biosorption of Textile Dye by Aspergillus Lentulus Pellets: Process Optimization and Cyclic Removal in Aerated Bioreactor." *Water, Air, & Soil Pollution* 225 (6): 1978.

Kharat, D. S. 2015. "Preparing Agricultural Residue Based Adsorbents for Removal of Dyes from Effluents – A Review." *Brazilian Journal of Chemical Engineering* 32: 1–12.

Kim, S. Y., M. R. Jin, C. H. Chung, Y. S. Yun, K. Y. Jahng, and K. Y. Yu. 2015. "Biosorption of Cationic Basic Dye and Cadmium by the Novel Biosorbent Bacillus

Catenulatus JB-022 Strain." *Journal of Bioscience and Bioengineering* 119 (4): 433–439.

Kuppusamy, S., P. Thavamani, M. Megharaj, and R. Naidu. 2015. "Bioremediation Potential of Natural Polyphenol Rich Green Wastes: A Review of Current Research and Recommendations for Future Directions." *Environmental Technology & Innovation* 4: 17–28.

Laufenberg, G., B. Kunz, and M. Nystroem. 2003. "Transformation of Vegetable Waste into Value Added Products: (A) the Upgrading Concept; (B) Practical Implementations." *Bioresource Technology* 87 (2): 167–198.

Lee, T., C. Ooi, R. Othman, and F. Y. Yeoh. 2014. Activated Carbon Fiber – The Hybrid of Carbon Fiber and Activated Carbon. *Reviews on Advanced Materials Science* 36: 118–136.

Li, Y., and J. Yu. 2014. "New Stories of Zeolite Structures: Their Descriptions, Determinations, Predictions, and Evaluations." *Chemical Reviews* 114 (14): 7268–7316.

Liu, M., C. Yu, Z. Dong et al. 2019. "Improved Separation Performance and Durability of Polyamide Reverse Osmosis Membrane in Tertiary Treatment of Textile Effluent through Grafting Monomethoxy-Poly(Ethylene Glycol) Brushes." *Separation and Purification Technology* 209: 443–451.

Liu, W., L. Liu, C. Liu et al. 2016. "Methylene Blue Enhances the Anaerobic Decolorization and Detoxication of Azo Dye by Shewanella Onediensis MR-1." *Biochemical Engineering Journal* 110: 115–124.

Lu, T., Q. Zhang, and S. Yao. 2017. "Efficient Decolorization of Dye-Containing Wastewater Using Mycelial Pellets Formed of Marine-Derived Aspergillus Niger." *Chinese Journal of Chemical Engineering* 25 (3): 330–337.

Mafra, M. R., L. Igarashi-Mafra, D. R. Zuim, É. C. Vasques, and M. A. Ferreira. 2013. "Adsorption of Remazol Brilliant Blue on an Orange Peel Adsorbent." *Brazilian Journal of Chemical Engineering* 30: 657–665.

Mahmoud, M. S. 2016. "Decolorization of Certain Reactive Dye from Aqueous Solution Using Baker's Yeast (Saccharomyces Cerevisiae) Strain." *HBRC Journal* 12 (1): 88–98.

Martorell, M. M., M. M. R. Soro, H. F. Pajot, and L. I. C. de Figueroa. 2017. "Optimization and Mechanisms for Biodecoloration of a Mixture of Dyes by Trichosporon Akiyoshidainum HP 2023." *Environmental Technology* 1–12.

Maruyama, T., H. Matsushita, Y. Shimada et al. 2007. "Proteins and Protein-Rich Biomass as Environmentally Friendly Adsorbents Selective for Precious Metal Ions." *Environmental Science & Technology* 41 (4): 1359–1364.

Mathur, M., D. Gola, R. Panja, A. Malik, and S. Z. Ahammad. 2018. "Performance Evaluation of Two Aspergillus Spp. for the Decolourization of Reactive Dyes by Bioaccumulation and Biosorption." *Environmental Science and Pollution Research* 25 (1): 345–352.

Mechati, F., C. Bouchelta, M. S. Medjram, R. Benrabaa, and N. Ammouchi. 2015. ""Effect of Hard and Soft Structure of Different Biomasses on the Porosity Development of Activated Carbon Prepared under N_2/Microwave Radiations."." *Journal of Environmental Chemical Engineering* 3 (3): 1928–1938.

Mella, B., B. S. C. Barcellos, D. E. S. Costa, and M. Gutterres. 2017a. "Treatment of Leather Dyeing Wastewater with Associated Process of Coagulation-Flocculation/Adsorption/Ozonation." *Ozone: Science & Engineering* 40 (2): 133–140.

Mella, B., M. J. Puchana-Rosero, D. E. S. Costa, and M. Gutterres. 2017b. "Utilization of Tannery Solid Waste as an Alternative Biosorbent for Acid Dyes in Wastewater Treatment." *Journal of Molecular Liquids* 242: 137–145.

Melo, P. S., A. P. Massarioli, C. Denny et al. 2015. "Winery By-Products: Extraction Optimization, Phenolic Composition and Cytotoxic Evaluation to Act as a New Source of Scavenging of Reactive Oxygen Species." *Food Chemistry* 181: 160–169.

Michalak, I., K. Chojnacka, and A. Witek-Krowiak. 2013. "State of the Art for the Biosorption Process – A Review." *Applied Biochemistry and Biotechnology* 170 (6): 1389–1416.

Mohan, D., V. K. Gupta, S. K. Srivastava, and S. Chander. 2001. "Kinetics of Mercury Adsorption from Wastewater Using Activated Carbon Derived from Fertilizer Waste." *Colloids and Surfaces A: Physicochemical and Engineering Aspects* 177 (2–3): 169–181.

Montazer-Rahmati, M. M., P. Rabbani, A. Abdolali, and A. R. Keshtkar. 2011. "Kinetics and Equilibrium Studies on Biosorption of Cadmium, Lead, and Nickel Ions from Aqueous Solutions by Intact and Chemically Modified Brown Algae." *Journal of Hazardous Materials* 185 (1): 401–407.

Mullai, P., M. K. Yogeswari, S. Vishali et al. 2017. "Aerobic Treatment of Effluents from Textile Industry." In *Current Developments in Biotechnology Pandey, and Bioengineering: Biological Treatment of Industrial Effluents* edited by D.-J. Lee, V. Jegatheesan, H. H. Ngo, P. C. Hallenbeck, A. Pandey, 3–34. Amsterdam: Elsevier.

Mustafa, M. M., P. Jamal, M. F. Alkhatib, S. S. Mahmod, D. N. Jimat, and N. N. Ilyas. 2017. "Panus Tigrinus as a Potential Biomass Source for Reactive Blue Decolorization: Isotherm and Kinetic Study." *Electronic Journal of Biotechnology* 26: 7–11.

Naja, G. M., and B. Volesky. 2011. "Biosorption for Industrial Applications." In *Comprehensive Biotechnology* edited by M. Moo-Young, 2nd ed., 685–700. Burlington: Academic Press.

Nayak, A., B. Bhushan, V. Gupta, and L. Rodriguez-Turienzo. 2016. "Development of a Green and Sustainable Clean up System from Grape Pomace for Heavy Metal Remediation." *Journal of Environmental Chemical Engineering* 4 (4): 4342–4353.

Neumann, H. G. 2010. "Aromatic Amines: Mechanisms of Carcinogenesis and Implications for Risk Assessment." *Frontiers in Bioscience (Landmark Edition)* 15: 1119–1130.

Ngulube, T., J. R. Gumbo, V. Masindi, and A. Maity. 2017. "An Update on Synthetic Dyes Adsorption onto Clay Based Minerals: A State-of-Art Review." *Journal of Environmental Management* 191: 35–57.

Nguyen, T. A., C. C. Fu, and R. S. Juang. 2016. "Biosorption and Biodegradation of a Sulfur Dye in High-Strength Dyeing Wastewater by Acidithiobacillus Thiooxidans." *Journal of Environmental Management* 182: 265–271.

Northup, R. R, R. A. Dahlgren, and J. G. McColl. 1998. "Polyphenols as Regulators of Plant-Litter-Soil Interactions in Northern California's Pygmy Forest: A Positive Feedback?." In *Plant-Induced Soil Changes: Processes and Feedbacks* edited by Nico Van Breemen, 189–220. Dordrecht: Springer Netherlands.

Oliveira, D. Q. L., M. Gonçalves, L. C. A. Oliveira, and L. R. G. Guilherme. 2008. "Removal of As(V) and Cr(VI) from Aqueous Solutions Using Solid Waste from Leather Industry." *Journal of Hazardous Materials* 151 (1): 280–284.

Olivella, M. A., N. Fiol, F. de la Torre, J. Poch, and I. Villaescusa. 2012. "A Mechanistic Approach to Methylene Blue Sorption on Two Vegetable Wastes: Cork Bark and Grape Stalks." BioResources 7 (3): 3340–3354.

Ortiz-Monsalve, S., J. Dornelles, E. Poll, M. Ramirez-Castrillón, P. Valente, and M. Gutterres. 2017. "Biodecolourisation and Biodegradation of Leather Dyes by a Native Isolate of Trametes Villosa." *Process Safety and Environmental Protection* 109: 437–451.

Park, D., Y.-S. Yun, and J. M. Park. 2010. "The Past, Present, and Future Trends of Biosorption." *Biotechnology and Bioprocess Engineering* 15 (1): 86–102.

Pereira, L., and M. Alves. 2012. "Dyes: Environmental Impact and Remediation." In *Environmental Protection Strategies for Sustainable Development* edited by A. Malik and E. Grohmann, 111–162. Dordrecht: Springer Netherlands.

Perez-Ameneiro, M., X. Vecino, L. Barbosa-Pereira, J. M. Cruz, and A. B. Moldes. 2014. "Removal of Pigments from Aqueous Solution by a Calcium Alginate–grape Marc Biopolymer: A Kinetic Study." *Carbohydrate Polymers* 101: 954–960.

Petruzzelli, D., R. Passino, and G. Tiravanti. 1995. "Ion Exchange Process for Chromium Removal and Recovery from Tannery Wastes." *Industrial & Engineering Chemistry Research* 34 (8): 2612–2617.

Piccin, J. S., C. S. Gomes, L. A. Feris, and M. Gutterres. 2012. "Kinetics and Isotherms of Leather Dye Adsorption by Tannery Solid Waste." *Chemical Engineering Journal* 183: 30–38.

Piccin, J. S., C. S. Gomes, B. Mella, and M. Gutterres. 2016a. "Color Removal from Real Leather Dyeing Effluent Using Tannery Waste as an Adsorbent." *Journal of Environmental Chemical Engineering* 4 (1): 1061–1067.

Piccin, J. S., M. Guterres, N. P. G. Salau, and G. L. Dotto. 2016b. "Mass Transfer Models for the Adsorption of Acid Red 357 and Acid Black 210 by Tannery Solid Wastes." *Adsorption Science & Technology* 35: 300–316.

Puchana-Rosero, M. J., M. A. Adebayo, E. C. Lima et al. 2016a. "Microwave-Assisted Activated Carbon Obtained from the Sludge of Tannery-Treatment Effluent Plant for Removal of Leather Dyes." *Colloids and Surfaces A: Physicochemical and Engineering Aspects* 504: 105–115.

Puchana-Rosero, M. J., E. C Lima, S. Ortiz-Monsalve et al. 2016b. Fungal Biomass as Biosorbent for the Removal of Acid Blue 161 Dye in Aqueous Solution. *Environmental Science and Pollution Research* 24 (4): 4200–4209.

Rangabhashiyam, S., N. Anu, and N. Selvaraju. 2013. "Sequestration of Dye from Textile Industry Wastewater Using Agricultural Waste Products as Adsorbents." *Journal of Environmental Chemical Engineering* 1 (4): 629–641.

Rawat, D., V. Mishra, and R. S. Sharma. 2016. "Detoxification of Azo Dyes in the Context of Environmental Processes." *Chemosphere* 155: 591–605.

Rosa, A. L. D. da, E. Carissimi, G. L. Dotto, H. Sander, and L. A. Feris. 2018. "Biosorption of Rhodamine B Dye from Dyeing Stones Effluents Using the Green Microalgae Chlorella Pyrenoidosa." *Journal of Cleaner Production* 198: 1302–1310.

Rosa, J. M., A. M. F. Fileti, E. B. Tambourgi, and J. C. C. Santana. 2015. "Dyeing of Cotton with Reactive Dyestuffs: The Continuous Reuse of Textile Wastewater Effluent Treated by Ultraviolet/Hydrogen Peroxide Homogeneous Photocatalysis." *Journal of Cleaner Production* 90: 60–65.

Rosas-Castor, J. M., M. T. Garza-González, R. B. García-Reyes et al. 2014. "Methylene Blue Biosorption by Pericarp of Corn, Alfalfa, and Agave Bagasse Wastes." *Environmental Technology* 35 (9): 1077–1090.

Ruthven, D. M. 1984. *Principles of Adsorption and Adsorption Processes.* New York: Wiley & Sons.

Saeed, A., M. Sharif, and M. Iqbal. 2010. "Application Potential of Grapefruit Peel as Dye Sorbent: Kinetics, Equilibrium and Mechanism of Crystal Violet Adsorption." *Journal of Hazardous Materials* 179 (1–3): 564–572.

Safa, Y., and H. N. Bhatti. 2010. "Factors Affecting Biosorption of Direct Dyes from Aqueous Solution." *Asian Journal of Chemistry* 22 (9): 6625–6639.

Sangon, S., A. J. Hunt, T. M. Attard, P. Mengchang, Y. Ngernyen, and N. Supanchaiyamat. 2018. "Valorisation of Waste Rice Straw for the Production of

Highly Effective Carbon Based Adsorbents for Dyes Removal." *Journal of Cleaner Production* 172: 1128–1139.

Saratale, R. G., G. D. Saratale, J. S. Chang, and S. P. Govindwar. 2011. "Bacterial Decolorization and Degradation of Azo Dyes: A Review." *Journal of the Taiwan Institute of Chemical Engineers* 42 (1): 138–157.

Sarvajith, M., G. K. K. Reddy, and Y. V. Nancharaiah. 2018. "Textile Dye biodecolourisation and Ammonium Removal over Nitrite in Aerobic Granular Sludge Sequencing Batch Reactors." *Journal of Hazardous Materials* 342: 536–543.

Sekaran, G., K. A. Shanmugasundaram, and M. Mariappan. 1998. "Characterization and Utilisation of Buffing Dust Generated by the Leather Industry." *Journal of Hazardous Materials* 63 (1): 53–68.

Sen, S. K., S. Raut, P. Bandyopadhyay, and S. Raut. 2016. "Fungal Decolouration and Degradation of Azo Dyes: A Review." *Fungal Biology Reviews* 30 (3): 112–133.

Sharma, S., A. Hasan, N. Kumar, and L. M. Pandey. 2018. "Removal of Methylene Blue Dye from Aqueous Solution Using Immobilized Agrobacterium Fabrum Biomass along with Iron Oxide Nanoparticles as Biosorbent." *Environmental Science and Pollution Research* 25 (22): 21605–21615.

Shi, X., A. Tian, J. You, H. Yang, Y. Wang, and X. Xue. 2018. "Degradation of Organic Dyes by a New Heterogeneous Fenton Reagent - Fe2GeS4 Nanoparticle." *Journal of Hazardous Materials* 353: 182–189.

Si, J., T.-Q. Yuan, and B.-K. Cui. 2015. "Exploring Strategies for Adsorption of Azo Dye Congo Red Using Free and Immobilized Biomasses of Trametes Pubescens." *Annals of Microbiology* 65 (1): 411–421.

Solís, M., A. Solís, H. I. Pérez, N. Manjarrez, and M. Flores. 2012. "Microbial Decolouration of Azo Dyes: A Review." *Process Biochemistry* 47 (12): 1723–1748.

Speight, J. G. 2018. "Sorption, Dilution, and Dissolution." In *Reaction Mechanisms in Environmental Engineering* edited by J. G. Speight, 165–201. Laramie: Butterworth-Heinemann.

Sposito, G. 2008. *The Chemistry of Soils*, 2th ed. New York: Oxford University Press.

Srinivasan, A., and T. Viraraghavan. 2010. "Decolorization of Dye Wastewaters by Biosorbents: A Review." *Journal of Environmental Management* 91 (10): 1915–1929.

Subramaniam, R., and S. K. Ponnusamy. 2015. "Novel Adsorbent from Agricultural Waste (Cashew NUT Shell) for Methylene Blue Dye Removal: Optimization by Response Surface Methodology." *Water Resources and Industry* 11: 64–70.

Tavlieva, M. P., S. D. Genieva, V. G. Georgieva, and L. T. Vlaev. 2013. "Kinetic Study of Brilliant Green Adsorption from Aqueous Solution onto White Rice Husk Ash." *Journal of Colloid and Interface Science* 409: 112–122.

Tran, V. S., H. H. Ngo, W. Guo, J. Zhang, S. Liang, C. Ton-That, and X. Zhang. 2015. "Typical Low Cost Biosorbents for Adsorptive Removal of Specific Organic Pollutants from Water." *Bioresource Technology* 182: 353–363.

Vijayaraghavan, K., and Y. S. Yun. 2008. "Bacterial Biosorbents and Biosorption." *Biotechnology Advances* 26 (3): 266–291.

Vikrant, K., B. S. Giri, N. Raza et al. 2018. "Recent Advancements in Bioremediation of Dye: Current Status and Challenges." *Bioresource Technology* 253: 355–367.

Wang, D., J. Zhang, X. Tian, D. Liu, and K. Sumathy. 2014. "Progress in Silica Gel–water Adsorption Refrigeration Technology." *Renewable and Sustainable Energy Reviews* 30: 85–104.

Wang, M. X., Q. L. Zhang, and S. J. Yao. 2015. "A Novel Biosorbent Formed of Marine-Derived Penicillium Janthinellum Mycelial Pellets for Removing Dyes from Dye-Containing Wastewater." *Chemical Engineering Journal* 259: 837–844.

Wang, S. N., Q. J. Chen, M. J. Zhu, F. Y. Xue, W. C. Li, T. J. Zhao, G. D. Li, and G. Q. Zhang. 2018. "An Extracellular Yellow Laccase from White Rot Fungus Trametes Sp. and Its Mediator Systems for Dye Decolorization." *Biochimie* 148: 46–54.

Wei, D., B. Wang, H. H. Ngo et al. 2015. "Role of Extracellular Polymeric Substances in Biosorption of Dye Wastewater Using Aerobic Granular Sludge." *Bioresource Technology* 185: 14–20.

Witt, O. N. 1876. "Zur Kenntniss Des Baues Und Der Bildung Färbender Kohlenstoffverbindungen." *Berichte Der Deutschen Chemischen Gesellschaft* 9 (1) Wiley-Blackwell: 522–527.

WWAP (United Nations World Water Assessment Programme). 2014. The United Nations World Water Development Report 2014: *Water and Energy*. Paris: UNESCO.

WWAP (United Nations World Water Assessment Programme). 2017. The United Nations World Water Development Report 2017. *Wastewater: The Untapped Resource*. Paris: UNESCO.

Yahya, M. A., Z. Al-Qodah, and C. W. Z. Ngah. 2015. "Agricultural Bio-Waste Materials as Potential Sustainable Precursors Used for Activated Carbon Production: A Review." *Renewable and Sustainable Energy Reviews* 46: 218–235.

Yang, R. T. 2003. *Adsorbents - Fundamentals and Applications*. Hoboken: John Wiley & Sons, Inc.

Yılmaz, O., I. C. Kantarli, M. Yuksel, M. Saglam, and J. Yanik. 2007. "Conversion of Leather Wastes to Useful Products." *Resources, Conservation and Recycling* 49 (4): 436–448.

Zhang, M., and B. Shi. 2004. "Adsorption of Dyes from Aqueous Solution by Chromium Containing Leather Waste." *Journal of the Society of Leather Technologists and Chemists* 88: 236–241.

8 Use of Bioremediation in Treatment of Industrial Effluents

A. Gürses and K. Güneş
Department of Chemistry Education, K. K. Education Faculty, Atatürk University, Erzurum, Turkey

8.1 INTRODUCTION

Nowadays, most of the environmental problems and their potential solutions are intertwined with the microbial components of the global ecosystem. Scientific studies have shown that microorganisms play an important role in the continuous and regular flow of matter and energy from the global ecosystem through their metabolic activities to transform organic and inorganic substances. The bioactive potentials of natural microorganisms are very encouraging in terms of degradation and recycling of wastes arising from industrial and especially chemical processes and negatively affecting the environment (Iihan et al. 2004; Ertugrul et al. 2009; Samantaray et al. 2014).

Bioremediation, which can provide decomposition or immobilization by exploiting the present metabolic potentials of microorganisms with novel catabolic functions derived from selection, is a green sustainable process using the metabolic potential of microorganisms for decontamination of environment. Bioremediation, which is more economical than thermal (combustion) and many other physicochemical techniques and a well-established technique for treating domestic and industrial effluents, is an effective biodegradation process that is mainly applied to convert organic pollutants to harmless metabolites or carbon dioxide and water (Alexander 1999; Cho 2000; Kamat and Kamat 2015; Porwal et al. 2015).

8.2 INDUSTRIAL WASTES AND THEIR CLASSIFICATION

The wastes, which often emerge as an inevitable by-product of industrial production and human activities, are rapidly increasing in parallel with the industrial development in the world, threatening human health as well as adversely affecting the environment (Kunal 2011). A wide variety of chemical, physical and biological industrial processes commonly use non-renewable (petroleum, coal, ore, minerals, etc.) and renewable resources (plants, chemicals, feed, etc.)

to produce many raw materials or final products (fuels, chemicals, plastics, pharmaceuticals, food, feed, etc.). In particular, industrial wastes produced by factories and industrial facilities have been much more dominant in increasing environmental pollution, resulting in radical changes in terms of pollution control and the sustainability of environmental quality (Wei and Huang 2001; Wen 2009). Industrial wastes are traditionally defined as unwanted residues produced by industrial processes or derived from manufacturing processes (Tchobanoglous et al. 1993; Abduli 1996; Casares et al. 2005; Aivalioti et al. 2014). They emerge as hazardous and non-hazardous solid, liquid, and gaseous materials containing hazardous substances, toxic chemicals, medical wastes, mineral and mineral processing wastes, wood and paper wastes, organic wastes, and inorganic deposits. Industrial wastes (solid and liquid) are classified as in Figure 8.1 by the Protection of the Environmental Protection Act of 1997 (POEO Act). Other classifications for waste types are also made by EPA as hazardous waste, restricted solid waste, and general solid waste—putrescible and non-putrescible (EPA 2014).

Special waste, including clinical wastes (laboratory specimens or cultures), cytotoxic wastes, drugs and medicines, or pharmaceutical wastes, is classified as non-hazardous waste with special regulatory restrictions. Asbestos waste and waste tires are also considered special waste. Asbestos corresponds to the fibrous form of the silicate minerals that belong to the serpentine or amphibole

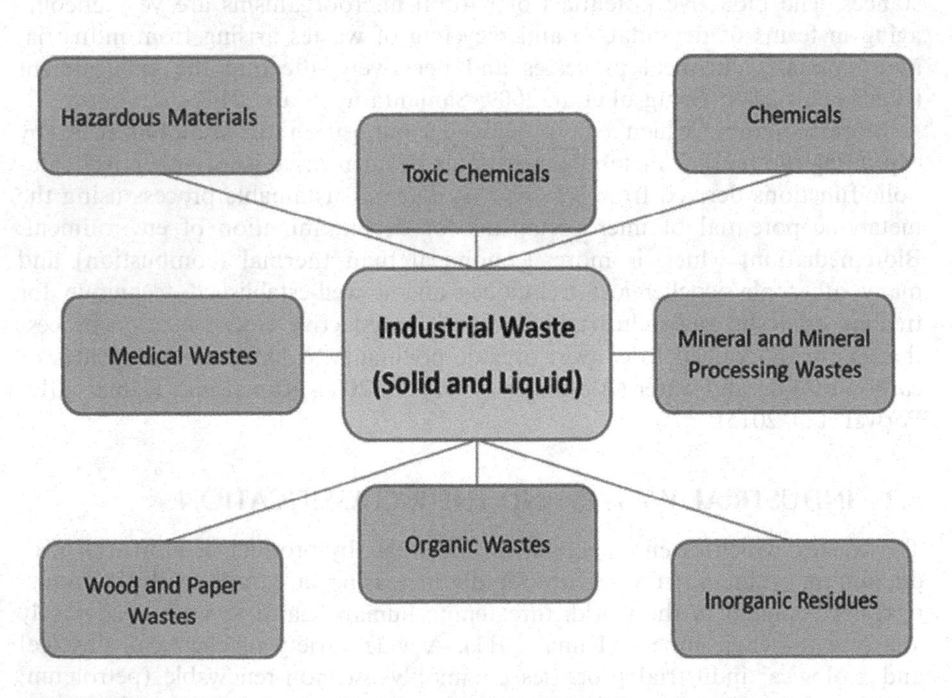

FIGURE 8.1 Schematic representation of industrial (solid and liquid) wastes.

groups of rock-forming minerals, including actinolite, amosite (brown asbestos), anthophyllite, chrysotile (white asbestos), crocidolite (blue asbestos), and tremolite. Waste tire category include the used and unwanted tires, including casings, shredded tires, or tire pieces.

Potentially hazardous wastes for humans or other organisms are defined as hazardous wastes, and waste types other than special wastes are also included in this category by the EPA (for example, containers used for coal tar or bitumen storage, lead-acid or nickel-cadmium batteries, and lead dyes).

Wastes that do not exceed the maximum possible contaminant content, specific contaminant concentration (SCC), and/or toxicity characteristics leaching procedure (TCLP) are classified as *restricted solid wastes.*

Wastes other than special waste, liquid waste, hazardous waste, or restricted solid waste have been classified as *general (putrescible) solid waste* by EPA (e.g., household waste with putrescible organic substances, waste from litter-bins, and food waste).

Also, wastes other than special waste, liquid waste, hazardous waste, restricted solid waste, and general (putrescible) solid waste are defined as *non-putrescible solid waste.* Typical examples of these wastes are glass, plastic, rubber, gypsum board, ceramic, brick, concrete, metal, and paper or cardboard (Gürses et al. 2016).

Wastes can be minimized, rehabilitated, or remedied and disposed using applicable strategies and techniques, and the energy supplied or produced can be evaluated through as energy-efficient processes with the appropriate conversions, right catalyst and reactor selections (Al-Dahhan 2016).

8.3 INDUSTRIAL EFFLUENTS

Rapid population growth and increasing industrialization lead to serious environmental problems (soil, water, and air pollution). Especially, the discharge of industrial effluents, as well as chemical spills, domestic wastes, sewage, and pesticide residues, is of critical importance in terms of environmental pollution. Major examples of manufacturing and processing industries that can be listed as the main source of environmental pollution are tanneries, fertilizer factories, refineries, chemical production plants, and petrochemical, vegetable oil, paper, textile, paint, sugar, and food industries. On the other hand, it is well known that wastewater released by refineries, petrochemical industries, and oil storage facilities contains large quantities of crude petroleum products, polycyclic and aromatic hydrocarbons, phenols, metal derivatives, surfactants, sulfites, naphthyl acids, and other chemicals (Bako et al. 2008).

Methods for the treatment of wastewater are generally classified as physical, chemical, and biological methods. Sedimentation, floatation, screening, adsorption, coagulation, oxidation, ozonation, electrolysis, reverse osmosis, ultrafiltration, and nanofiltration techniques are widely used to remove floating, colored and toxic substances as well as colloidal particles (Pokhrel and Viraraghavan 2004). Physical and chemical methods have many disadvantages in practice and in particular, the need for re-treatment of post-treatment products leads to higher investing and operating costs. Biological processes can produce a smaller quantity of final product, especially by converting large quantities of organic waste components to carbon dioxide and water, which are harmless products, or by converting residues of organic substances in waste water into methane gas. In biological treatment processes, contaminants in wastewater can be dissolved, detoxified, and separated using a number of known microorganisms and especially activated sludge, which is a combination of different microbial combinations. The microbial consortia formed represent a broad selection of microbial populations. The formulated composition can degrade organic substance and present a wide range of substrates in a reproducible manner, as well as reduce the time required for the microbial process to take place (Priya darshini and Sharpudin 2016). Biological methods are widely used worldwide due to their relatively low cost and application diversity (Kaushik 2015).

Although many physical and physicochemical techniques are used to remove residues in effluents, bio-remediation is becoming much more important as an effective treatment technology because of the biodegradability of the majority of the particles in the wastewater and also the abundance of microorganisms able to degrade waste (Chaîneau et al. 2000; Prince 2002; Joo et al. 2008; Mcheik et al. 2013).

An effective and reliable method for the treatment of industrial wastewater has to meet a number of basic requirements, such as prevention of water pollution, protection of natural resources, protection of human health and ecosystem, and compliance with environmental protection regulations.

The most commonly used methods for decontamination of toxic wastes are activated carbon adsorption, chemical oxidation, incineration, direct photolysis, and biodegradation-bioremediation.

8.4 BIOREMEDIATION TECHNOLOGY

Bioremediation, which primarily uses living organisms consisting of microorganisms to convert pollutants into less toxic forms, is defined as a process in which biological degradation is performed to reduce organic wastes to harmless levels below the concentration limits determined by environmental regulations (Mueller et al. 1996). This process can be applied *in situ* or *ex situ*, mediated by mixed microbial consortia and/or pure microbial strains. The most common bioremediation techniques are schematically shown in Figure 8.2.

In situ bioremediation involves techniques based on in situ processing of contaminant materials (on-site treating) without requiring any excavation, and therefore this causes almost no structural change in the soil structure to be treated and construction and application costs are lower than other techniques.

- *Bioventing* is an in situ remediation technique that uses microorganisms to biodegrade organic compounds that adsorbed on the unsaturated zones of the soil, improving the activity of indigenous bacteria and enhancing in situ biodegradation of hydrocarbon compounds in the

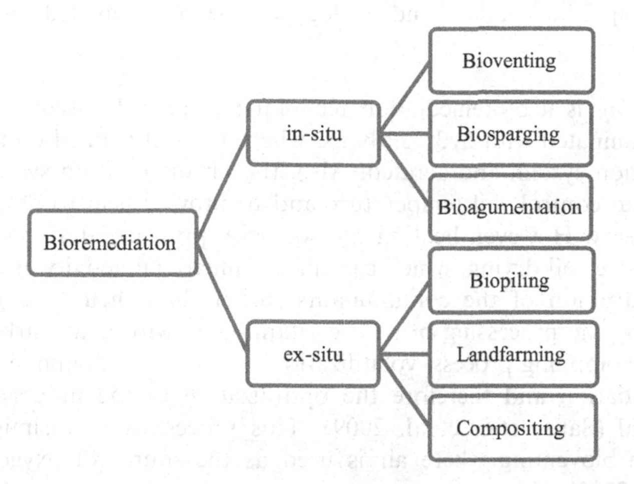

FIGURE 8.2 Schematic representation of main bioremediation techniques.

soil by facilitating air or oxygen flow. In situ bioventing systems with components very similar to soil ventilation systems consist essentially of a blower and a series of air induction (influent) and venting (effluent) wells (Lee and Swindoll 1993).

- *Biosparging* is an in situ remediation technique based on the injection of air under pressure below the water table to enrich the oxygen concentration of groundwater, as well as to increase the rate of biological degradation of contaminants using naturally occurring bacteria. This technique can improve the interaction between soil and groundwater by improving the mixing regime in the saturated zone. Also, smaller-diameter air injection systems, which can be easily installed and are of low cost, provide significant advantages in the design and construction of the system (Vidali 2001).
- *Bioaugmentation,* a technique in which microorganisms such as bacteria containing the necessary catabolic genes are inoculated to the soil to increase the rate of contaminant degradation, has a great potential for bioremediation of soil contaminated with organic wastes. This technique can be applied in two different ways as *cell bio augmentation* based on the survival and growth of inoculated strains to achieve target contaminants degradation, and *genetic bioaugmentation* based on the spread of catabolic genes located in mobile genetic elements to native microbial populations (Garbisu et al. 2017). Bioaugmentation can significantly improve the biodegradation capacities of contaminated sites by using single strains or microorganisms with the desired catalytic properties (Mrozik and Piotrowska-Seget 2010).

In *ex situ* bioremediation techniques, which require excavation and transport of contaminants from contaminated sites, the treatment costs are generally assessed taking into consideration such factors as pollution level, pollutant type, geographical location, and geology of the contaminated area (Azubuike et al. 2016).

- *Biopiling* is a bioremediation technique commonly used to treat soil contaminated with hydrocarbons, where the soil is piled on an air distribution system and aerated. Also, the air distribution system can be used to control soil temperature and to provide heat to the soil when necessary. However, heating the soil with pressurized air may lead to excessive soil drying, which can inhibit microbial activity and promote volatilization of the contaminants rather than their biodegradation. During the processing of soils contaminated with hydrocarbon wastes in the biopiling process, volatilization may become dominant over biodegradation and therefore the optimization of the process is highly critical (Sanscartier et al. 2009). This process is very similar to the active bioventing where air is used as the source of oxygen (Hazen et al. 2003).
- *Landfarming* is one of the oldest and most well-known ex situ biological soil remediation treatments, and the US Environmental Protection Agency (US EPA) defines it as a natural treatment technique, in which hazardous wastes accumulated on or in the soil are degraded by microorganisms. In this process, the soil regime should be controlled by monitoring characteristics such as moisture content, nutrient content, aeration frequency, and soil pH value to optimize the degradation rate and efficiency of the pollutants (Lukić et al. 2017). In addition, contaminated soil is occasionally spread over with waste materials or mixed with soil amendments, such as bulking agents and nutrients, for improving the oxidation and degradation process with the current microbial population (Gan et al. 2009).
- *Composting*, which is a process in which organic materials are degraded by microorganisms and thus produce organic and/or inorganic by-products and energy in the form of heat, is particularly suitable for the destruction of biodegradable harmful pollutants. The potential for bioremediation by composting techniques of environmental matrices contaminated is extremely high due to the intensity of microbial activity in such a composting matrix (Bavarva 2015). This activity is often increased in the presence of a warm, moist, aerobic, and nutrient- and carbon-rich environment, and metabolic heat production and isolation properties of the physical matrix can create a self-heating environment that can further increase microbial activity. This metabolism-induced temperature increase can improve the development of microbial communities. In terms of bioremediation potential, three general levels of composting technology (windrow, static pile, and in-vessel) are considered (Williams and Myler 1990).

Bioremediation in which naturally occurring bacteria and fungi or plants are used to degrade or detoxify substances harmful to human health and/or the environment is mainly based on the stimulation of microorganisms with nutrients and other chemicals that can lead to the destruction of contaminants (Diaz 2008). Microorganisms may be indigenous to the contaminated area or may be isolated from other places and brought to the contaminated area. Contaminant compounds are transformed by living organisms through reactions in metabolic processes, and biodegradation of a compound can occur only as a consequence of the actions of multiple organisms (Vidali 2001). Therefore, it can be argued that the bioremediation systems are limited to the capabilities of native microbes. However, there are intensive investigations on ways to augment contaminated sites with nonnative microorganisms, including genetically engineered microorganisms, especially those suitable for degrading contaminants in particular sites. The bio augmentation technique, an ex situ process, is promising in this sense in terms of usability for future bioremediation systems (Krishnaveni et al. 2013). It is also known that bioremediation is a very convenient process for processing water efficiently, economically, versatile and environmentally friendly, especially for reuse in many different applications (Norris 1994).

Bioremediation, a pollution control technology that uses biological systems to convert various toxic chemicals to less harmful forms or catalyze the conversion, is one of the most popular among all known technologies in waste cleaning to clear many environmental contaminants (Murugesan 2003; Asamudo et al. 2005; Yadav et al. 2017).

Bioreactors, which are units designed for optimal growth and metabolic activity of the organism through the action of biocatalysts, enzymes or microorganisms, and animal or plant cells, are technologically the best environmental bioremediation tools. The raw materials processed in these reactors can be organic or inorganic chemical or more complex substances. The reaction engineering of biocatalytic processes requires not only a number of specific features but also that the bioreactor conditions should be adjusted to suit the conditions under which live microorganisms can exhibit their activities under certain conditions (Singh et al. 2014). These reactors, unlike the common chemical reactors, have distinctly different construction and processing characteristics because they use biological species and require controlled conditions.

Since microorganisms are more sensitive and less stable than chemicals, bioreactor systems are being carried out through processes that maintain the desired biological activity and minimize undesired activity. Contaminants that can control biological activity can directly affect the nature of the process, as microorganisms can alter the biochemistry and other properties of organisms due to mutation. Also, analogous to heterogeneous catalysis, deactivation or mortality may occur and promoters or coenzymes influence the kinetics of the bioreactor (Luka et al. 2018).

Bioreactors that may be much faster than land treatment in the biodegradation of wastes are more advantageous than most biofilters in terms of control of reaction conditions and effluent quality. Although land treatment requires a very long time, bioreactors require a relatively short run-time of only a few days or weeks to effectively degrade certain pollutants. Slurry-phase bioreactors suitable for the bioremediation of soluble organic wastes at high concentrations in soil or sludge have a high processing capacity up to organic wastes at the level of 250 g/kg (Bonaventura and Johnson 1997).

Bioremediation uses microorganisms such as bacteria (aerobic and anaerobic), fungi, algae, and actinomycetes (filamentous bacteria) to degrade or immobilize waste materials (Shanatan 2004; Adams et al. 2015).

Microorganisms cannot mineralize most of the harmful substances individually, and complete mineralization can only occur as a result of a subsequent degradation by a consortium of microorganisms, including synergism and co-metabolism actions. The natural communities of microorganisms in various habitats, which can metabolize and even mineralize many organic molecules, have a huge physiological versatility. Most bioremediation processes occur under aerobic conditions, but a system under anaerobic conditions can allow microbial organisms to degrade the recalcitrant molecules (Colberg and Young 1995).

- *Bacteria* whose metabolism is dependent on the presence of electron donors and acceptors and essential nutrients and the ions necessary for growth are highly effective in removing contaminants in both organic and inorganic media. Bacteria can carry out the uptake and transformation of pollutants, as well as their immobilization or mobilization, over very different mechanisms (Antizar-Ladislao and Galil 2004, 2006; Antizar-Ladislao 2010). They are usually divided into two categories, depending on the running conditions: aerobic bacteria, which are normally found in healthy or oxygenated environments, and anaerobic bacteria, which are found in low oxygen environments such pond muck, poorly maintained compost piles, and intestinal tracts. While some anaerobic bacteria are pathogenic organisms, the many bacteria such as acidophilus and lactobacillus are effective and necessary for human digestion (Bavarva 2015).
- *Fungi*, which have been used in many fields from food fermentation to production of pharmaceuticals, have long been known microorganisms with their biological properties, economic values, and pathogenic abilities. They are active even in habitats with environmental extremes due to the enzyme system (Cooke 1979). They may biodegrade many hazardous materials or compounds or convert them to harmless, nontoxic, or useful products. Many organisms can be involved in the biodegradation of organic wastes and the production of many new and biotechnologically important substances (Tripathi et al. 2007). The fungi, which are known for their superior ability to produce a wide variety of extracellular proteins, organic acids and other metabolites, and the capacities to adapt to extreme environmental constraints, are particularly effective in treating colored and metallic effluents (Lilly and Barnett 1951; Cochrane 1958; Ezeronye and Okerentugba 1999). White rot fungi are another group of widely exploited microorganism in distillery effluent bioremediation. They produce various isoforms of

extracellular oxidases, including laccases, manganese peroxidases, and lignin peroxidase, which are involved in the degradation of various xenobiotic compounds and dyes. Adsorption is another important mechanism involved in removing color of the distillery mill effluents using fungi (Kaushik 2015).

- *Algae* are the important bioremediation substances currently used in wastewater treatment, and their potential in wastewater treatment is higher than their current level (Volesky 1990; Wase and Forster 1997). Blue green algae (Cyanobacteria), which are considered to be the most primitive photosynthetic prokaryotes, are a unique assemblage of organisms that occupy a vast array of habitats (Abd Allah 2006; Ash and Jerkins 2006; Prabha 2012).
- *The filamentous bacteria (Actinomycetes)*, which are mainly an important component of soil and sometimes live in marine and aquatic sediments, have an important role in removing heavy metals and other different pollutants (Siñeriz et al. 2005; Aburas 2016). They are used for biotransformation, biodegradation, and many other purposes besides their various applications in bioremediation. Actinomycetes, which can induce degradation of herbicides, pesticides, chromium ions, petrochemicals, and nitro-aromatics, are toxic and most of them are pathogenic and carcinogenic and can cause respiratory diseases and mental illness. However, despite their disadvantages, they are extremely important in the biotechnological field due to their wide applications.

8.5 BASIC PRINCIPLES OF BIOREMEDIATION AND RELATED PROCESS VARIABLES

Bioremediation is basically a process in which organisms decompose organic pollutants into CO_2, water, salt, and other harmless products to improve their growth and metabolism. Moreover, bioremediation capacity can be improved by genetic modifications in microorganisms and plants. Bioremediation, which usually requires the use of biological techniques, is very low cost and can be coupled with physical and chemical waste processing technologies when necessary. It is also a non-invasive process that conserves the ecosystem and thus does not disturb the ecological equilibrium (Mani and Kumar 2014). The ability of economically producing sufficient quantity of biomass of suitable microorganisms in a much shorter period, although it naturally requires very long periods, is a distinctive feature for this process (Chen et al. 2003). The process is generally carried out by applying the microorganisms directly to the contaminated area or by using sufficient concentrations of various chemicals that promote the rapid growth of appropriate microorganisms (Galadima 2012).

The main control variables in bioremediation of harmful chemicals in wastewater are microorganism type, temperature, pH, dissolved oxygen concentration, and inorganic nutrient content. Microorganisms can only interact with certain pollutants when they can access a variety of compounds that can help them produce energy and nutrients to form more cells. Therefore, the selection of the appropriate microorganism is critical in terms of bioremediation efficiency and yield (El Fantroussi and Agathos 2005).

Microbial degradation of a specific pollutant requires an optimal temperature as it controls the rate of bioremediation based on the high dependence of the physiological properties on the temperature (Abatenh et al. 2017). Since metabolic processes are highly susceptible to slight changes in pH, the pH of the medium can directly affect microbial growth potential and metabolic activity and thus control bioremediation efficiency (Wang et al. 2011; Asira 2013).

Dissolved oxygen concentration is a critical variable for most living organisms. Biological degradation takes place under aerobic and anaerobic conditions, but in most cases, the presence of oxygen can improve hydrocarbon metabolism (Macaulay 2014).

Nutritional supplementation is essential in terms of nutrient balance in the context of microbial growth and reproduction, as well as being critical for the biodegradation yield and effectiveness. Nutritional balancing based on the supplementation of essential elements such as N and P may improve the biodegradation efficiency by controlling the bacterial C, N, and P ratio. Couto et al. 2014 Moreover, carbon, nitrogen, and phosphorus are the essential elements for microorganisms to maintain their microbial activity and vitality. Optimal nutritional supplementation can improve the biodegradation yields of microorganisms, especially by improving their metabolic activities in cold environments.

Proper control of process variables is critical to the growth and reproduction of microorganisms as well as biodegradation potentials and, if the optimum conditions are not met, they can grow and multiply more slowly, or they may die and even produce more harmful chemicals.

8.6 BIOREMEDIATION OF INDUSTRIAL EFFLUENTS

The primary advantage of bioremediation, a sustainable, environmentally friendly green technology for a wide variety of industrial wastes worldwide, is its applicability in a wide range of environments for a wide variety of pollutants. However, industrial scale applications of this process still require more intensive pre-tests and feasibility studies at the laboratory scale (Godoy-Faúndez et al. 2014).

Bioremediation should be considered as a general concept that covers all processes carried out to transform a contaminated environment in its current state using biological pathways into its original state. Even if the procedures for achieving the set goals differ, the basic principles are the same because of using microorganisms and their enzymes. Although bioremediation practices have started earlier, especially in the food industry, developments in microbiological and genetic fields have further expanded the scope of applications. Pesticides, herbicides, cleaning chemicals, and chemicals used in food production processes are among the new pollutants entering biogeochemical cycles. Bioremediation can convert these pollutants into products that can be absorbed and used with autotrophic organisms that do not have toxic effects on them (Thassitou and Arvanitoyannis 2001).

The efficient use of raw materials in the food sector, which is one of the industrial areas that use the most raw materials, is extremely important. Since the real value of the products in agricultural production is firmly dependent on material and energy costs, the use of materials and energy expenditures are becoming increasingly important. This can only be achieved through the feasible use of non-waste technologies, raw materials, and recycling resources. Also, it is critical to eliminate the negative effects of wastes on the environment, in particular to clean effluents from food industries and ensure the ecological sustainability of plants used in food production (Dhanalakshmi et al. 2016). Effluents from meat, vegetable, and food raw materials and from cooking utensils, dishwashers, and kitchens can also be included in the food waste category (Zulaikha et al. 2014). In recent years, the physical and chemical methods used for the treatment of food wastewater have been replaced by biological methods and researches have begun to focus on bioremediation (Xue et al. 2016). The main microorganism types available for food wastewater bioremediation are given in Table 8.1.

The textile industry has the potential to produce hazardous wastes that could threaten the ecosystem from different operations. These pollutants not only emerge as a result of the actions taken to meet human expectations of fashion and improve living standards but can also lead to serious potential for environmental pollution. Textile effluents can contaminate both surface water and groundwater, as it can be released into access waters without any treatment that can improve the water quality (Asamudo et al. 2005; Ajao et al. 2011). These effluents containing dyes and similar compounds with high molar mass and complex chemical structures have unfortunately low biodegradability and thus their direct discharge to the sewage networks can lead to serious performance losses for biological treatment processes (Babu et al. 2000; Olayinka and Alo 2004). Moreover, since such effluents can lead to high concentrations of salt, acid, and bases during biological processes and also because the dyes are stable at light and high temperature, higher costs for biological treatment can be expected. Classical and conventional treatment methods for textile effluents are generally based on processes such as chemical precipitation, chlorination, and activated sludge and activated carbon adsorption (Babu et al. 2000). Although microbial degradation mechanisms of dyes are not well known, most dyes were found to be sensitive to reductive degradation processes. Some bacteria *(Bacillus spp., Alcaligenes spp., Acinetobacter spp.)* are reported to be effective in the bioremediation of halogenated aromatic compounds and textile effluents (Olayinka and Alo 2004).

Generally, the biodegradability of the natural compounds is higher. In addition, compounds with high molar mass containing complex ring structure and halogen substituents may degrade more slowly than complex straight chain hydrocarbon or low molar mass compounds (Olayinka and Alo 2004). The degree to which the synthetic compounds can be metabolized by microorganisms relates to the various structural properties of the compound similar

TABLE 8.1

Major types of microorganisms available for bioremediation of food wastewater

Types of microorganisms	Genus/species	References
Bacteria	*Acinetobacter sp.*	Fatajeva et al. 2014
	A. junii	Kielak et al. 2017
	A. radioresistens	Al Atrouni et al. 2016
	Bacillus sp.	Parthipan et al. 2017
	B. subtilis	Prasad and Manjunath 2011
	B. licheniformis	Xue et al. 2016
	B. amyloliquefaciens	Tzirita et al. 2018
	B. cereus	Amin et al. 2013
	Pseudomonas sp.	Shah Maulin 2017
	P. aeruginosa	Pérez Silva et al. 2006
	P. fluorescens	Khan and Ahmad 2006
	Aeromonas sp.	Divya et al. 2015
	Arthrobacter sp.	See-Too et al. 2017
	Staphylococcus aureus	Kanmani et al. 2015
	Clostridium sp.	Nigam 2013
	Enterococcus sp.	Tayabali et al. 2017
	Klebsiella sp.	Peil et al. 2016
	Nitrobacter sp.	Antony and Philip 2006
	Raoultella planticola	Peil et al. 2016
	Zoogloea sp.	Akpor and Muchie 2010
Fungal	*Cunninghamella bertholletiae*	Raja et al. 2017
	Aspergillus sp.	Deshmukh et al. 2016
	Fusarium moniliforme	AI-Jawhari 2016
	Penicillium sp.	Chávez et al. 2011
	Rhizopus spp.	El-Shiekh et al. 2007
	Alternaria gaisen	Saranraj and Stella 2014
	Trichoderma sp.	Hasan 2016
	Thermoactinomyces sp.	Verma et al. 2016
Yeast	*Candida*	Sáenz-Marta et al. 2015
	Saccharomyces sp.	Türker 2014
	Saccharomyces cerevisiae	Mcheik et al. 2013

to the natural compounds. The degree and rate of metabolism of any compound in the environment usually depends on the availability of electron acceptors and other nutrients (Chen 2002). Furthermore, decolorization and degradation can effectively detoxify effluent without leaving any residues (Galadima 2012).

The paper pulp and paper industry, which creates serious environmental problems with wastewaters containing high levels of halogenated organic matter, is one of the largest industrial waste producing industries in the world (Christov and Van Driessel 2003; Murugesan 2003).

It is known that individual isolates and isolate combinations of some bacterial species (*Bacillus subtilis, Citrobacter freundii, Alcaligenes, Burkholderia, Pseudomonas asaeruginosa,* etc.) and fungus (*Phanerochaete chrysosporium, Rhizopus stolonifer, Pleurotus eryngii, Pleurotus ostreatus,* etc.) commonly found in paper mill waste waters significantly affect the degradability of wastewater. (Shanthi et al. 2012; Olawale 2014; Jeenathunisa et al. 2017). Due to the presence of extracellular enzymes, fungi that are naturally present in pulp and paper mill effluents can survive longer in higher effluent loads, compared to bacteria (Singhal and Thakur 2009; Yang et al. 2011; Kamali and Khodaparast 2015). White rot fungi which can degrade the wood can effectively degrade lignin. In addition, *Tinctoria borbonica, Phanerochaete chrysosporium,* and *Trametes versicolor* have been determined to degrade lignin and metabolize together with carbohydrates. On the other hand, *Aspergillus niger* and *Trichoderma spp.* are capable of decolorizing hard wood bleaching effluent and also degrading lignin (Dashtban et al. 2010; Kamali and Khodaparast 2015). White rot fungi such as *Schizophyllum commune* can decolorize the effluent from bagasse based pulp and paper mills but cannot degrade lignin unless the metabolizable carbon source is provided simultaneously. (Dashtban et al. 2010). It is reported that *Tinctoria borbonica* can decolorize kraft waste liquid to light yellow and provide a color reduction of approximately 90–99% after four days of cultivation (Abd El-Rahim and Zaki 2005). *Gliocladium virens,* a soil saprophyte, is a natural and common microorganism and is used for bioremediation of pulp and paper mill wastes. It has been found that fungus present in wastewater can provide approximately 50% color removal, with significant reductions in the lignin, cellulose, and biological oxygen demand contents of the waste water (Kamali and Khodaparast 2015). *Coriolus versicolor,* which has a white rot fungus and can produce extracellular laccase and extracellular enzyme, performed well in lignin biodegradation (Hossain and Ismail 2015). The gasteromycetes *Cyrus stercoreus,* which is particularly effective in litter decomposition, can effectively degrade lignin like other white rot fungi (Achoka 2002; Saritha et al. 2010; Kamali and Khodaparast 2015). *Sordaria fimicola, Halosatpheia spp.*, and *Basidiomycetes spp.* can produce lignin-modifying enzymes such as laccase, manganese peroxidase, and lignin peroxidase. These marine fungi were used for color removal in the paper mill bleach kraft effluents (Kamali and Khodaparast 2015).

Pharmaceuticals, which can be found in wastewater, surface water, groundwater, and soil, interact with living organisms and can cause acute effects on fauna and flora. These are also considered as a significant group of environmental pollutants (Jorgensen and Halling Sorensen 2000, Heberer 2002; Cahill et al. 2004; Debska et al. 2004; Hernando et al. 2004; Stackelberg et al. 2004; Gomez et al. 2006; Mansour et al. 2012). Although the pharmaceuticals and other by-products released into the environment in large quantities by the pharmaceutical industry are biologically active, they are not biodegradable and are resistant to degradation. Therefore, it is extremely difficult to eliminate pharmaceutical contaminants by conventional methods (Kapoor 2015). Moreover, pharmaceutical products such as antibiotics and hormones can affect biological diversity and the natural microbiota of the soil, as well as deteriorate ecological equilibrium and soil fertility (Ding and He 2010; Shalini et al. 2010; Majumder and Kapagunta 2017). Furthermore, since pharmaceutical pollutants can also enter in the food chain, they may adversely affect aquatic and terrestrial organisms, including humans (Randhawa and Kullar 2011). Biological processes can provide the converting or complete mineralization of recalcitrant and xenobiotic pharmaceuticals into less toxic forms. On the other hand, some microorganism strains are reported to have the potential to use pharmaceutical waste as a carbon source (Kartheek et al. 2011). Biodegradation of pharmaceutical waste, which has vital importance, depends on many factors such as compound structure, toxicity, concentration, environmental conditions during degradation, the efficiency of microbes to be used, and the presence of other compounds and their concentrations (Misal et al. 2011). The metabolizing capability of naturally present microbes that can metabolize pharmaceutical compounds are directly related to their degree of access to them and nutrients. The relatively low abundance of such strains can be mentioned as another important limitation in terms of their remediation capabilities (Edwards and Kjellerup 2013).

Bioremediation is one of the newest and most common techniques used in the treatment of effluents from pharmaceutical industries (Rana et al. 2017). The types of microorganisms commonly used for pharmaceutical bioremediation are given in Table 8.2.

As heavy metals accumulating in water, sediments, and soil cause serious environmental problems, intensive researches are being carried out to remove and recover heavy metals from contaminated environments, especially in recent years (Akıncı and Guven 2011; Kang et al. 2016). Among the various physical, chemical, and biological methods for the removal of heavy metals from contaminated soils and water, bacterial-based bioremediation appears to be the most promising and most widely applicable method (Shi et al. 2009; Govarthanan et al. 2015a). However, bioremediation is greatly influenced by a variety of factors, including the survival of bacteria in contaminated soil or water, the effect of abiotic factors on the growth of bacteria, the metal detoxification mechanism, the expression rate of the metal detoxification genes, and the effect of pollutants on bacterial activity (Suja et al. 2014; Loganathan et al. 2015; Govarthanan et al. 2015b, 2016). The interaction of some heavy metals, such as Co, Cu, Fe, Mn, Mo, Ni, V, and Zn, with organisms can occur in a very short time, but their excess amounts can be harmful to microorganisms. On the other hand, metalloids such as Pb, Cd, Hg, and As have no beneficial effect on organisms and are considered to be the most harmful heavy metals because they are very harmful for plants and animals (Chibuike and Obiora 2014). Also, high concentrations of heavy metals in the environment can cause many toxic effects, such as cell destruction and inhibition of enzyme activity. It is known that metal toxins can enter the human body through the skin, lungs, and gastrointestinal tract, and some heavy metals are immunosuppressive, causing neurological disorder, anemia, gout, kidney damage,

TABLE 8.2
The types of microorganisms commonly used for pharmaceutical bioremediation

Types of microorganisms	Examples	References
Bacteria	*Pseudomonas* sp.	Aissaoui et al. 2017
	P. fluorescens	Das et al. 2012
	P. putida	Mansour et al. 2012
	P. aeruginosa	Surti 2016
	Enterobactor sp.	Muthulakshmi et al. 2017
	E. hormaechei	Sivagamasundari and Jayakumar 2017
	E. cloacae	Aissaoui et al. 2017
	E. aerogenes	Davin-Regli and Pagès 2015
	Stenotrophomonas sp.	Liu et al. 2007
	S. maltophilia	Zhang et al. 2013
	Aeromonas sp. *A. punctata*	Lakshmi Priya 2013
	Acinetobacter sp.	Wang et al. 2018
	A. calcoaceticus	Liu et al. 2016
	Arthrobacter sp.	Wang et al. 2015
	A. nicotianae	Aissaoui et al. 2017
	Rhodococcus sp.	De Carvalho et al. 2014
Fungi	*Bjerkandera* sp.	Cruz Morató 2013
	B. adusta	Pinedo-Rivilla et al. 2009
	Aspergillus sp.	Kartheek et al. 2011
	A. niger	Kalyani et al. 2016
	A. flavus	Lalitha et al. 2011
	Rhizopus sp.	Kedia and Sharma 2015
	Penicillium sp.	Leitão 2009
	P. funiculosum	Nicoletti et al. 2012
	P. chrysogenum	Pereira et al. 2014
Plants	*Aquatic plants*	Singh et al. 2012
	Eichhornia crassipes	Mangunwardoyo et al. 2013
	Tomato hairy root cultures	Gonzalez et al. 2006
	Phragmites karka	Raza et al. 2015
	Typha latifolia	Calheiros et al. 2008

gastroenteritis, and high blood pressure (Iye 2015). However, the deadly effect of metallic toxins on health is also directly related to absorption, distribution, excretion, and metabolic rates in living organisms.

Various bioremediation mechanisms such as biosorption, bioaccumulation, biomineralization, biotransformation, and bioleaching have been proposed.

Microorganisms capable of reducing or oxidizing metals can extract heavy metals from soil to use for growth and development. The following mechanisms have been proposed for the removal of heavy metals (Ojuederie and Babalola 2017):

- Sequestration of toxic metals by cell wall components or by intracellular metal binding proteins and peptides, such as metallothioneins and phytochelatins along with compounds such as bacterial *siderophores*, which are mostly catecholates compared to fungi that produce *hydroxamate siderophores*.
- Changing biochemical pathways to block metal uptake.
- Conversion of metals into innocuous forms by enzymes.
- Reduction of intracellular concentration of metals using precision efflux systems.

Bioremediation, which is a serious alternative to high-cost and low-efficiency traditional methods for heavy metal removal, can be applied successfully, taking into account the mechanisms controlling the growth and activity of microorganisms. (Dixit et al. 2015; Verma and Sharma 2017). Although organic solvents are contaminants that can destroy cell membranes, cells often develop appropriate defense mechanisms with substances that can protect the outer cell membrane (Sikkema et al. 1995; Roane and Pepper 2000). Some microalgae, macroalgae, and cyanobacteria used in the removal of heavy metals are given in Table 8.3.

Pesticides, developed to eliminate or control biological organisms such as weeds, insects, and rodents, are a common class of chemical substance. Approximately, 2.4 million metric tons of pesticide active ingredients are used annually worldwide to control the occurrence of such unwanted organisms in agricultural and urban environments (Grube et al. 2011; Rani and Dhania 2014). The presence of pesticide residues, even at trace concentrations in water sources around the world, has been identified by long-term studies and it has been determined that their access to people is predominantly with drinking water. For this reason, even waters containing pesticides at trace concentrations above the permissible concentration threshold may need to be remediated in order to protect human health (Benner et al. 2013; Helbling 2015). Biodegradation and bioremediation, which are highly effective for the elimination of pesticides, are eco-friendly and low-cost processes based on the conversion or metabolism of pesticides by microorganisms. The main difference between these two processes is that biodegradation is a natural and in situ process, whereas bioremediation is one based on a technological basis. The effectiveness of bioremediation depends on the availability of an effective bacterial strain that can metabolize contaminants (e.g., pesticide residues) to allowable levels in a short period (Singh 2008). Moreover, the effective bioremediation of pesticides is critically important because contaminants such as pesticides can not only affect the soil but also affect the groundwater (Gavrilescu 2005). The low level of biodegradability of chemical pesticides puts them into a permanent class of toxic substances. The most critical impacts on soil microbiota are loss of biodiversity and functional problems in the nutrient cycle. Moreover, high toxicity levels caused by pesticides may require further bioremediation (Tayade et al. 2013). Microorganisms that are already present in contaminated ecosystems may lead to insufficient intrinsic bioremediation

TABLE 8.3
Some microalgae, macroalgae, and cyanobacteria used in the removal of heavy metals

Microorganisms	Metal	References
	Microalgae	
Chlorella vulgaris	Pb, Cd, Cu	Dewi and Nuravivah 2018; Mallick 2003; Al-Qunaibit 2009
Spirogyra hyaline	Hg, As, Cd, Pb, Co	Kumar and Oommen 2012
Pseudochlorococcum typicum	Pb, Cd, Hg	Hajdu-Rahkama 2014.
Scenedesmus intermedius	Pb, Cd, As	Baos et al. 2002
Scenedesmus quadricauda var quadrispina	Pb, Cd, Hg	Shanab et al. 2012
Rhizoclonium	As	Saunders et al. 2012
Schizomeris leibleinii	Pb	Özer et al. 1999
Desmodesmus pleiomorphus	Zn	Monteiro et al. 2009
Scenedesmus obliquus	Cd	Gürbüz et al. 2009
Nannochloropsis oculata	Cd	Torres 2016
Dunaliella	Hg, Pb, Cd	Imani et al. 2011
	Macroalgae	
Laminaria japonica	Cd, Zn	Fourest and Volesky 1997
Ascophyllum nodosum	Cd, Pb	Baumann et al. 2009
Bifurcaria bifurcata	Cu	Shobha et al. 2014
Sargassum muticum	Pb	Rubín Carrero et al. 2006
Fucus spiralis	Cd, Cu, Ni, Pb, Zn	Romera et al. 2008
Ecklonia radiata	Pb, Cd, Cu	Matheickal and Yu 1996
Ecklonia maxima	Zn, Cd, Cu	Stirk and Van Staden 2000
Palmaria palmata	Cu	Prasher et al. 2004
Padina pavonica	Cd, Ni	Ofer et al. 2003
	Cyanobacteria	
Anabaena subcylindrica	Pb, Cu, Co, Mn	El-Sheekh et al. 2005
Spirulina platensis	Pb	Murali and Mehar 2014
Anabaena nodusum	Cd	Sandau et al. 1996
Phormidium ambiguum	Pb, Cd, Hg	Shanab et al. 2012
Spirulina maxima	Pb, Cu	Biswas 2017
Lyngbya taylorii	Pb, Cd, Zn, Ni	Klimmek et al. 2001
Synechococcus sp.	Pb	Sakaguchi et al. 1978

(Uqab et al. 2016). The bacterial species belonging to the genera *Flavobacterium, Arthobacter, Azotobacter, Burkholderia,* and *Pseudomonas,* as well as bacterium *Raoultella* sp., can degrade pesticides (Glazer and Nikaido 2007). However, the mechanism of degradation may vary depending on bacterial species and target compounds. For example, *Pseudomonas* sp. and *Klebsiella pneumoniae* have hydrolytic enzymes capable of degrading s-triazine herbicides such as atrazine. Generally, various fungi such as *Flammulina velutipes, Stereum hirsutum, Coriolus versicolor, Dichomitus squalens, Hypholoma fasciculare, Auricularia auricula, Pleurotus ostreatus, Avatha discolor,* and *Agrocybe semiorbicularis* can also degrade pesticides (Odukkathil and Vasudevan 2013).

Herbicides, a class of compounds, which are a potential threat to the environment because of their wide variety of toxic effects, are among those responsible for surface and groundwater contamination due to their widespread use (Kanissery and Sims 2011). *Alachlor* (an amide herbicide), *atrazine* (a triazine herbicide), *2,4-D* (2,4-dichlorophenoxyacetic acid, a phenoxy herbicide), *Metolachlor* (an amide herbicide), Trifluralin (a dinitroaniline herbicide), *simazine* (a triazine herbicide), and *metribuzin* (an organophosphate insecticide) are the most widely used herbicides worldwide. Bioremediation is an attractive technology that can be used to remove herbicides from contaminated environment (Seeger et al. 2010).

Bioremediation, which is a natural process and produces harmless residues and by-products such as carbon dioxide, water, and cell biomass, is considered to be a very attractive and applicable process, especially for the treatment of contaminated soil (Vidali 2001). The most important advantage of bioremediation compared to other biological technologies is the potential to utilize enzymatic metabolic pathways that have evolved in nature in a very long period. Thus, the combined enzymatic metabolic pathways can make possible the degradation of a wide range of hazardous contaminants (Hatzikioseyian 2010).

REFERENCES

Abatenh, E., Gizaw, B., Tsegaye, Z., Wassie, M. 2017. The role of microorganisms in bioremediation-A review. *Open Journal of Environmental Biology* 2(1): 038–046.

Abd Allah, L. S. 2006. Metal-binding ability of cyanobacteria: The responsible genes and optimal applications in bioremediation of polluted water for agriculture use. PhD diss., Alexandria Univ. Alexandria, Egypt.

Abd El-Rahim, W. M., Zaki, E. A. 2005. Functional and molecular characterization of native Egyptian fungi capable of removing textile dyes. *Arab Journal of Biotechnology* 8: 189–200.

Abduli, M. A. 1996. Industrial waste management in Tehran. *Environment International* 22: 335–341.

Aburas, M. M. A. 2016. Bioremediation of toxic heavy metals by waste water actinomycetes. *International Journal of Current Research* 8(1): 24870–24875.

Achoka, J. D. 2002. The efficiency of oxidation ponds at the Kraft pulp and paper mill at Webuye in Kenya. *Water Research* 36(5): 1203–1212.

Adams, G. O., Fufeyin, P. T., Okoro, S. E., Ehinomen, I. 2015. Bioremediation, biostimulation and bioaugmention: A review. *International Journal of Environmental Bioremediation & Biodegradation* 3(1): 28–39.

AI-Jawhari, I. F. H. 2016. Bioremediation of Anthracene by Aspergillus niger and Penicillium Funiculosu. *International Research Journal of Biological Sciences* 5(6): 1–11.

Aissaoui, S., Ouled-Haddar, H., Sifour, M., Beggah, C., Benhamada, F. 2017. Biological removal of the mixed pharmaceuticals: Diclofenac, Ibuprofen, and Sulfamethoxazole using a bacterial consortium. *Iranian Journal of Biotechnology* 15(2): 135–142.

Aivalioti, M., Cossu, R., Gidarakos, E. 2014. New opportunities in industrial waste management. *Waste Management* 34: 1737–1738.
Ajao, A. T., Adebayo, G. B., Yakubu, S. E. 2011. Bioremediation of textile industrial effluent using mixed culture of Pseudomonas aeruginosa and Bacillus subtilis immobilized on agaragar in a Bioreactor. *Journal of Microbiology and Biotechnology Research* 1(3): 50–56.
Akıncı, G., Guven, D. E. 2011. Bioleaching of heavy metals contaminated sediment by pure and mixed cultures of Acidithiobacillus spp. *Desalination* 268: 221–226.
Akpor, O. B., Muchie, M. 2010. Bioremediation of polluted wastewater influent: Phosphorus and nitrogen removal. *Scientific Research and Essays* 5(21): 3222–3230.
Al Atrouni, A., Joly-Guillou, M. L., Hamze, M., Kempf, M. 2016. Reservoirs of non-baumannii Acinetobacter species. *Frontiers in Microbiology* 7: 1–12.
Al-Dahhan, M. H. 2016. Trends in minimizing and treating industrial wastes for sustainable environment. *Procedia Engineering* 138: 347–368.
Alexander, M. 1999. Bioremediation in the rhizosphere. *Environmental Science Technology* 27: 2630–2636.
Al-Qunaibit, M. H. 2009. Divalent Cu, Cd, and Pb biosorption in mixed solvents. *Bioinorganic Chemistry and Applications* 2009: 5 pages.
Amin, A., Naik, A. T. R., Azhar, M., Nayak, H. 2013. Bioremediation of different waste waters – A review. *Continental Journal of Fisheries and Aquatic Science* 7(2): 7–17.
Antizar-Ladislao, B. 2010. Bioremediation: Working with bacteria. *Elements* 6(6): 389–394.
Antizar-Ladislao, B., Galil, N. I. 2004. Biosorption of phenol and chlorophenols by acclimated residential biomass under bioremediation conditions in a sandy aquifer. *Water Research* 38: 267–276.
Antizar-Ladislao, B., Galil, N. I. 2006. Biodegradation of 2,4,6-trichlorophenol and associated hydraulic conductivity in sand-bed columns. *Chemosphere* 64: 339–349.
Antony, S. P., Philip, R. 2006. Bioremediation in shrimp culture systems. *Naga the World Fish Center Quarterly* 29(3&4): 62–66.
Asamudo, N. U., Daba, A. S., Ezeronye, O. U. 2005. Bioremediation of textile effluent using Phanerochaete chrysosporium. *African Journal of Biotechnology* 4(13): 1548–1553.
Ash, N., Jerkins, M. 2006. Biodiversity and poverty reduction: The importance of biodiversity for ecosystem services. Final report prepared by the United Nations.
Asira, E. E. 2013. Factors that determine bioremediation of organic compounds in the soil. *Academic Journal of Interdisciplinary Studies* 2: 125–128.
Azubuike, C. C., Chikere, C. B., Okpokwasili, G. C. 2016. Bioremediation techniques–classification based on site of application: Principles, advantages, limitations and prospects. *World Journal of Microbiology and Biotechnology* 32(11): 180.
Babu, B. V., Rana, H. T., Ramakrishna, V., Sharma, M. 2000. COD reduction of reactive dyeing effluent from cotton textile industry. *Journal of the Institution of Public Health Engineers India* 4: 5–11.
Bako, S. P., Chukwunonso, D., Adamu, A. K. 2008. Bio-remediation of refinery effluents by strains of Pseudomonas aerugenosa and Penicillium janthinellum. *Applied Ecology and Environmental Research* 6(3): 49–60.
Baos, R., García-Villada, L., Agrelo, M., López-Rodas, V., Hiraldo, F., Costas, E. 2002. Short-term adaptation of microalgae in highly stressful environments; An experimental model analysing the resistance of Scenedesmus intermedius (Chlorophyceae) to the heavy metals mixture from the Aznalcóllar mine spill. *European Journal of Phycology* 37(4): 593–600.
Baumann, H. A., Morrison, L., Stengel, D. B. 2009. Metal accumulation and toxicity measured by PAM—chlorophyll fluorescence in seven species of marine macroalgae. *Ecotoxicology and Environmental Safety* 72(4): 1063–1075.

Bavarva, S. R. 2015. Bioremediation a secure and reverberation module for ground-water and soil. *International Journal of Innovative and Emerging Research in Engineering* 2: 32–38.
Benner, J., Helbling, D. E., Kohler, H. P. E., Wittebol, J., Kaiser, E., Prasse, C., Ternes, T. A., Albers, C. N., Aamand, J., Horemans, B., Springael, D., Walravens, E., Boon, N. 2013. Is biological treatment a viable alternative for micropollutant removal in drinking water treatment processes? *Water Research* 47(16): 5955–5976.
Biswas, A. 2017. Biosorption of lead and copper from contaminated solutions by the cyanobacteria–Spirulina maxim. PhD diss., California State Univ., USA.
"Bonaventura, C., & Johnson, F. M. 1997. Healthy environments for healthy people: bioremediation today and tomorrow. Environmental health perspectives, 105 (suppl 1): 5-20."
Cahill, J. D., Furlong, E. T., Burkhardt, M. R., Kolpin, D., Anderson, L. G. 2004. Determination of pharmaceutical compounds in surface- and ground-water samples by solid phase extraction and high-performance liquid chromatography-electrospray ionization mass spectrometry. *Journal of Chromatography A* 1041: 171–180.
Calheiros, C. S., Rangel, A. O., Castro, P. M. 2008. Evaluation of different substrates to support the growth of Typha latifolia in constructed wetlands treating tannery wastewater over long-term operation. *Bioresource Technology* 99(15): 6866–6877.
Casares, M. L., Ulierte, N., Mataran, A., Ramos, A., Zamorano, M. 2005. Solid industrial wastes and their management in Asegra (Granada, Spain). *Waste Management* 25(10): 1075–1082.
Chaineau, C. H., Morel, J. L., Oudot, J. 2000. Biodegradation of fuel oil hydrocarbons in the rhizosphere of maize. *Journal of Environmental Quality* 29(2): 569–578.
Chávez, R., Fierro, F., García-Rico, R. O., Laich, F. 2011. Mold-fermented foods: Penicillium spp. as ripening agents in the elaboration of cheese and meat products. In *Mycofactorie*, ed. A. L. Leitão, 73–98. Netherlands: Bentham Science Publishers Ltd.
Chen, B. 2002. Understanding decolourisations characteristic of reactive azo dyes by Pseudomonas luteola: Toxicity and kinetics. *Process Biochemistry* 38: 437–446.
Chen, K., Wua, J., Liou, D., Hwang, S. J. 2003. Decolourisation of the textile dyes by newly isolated bacterial strains. *Journal of Biotechnology* 101: 57–68.
Chibuike, G. U., Obiora, S. C. 2014. Heavy metal polluted soils: Effect on plants and bioremediation methods. *Applied and Environmental Soil Science* 2014: 1–12.
Cho, Y. G. 2000. Decolorization of textile dyes by newly isolated bacterial strains. *Journal of Biotechnology* 101(1): 57–68.
Christov, L., Van Driessel, B. 2003. Waste water bioremediation in the pulp and paper industry. *Indian Journal of Biotechnology* 2: 444–450.
Cochrane, V. W. 1958. *Physiology of the Fungi*. New York: John Wiley and Sons Inc.
Colberg, P. J. S., Young, L. Y. 1995. Anaerobic degradation of no halogenated homocyclic aromatic compounds coupled with nitrate, iron, or sulfate reduction. In *Microbial Transformation and Degradation of Toxic Organic Chemicals*, eds. L. Y. Young, and C. E. Cerniglia, 307–330. New York: Wiley-Liss.
Cooke, W. B. 1979. *The Ecology of Fungi*. Boca Raton, FL: CRC Press Inc.
Couto, N., Fritt-Rasmussen, J., Jensen, P. E., Højrup, M., Rodrigo, A. P., Riberio, A. B. 2014. Suitability of oil bioremediation in an Artic soil using surplus heating from an incineration facility. *Environmental Science and Pollution Research* 21: 6221–6227.
Cruz Morató, C. 2013. Biodegradation of pharmaceuticals by Trametes versicolor. PhD diss., Autonomous University of Barcelona, Spain.
Das, M. P., Bashwant, M., Kumar, K., Das, J. 2012. Control of pharmaceutical effluent parameters through bioremediation. *Journal of Chemical and Pharmaceutical Research* 4(2): 1061–1065.

Dashtban, M., Schraft, H., Syed, T. A., Qin, W. 2010. Fungal biodegradation and enzymatic modification of lignin. *International Journal of Biochemistry and Molecular Biology* 1(1): 36–50.

Davin-Regli, A., Pagès, J- M. 2015. Enterobacter aerogenes and Enterobacter cloacae; versatile bacterial pathogens confronting antibiotic treatment. *Frontiers in Microbiology* 6: 1–10.

De Carvalho, C. C., Costa, S. S., Fernandes, P., Couto, I., Viveiros, M. 2014. Membrane transport systems and the biodegradation potential and pathogenicity of genus Rhodococcus. *Frontiers in Physiology* 5: 1–13.

Debska, J., Kot-Wasik, A., Namiesnik, J. 2004. Fate and analysis of pharmaceutical residues in the aquatic environment. *Critical Reviews in Analytical Chemistry* 34: 51–67.

Deshmukh, R., Khardenavis, A. A., Purohit, H. J. 2016. Diverse metabolic capacities of fungi for bioremediation. *Indian Journal of Microbiology* 56(3): 247–264.

Dewi, E. R. S., Nuravivah, R. 2018. Potential of microalgae chlorella vulgaris as bioremediation agents of heavy metal Pb (Lead) on culture media. *E3S Web of Conferences* 31(05010): 1–4.

Dhanalakshmi, D., Maleeka Begum, S. F., Rajesh, G. 2016. Biodegradation and bioremediation of food industry waste effluents. *International Journal of Advanced Research in Science, Engineering and Technology* 3(1): 1195–1201.

Diaz, E. 2008. *Microbial Biodegradation: Genomics and Molecular Biology.* Wymondham, UK: Caister Academic Press.

Ding, C., He, J. 2010. Effect of antibiotics in the environment on microbial populations. *Applied Microbiology and Biotechnology* 87(3): 925–941.

Divya, M., Aanand, S., Srinivasan, A., Ahilan, B., Uma, A. 2015. Bioremediation of seafood processing plant effluents using indigenous bacterial isolates. *International Journal of Advanced Biotechnology and Research* 6(3): 443–449.

Dixit, R., Malaviya, D., Pandiyan, K., Singh, U. B., Sahu, A., Shukla, R., Singh, B. P., Rai, J. P., Sharma, P. K., Lade, H., Paul, D. 2015. Bioremediation of heavy metals from soil and aquatic environment: An overview of principles and criteria of fundamental processes. *Sustainability* 7(2): 2189–2212.

Edwards, S. J., Kjellerup, B. V. 2013. Applications of biofilms in bioremediation and biotransformation of persistent organic pollutants, pharmaceutical pollution/personal care products, and heavy metals. *Applied Microbiology and Biotechnology* 97(23): 9909–9921.

El Fantroussi, S., Agathos, S. N. 2005. Is bioaugmentation a feasible strategy for pollutant removal and site remediation? *Current Opinion in Microbiology* 8: 268–275.

El-Sheekh, M. M., El-Shouny, W. A., Osman, M. E., El-Gammal, E. W. 2005. Growth and heavy metals removal efficiency of Nostoc muscorum and Anabaena subcylindrica in sewage and industrial wastewater effluents. *Environmental Toxicology and Pharmacology* 19(2): 357–365.

El-Shiekh, H. H., Mahdy, H. M., El-Aaser, M. M. 2007. Bioremediation of aflatoxins by some reference fungal strains. *Polish Journal of Microbiology* 56(3): 215.

Environment Protection Authority (EPA). 2014. Waste classification guidelines – Part 1: Classification of waste. NSW Environment Protection Authority, Sydney.

Ertugrul, S., San, N. O., Donmez, G. 2009. Treatment of dye (Remazol Blue) and heavy metals using yeast cells with the purpose of managing polluted textile wastewaters. *Ecological Engineering* 35: 128–134.

Ezeronye, O. U., Okerentugba, P. O. 1999. Performance and efficiency of a yeast biofilter for the treatment of a Nigerian fertilizer plant effluent. *World Journal of Microbiology and Biotechnology* 15: 515–516.

Fatajeva, E., Gailiūtė, I., Paliulis, D., Grigiškis, S. 2014. The use of Acinetobacter sp. for oil hydrocarbon degradation in saline waters. *Biologija* 60(3): 126–133.

Fourest, E., Volesky, B. 1997. Alginate properties and heavy metal biosorption by marine algae. *Applied Biochemistry and Biotechnology* 67(3): 215–226.
Galadima, A. D. 2012. Bioremediation of textile industries effluents using selected bacterial species in Kano, Nigeria. Master diss., Ahmadu Bello Univ., Zaria, Nigeria.
Gan, S., Lau, E. V., Ng, H. K. 2009. Remediation of soils contaminated with polycyclic aromatic hydrocarbons (PAHs). *Journal of Hazardous Materials* 172(2–3): 532–549.
Garbisu, C., Garaiyurrebaso, O., Epelde, L., Grohmann, E., Alkorta, I. 2017. Plasmid-mediated bioaugmentation for the bioremediation of contaminated soils. *Frontiers in Microbiology* 8: 1966.
Gavrilescu, M. 2005. Fate of pesticides in the environment and its bioremediation. *Engineer in Life Science* 5: 497–526.
Glazer, A. N., Nikaido, H. 2007. *Microbial Biotechnology: Fundamentals of Applied Microbiology.* Cambridge, UK: Cambridge University Press.
Godoy-Faúndez, A., Reyez-Bozo, L., Montecinos-Bustamante, W., Quiroz-Valenzuela, S. 2014. Sustainable bioremediation, industrial ecology and public policies: New challenges for Chile. In *Bioremediation: Biotechnology, Engineering and Environmental Management*, ed. A. C. Mason, 99–132. New York, USA: Nova press.
Gomez, M. J., Petrovic, M., Fernandez-Alba, A. R., Barcelo, D. 2006. Determination of pharmaceuticals of various therapeutic classes by solid-phase extraction and liquid chromatographytandemmass spectrometry analysis in hospital effluent wastewaters. *Journal of Chromatography A* 1114: 224–233.
Gonzalez, P. S., Capozucca, C. E., Tigier, H. A., Milrad, S. R., Agostini, E. 2006. Phytoremediation of phenol from wastewater, by peroxidases of tomato hairy root cultures. *Enzyme and Microbial Technology* 39(4): 647–653.
Govarthanan, M., Lee, S. M., Kamala-Kannan, S., Oh, B. T. 2015b. Characterization, real-time quantification and in silico modeling of arsenate reductase (arsC) genes in arsenic- resistant Herbaspirillum sp. GW103. *Research in Microbiology* 166(3): 196–204.
Govarthanan, M., Mythili, R., Selvankumar, T., Kamala-Kannan, S., Rajasekar, A., Chang, Y. C. 2016. Bioremediation of heavy metals using an endophytic bacterium Paenibacillus sp. RM isolated from the roots of Tridax procumbens. *3 Biotechonolgy* 6(2): 242.
Govarthanan, M., Park, S. H., Park, Y. J., Myung, H., Krishnamurthy, R. R., Lee, S. H., Lovanh, N., Kamala-Kannan, S., Oh, B. T. 2015a. Lead biotransformation potential of allochthonous Bacillus sp. SKK11 with sesame oil cake extract in mine soil. *RSC Advances* 5(67): 54564–54570.
Grube, A., Donaldson, D., Kiely, T., Wu, L. 2011. Pesticides industry sales and usage 2006 and 2007 market estimates. U.S. Environmental Protection Agency.
Gürbüz, F., Ciftci, H., Akcil, A. 2009. Biodegradation of cyanide containing effluents by Scenedesmus obliquus. *Journal of Hazardous Materials* 162(1): 74–79.
Gürses, A., Güneş, K., Korucu, M. E. Açikyildiz, M. 2016. Industrial waste. In *Kirk-Othmer Encyclopedia of Chemical Technology*, eds. A. Seidel, and M. Bickford, 1–26. New York: John Wiley and Sons Inc.
Hajdu-Rahkama, R. 2014. Bioremediation of heavy metals by using the microalga desmodesmus subspicatus, Bachelor's thesis, Ostfalia University of Applied Sciences, Germany.
Hasan, S. 2016. Potential of Trichoderma sp. in bioremediation: A review. *Journal of Basic and Applied Engineering Research* 3(9): 776–779.
Hatzikioseyian, A. 2010. Principles of bioremediation processes. In *Trends in Bioremediation and Phytoremediation*, ed. G. Płaza, 23–54. India: Research Signpost.
Hazen, T. C., Tien, A. J., Worsztynowicz, A., Altman, D. J., Ulfig, K., Manko, T. 2003. Biopiles for remediation of petroleum-contaminated soils: A polish case

study. In *The Utilization of Bioremediation to Reduce Soil Contamination: Problems and Solutions*, eds. P. Baveye, J. A., Glaser, and V. Šašek, 229–246. Germany: Springer.

Heberer, T. 2002. Occurrence, fate, and removal of pharmaceutical residues in the aquatic environment: A review of recent research data. *Toxicology Letters* 131: 5–17.

Helbling, D. E. 2015. Bioremediation of pesticide-contaminated water resources: The challenge of low concentrations. *Current Opinion in Biotechnology* 33: 142–148.

Hernando, M. D., Petrovic, M., Fernandez-Alba, A. R., Barcelo, D. 2004. Analysis by liquid chromatography-electro spray ionization tandem mass spectrometry and acute toxicity evaluation for betablockers and lipid-regulating agents in wastewater samples. *Journal of Chromatography A* 1046: 133–140.

Hossain, K., Ismail, N. 2015. Bioremediation and detoxification of pulp and paper mill effluent: A review. *Research Journal of Environmental Toxicology* 9(3): 113–134.

Iihan, S., Nourbakhsh, N. M., Kilicarslan, S., Ozdag, H. 2004. Removal of chromium, lead and copper ions from industrial wastewaters by Staphylococcus saprophyticus. *Turkish Electronic Journal of Biotechnology* 2: 50–57.

Imani, S., Rezaei-Zarchi, S., Hashemi, M., Borna, H., Javid, A., Abarghouei, H. B. 2011. Hg, Cd and Pb heavy metal bioremediation by Dunaliella alga. *Journal of Medicinal Plants Research* 5(13): 2775–2780.

Iye, O. J. 2015. *Bioremediation of Heavy Metal Polluted Water using Immobilized Freshwater Green Microalga, Botryococcus sp*, PhD diss., Tun Hussein Onn University of Malaysia, Malaysia.

Jeenathunisa. N., Jeyabharathi. S, Arthi. J. 2017. Bioremediation of paper and pulp industrial effluent using bacterial isolates. *International Journal for Research in Applied Science & Engineering Technology* 5: 692–698.

Joo, H. S., Ndegwa, P. M., Shoda, M., Phae, C. G. 2008. Bioremediation of oil-contaminated soil using Candida catenulata and food waste. *Environmental Pollution* 156(3): 891–896.

Jorgensen, S. E., Halling-Sorensen, B. 2000. Drugs in the environment. *Chemosphere* 40: 691–699.

Kalyani, P., Geetha, S., Hemalatha, V., Chandana Vineela, K., Hemalatha, K. P. J. 2016. Bioremediation of Chromium by Using Aspergillus Niger (Mttc-961) and Aspergillus Flavus (Mttc-3396). *World Journal of Pharmacy and Pharmaceutical Sciences* 5 (10): 881–885.

Kamali, M., Khodaparast, Z. 2015. Review on recent developments on pulp and paper mill wastewater treatment. *Ecotoxicology and Environmental Safety* 114: 326–342.

Kamat, D. V., Kamat, S. D. 2015. Bioremediation of industrial effluent containing reactive dyes. *International Journal of Environmental Sciences* 5(6): 1078–1084.

Kang, C. H., Kwon, Y. J., So, J. S. 2016. Bioremediation of heavy metals by using bacterial mixtures. *Ecological Engineering* 89: 64–69.

Kanissery, R. G., Sims, G. K. 2011. Biostimulation for the enhanced degradation of herbicides in soil. *Applied and Environmental Soil Science* 2011: 10 pages.

Kanmani, P., Kumaresan, K., Aravind, J. 2015. Utilization of coconut oil mill waste as a substrate for optimized lipase production, oil biodegradation and enzyme purification studies in Staphylococcus pasteuri. *Electronic Journal of Biotechnology* 18 (1): 20–28.

Kapoor, D. 2015. Impact of pharmaceutical industries on environment, health and safety. *Journal of Critical Reviews* 2(4): 25–30.

Kartheek, B. R., Maheswaran, R., Kumar, G., Sharmila Banu, G. 2011. Biodegradation of pharmaceutical wastes using different microbial strains. *International Journal of Pharmaceutical & Biological Archive* 2(5): 1401–1404.

Kaushik, G. 2015. Bioremediation of industrial effluents: Distillery effluent. In *Applied Environmental Biotechnology: Present Scenario and Future Trends*, ed. G. Kaushik, 19–32. New Delhi: Springer.
Kedia, R., Sharma, A. 2015. Bioremediation of industrial effluents using arbuscular mycorrhizal fungi. *Biosciences Biotechnology Research Asia* 12(1): 197–200.
Khan, M. W. A., Ahmad, M. 2006. Detoxification and bioremediation potential of a Pseudomonas fluorescens isolate against the major Indian water pollutants. *Journal of Environmental Science and Health Part A* 41(4): 659–674.
Kielak, A. M., Castellane, T. C., Campanharo, J. C., Colnago, L. A., Costa, O. Y., Da Silva, M. L. C., van Veen, J. A., Lemos, E. G. M. Kuramae, E. E. 2017. Characterization of novel Acidobacteria exopolysaccharides with potential industrial and ecological applications. *Scientific Reports* 7: 41193.
Klimmek, S., Stan, H. J., Wilke, A., Bunke, G., Buchholz, R. 2001. Comparative analysis of the biosorption of cadmium, lead, nickel, and zinc by algae. *Environmental Science & Technology* 35(21): 4283–4288.
Krishnaveni, R., Pramiladevi, Y., Ramgopal Rao, S. 2013. Bioremediation of steel industrial effluents using soil microorganisms. *International Journal of Advanced Biotechnology and Research* 4: 914–919.
Kumar, J. N., Oommen, C. 2012. Removal of heavy metals by biosorption using freshwater alga Spirogyra hyalina. *Journal of Environmental Biology* 33(1): 27–31.
Kunal, A. R. 2011. Bio-medical waste incinerator ash: A review with special focus on its characterization, utilization and leachate analysis. *International Journal of Geology, Earth and Environmental Science* I(1): 48–58.
Lakshmi Priya, J. 2013. Biodegradation of diesel by Aeromonas hydrophila. *International Journal of Pharmaceutical Science Invention* 2(4): 24–36.
Lalitha, P., Reddy, N. N. R., Arunalakshmi, K. 2011. Decolorization of synthetic dyes by Aspergillus flavus. *Bioremediation Journal* 15(2): 121–132.
Lee, M. D., Swindoll, C. M. 1993. Bioventing for in situ remediation. *Hydrological Sciences Journal* 38(4): 273–282.
Leitão, A. L. 2009. Potential of Penicillium species in the bioremediation field. *International Journal of Environmental Research and Public Health* 6(4): 1393–1417.
Lilly, V. M., Barnett, H. L. 1951. *Physiology of the Fungi*. New York: McGraw-Hill.
Liu, Z., Xie, W., Li, D., Peng, Y., Li, Z., Liu, S. 2016. Biodegradation of phenol by bacteria strain Acinetobacter calcoaceticus PA isolated from phenolic wastewater. *International Journal of Environmental Research and Public Health* 13(3): 300–308.
Liu, Z., Yang, C., Qiao, C. 2007. Biodegradation of p-nitrophenol and 4-chlorophenol by Stenotrophomonas sp. *FEMS Microbiology Letters* 277(2): 150–156.
Loganathan, P., Myung, H., Muthusamy, G., Lee, K. J., Seralathan, K. K., Oh, B. T. 2015. Effect of heavy metals on acdS gene expression in Herbaspirillium sp. GW103 isolated from rhizosphere soil. *Journal of Basic Microbiology* 55(10): 1232–1238.
Luka, Y., Highina, B. K., Zubairu, A. 2018. Bioremediation: A solution to environmental pollution-A review. *American Journal of Engineering Research* 7(2): 101–109.
Lukić, B., Panico, A., Huguenot, D., Fabbricino, M., van Hullebusch, E. D., Esposito, G. 2017. A review on the efficiency of landfarming integrated with composting as a soil remediation treatment. *Environmental Technology Reviews* 6(1): 94–116.
Macaulay, B. M. 2014. Understanding the behavior of oil-degrading microorganisms to enhance the microbial remediation of spilled petroleum. *Applied Ecology and Environmental Research* 13: 247–262.

Majumder, A., Kapagunta, C. 2017. A bio-remediation solution for pharmaceutical pollution. Project Guru. www.projectguru.in/publications/bio-remediation-solution-pharmaceutical-pollution/ (accessed July 12, 2018).

Mallick, N. 2003. Biotechnological potential of Chlorella vulgaris for accumulation of Cu and Ni from single and binary metal solutions. *World Journal of Microbiology and Biotechnology* 19(7): 695–701.

Mangunwardoyo, W., Sudjarwo, T., Patria, M. P. 2013. Bioremediation of effluent wastewater treatment plant Bojongsoang Bandung Indonesia using consortium aquatic plants and animals. *International Journal of Research and Reviews in Applied Sciences* 14(1): 150–160.

Mani, D., Kumar, C. 2014. Biotechnological advances in bioremediation of heavy metals contaminated ecosystems: An overview with special reference to phytoremediation. *International Journal of Environmental Science and Technology* 11(3): 843–872.

Mansour, H. B., Mosrati, R., Barillier, D., Ghedira, K., Chekir-Ghedira, L. 2012. Bioremediation of industrial pharmaceutical drugs. *Drug and Chemical Toxicology* 35 (3): 235–240.

Matheickal, J. T., Yu, Q. 1996. Biosorption of lead from aqueous solutions by marine algae Ecklonia radiata. *Water Science and Technology* 34(9): 1–7.

Mcheik, A., Fakih, M., Olama, Z., Holail, H. 2013. Bioremediation of four food industrial effluents. *American Journal of Agriculture and Forestry* 1(1): 12–21.

Misal, S. A., Lingojwar, D. P., Shinde, R. M., Gawai, K. R. 2011. Purification and characterization of azoreductase from alkaliphilic strain Bacillus badius. *Process Biochemistry* 46(6): 1264–1269.

Monteiro, C. M., Marques, A. P., Castro, P. M., Malcata, F. X. 2009. Characterization of Desmodesmus pleiomorphus isolated from a heavy metal-contaminated site: Biosorption of zinc. *Biodegradation* 20(5): 629–641.

Mrozik, A., Piotrowska-Seget, Z. 2010. Bioaugmentation as a strategy for cleaning up of soils contaminated with aromatic compounds. *Microbiological Research* 165(5): 363–375.

Mueller, J. G., Cerniglia, C. E., Pritchard. P. H. 1996. Bioremediation of environments contaminated by polycyclic aromatic hydrocarbons. In *Bioremediation: Principles and Applications*, eds. R. L., Crawford, and D. L. Crawford, 125–194. Cambridge, UK: Cambridge University Press.

Murali, O., Mehar, S. K. 2014. Bioremediation of heavy metals using spirulina. *International Journal of Geology, Earth and Environmental Sciences* 4(1): 244–249.

Murugesan, K. 2003. Bioremediation of paper and pulp mill effluents. *Indian Journal of Experimental Biology* 41: 1239–1248.

Muthulakshmi, L., Nellaiah, H., Kathiresan, T., Rajini, N., Christopher, F. 2017. Identification and production of bioflocculants by Enterobacter sp. and Bacillus sp. and their characterization studies. *Preparative Biochemistry and Biotechnology* 47(5): 458–467.

Nicoletti, R., Buommino, E., Antonietta Tufano, M. 2012. Patenting Penicillium strains. *Recent Patents on Biotechnology* 6(2): 81–96.

Nigam, P. S. 2013. Microbial enzymes with special characteristics for biotechnological applications. *Biomolecules* 3(3): 597–611.

Norris, D. 1994. *Handbook of Bioremediation*. Boca Raton, FL: CRC Press.

Odukkathil, G., Vasudevan, N. 2013. Toxicity and bioremediation of pesticides in agricultural soil. *Reviews in Environmental Science and Bio/Technology* 12(4): 421–444.

Ofer, R., Yerachmiel, A., Shmuel, Y. 2003. Marine macroalgae as biosorbents for cadmium and nickel in water. *Water Environment Research* 75(3): 246–253.

Ojuederie, O. B., Babalola, O. O. 2017. Microbial and plant-assisted bioremediation of heavy metal polluted environments: A review. *International Journal of Environmental Research and Public Health* 14(12): 1504–1530.

Olawale, A. M. 2014. Bioremediation of waste water from an industrial effluent system in Nigeria using Pseudomonas aeruginosa: Effectiveness tested on albino rats. *Journal of Petroleum & Environmental Biotechnology* 5(1): 166.

Olayinka, K. O., Alo, B. I. 2004. Studies on industrial pollution in Nigeria: The effect of textile effluents on the quality of groundwater in some parts of Lagos. *Nigerian Journal of Health and Biomedical Sciences* 3: 44–50.

Özer, A., Özer, D., Ekiz, H. I. 1999. Application of freudlich and langmuir models to multistage purification process to remove heavy metal ions using Schizomeris leibleinni. *Process Biochemistry* 34: 919–927.

Parthipan, P., Preetham, E., Machuca, L. L., Rahman, P. K., Murugan, K., Rajasekar, A. 2017. Biosurfactant and degradative enzymes mediated crude oil degradation by bacterium Bacillus subtilis A1. *Frontiers in Microbiology* 8: 1–14.

Peil, G. H., KuSS, A. V., Rave, A. F., VillArreAl, J. P., Hernandes, Y. M., Nascente, P. S. 2016. Bioprospecting of lipolytic microorganisms obtained from industrial effluents. *Anais da Academia Brasileira de Ciências* 88(3): 1769–1779.

Pereira, P., Enguita, F. J., Ferreira, J., Leitão, A. L. 2014. DNA damage induced by hydroquinone can be prevented by fungal detoxification. *Toxicology Reports* 1: 1096–1105.

Pérez Silva, R. M., Ábalos Rodríguez, A., Gómez Montes de Oca, J. M., Cantero Moreno, D. 2006. Biodegradation of crude oil by Pseudomonas aeruginosa AT18 strain. *Tecnologia Química* 26(1): 70–77.

Pinedo-Rivilla, C., Aleu, J., Collado, I. G. 2009. Pollutants biodegradation by fungi. *Current Organic Chemistry* 13(12): 1194–1214.

Pokhrel, D., Viraraghavan, T. 2004. Treatment of pulp and paper mill wastewater—a review. *Science of the Total Environment* 333(1–3): 37–58.

Porwal, H. J., Mane, A. V., Velhal, S. G. 2015. Biodegradation of dairy effluent by using microbial isolates obtained from activated sludge. *Water Resources and Industry* 9: 1–15.

Prabha, Y. 2012. Potential of algae in bioremediation of waste water, PhD diss., Deemed University, India.

Prasad, M. P., Manjunath, K. 2011. Comparative study on biodegradation of lipid-rich wastewater using lipase producing bacterial species. *Indian Journal of Biotechnology* 10: 121–124.

Prasher, S. O., Beaugeard, M., Hawari, J., Bera, P., Patel, R. M., Kim, S. H. 2004. Biosorption of heavy metals by red algae (Palmaria palmata). *Environmental Technology* 25(10): 1097–1106.

Prince, C. 2002. Bioremediation of petroleum and other hydrocarbons. In *Encyclopedia of Environmental Microbiology*, ed. G. Bitton, 2402–2416. New York: John Wiley & Sons.

Priya darshini, P., Sharpudin, J. 2016. Bioremediation of industrial and municipal waste water using bacterial isolates. *International Journal of Engineering Sciences & Research Technology* 5(5): 173–177.

Raja, M., Praveena, G., William, S. J. 2017. Isolation and identification of fungi from soil in Loyola college campus, Chennai, India. *International Journal of Current Microbiology and Applied Sciences* 6(2): 1789–1795.

Rana, R. S., Singh, P., Kandari, V., Singh, R., Dobhal, R., Gupta, S. 2017. A review on characterization and bioremediation of pharmaceutical industries' wastewater: An Indian perspective. *Applied Water Science* 7(1): 1–12.

Randhawa, G. K., Kullar, J. S. 2011. Bioremediation of pharmaceuticals, pesticides, and petrochemicals with gomeya/cow dung. *ISRN Pharmacology* 2011: 1–7.

Rani, K., Dhania, G. 2014. Bioremediation and biodegradation of pesticide from contaminated soil and water–a novel approach. *International Journal of Current Microbiology and Applied Sciences* 3(10): 23–33.

Raza, M. H., Sadiq, A., Farooq, U., Athar, M., Hussain, T., Mujahid, A., Salman, M. 2015. Phragmites karka as a biosorbent for the removal of mercury metal ions from aqueous solution: Effect of modification. *Journal of Chemistry* 2015: 12 pages.

Roane, T. M., Pepper, I. L. 2000. Microorganisms and metal pollution. In *Environmental Microbiology*, eds. I. L. Maier, and C. B. Pepper, 387–442. London, UK: Academic Press.

Romera, E., Gonzalez, F., Ballester, A., Blazquez, M. L., Munoz, J. A. 2008. Biosorption of heavy metals by Fucus spiralis. *Bioresource Technology* 99(11): 4684–4693.

Rubín Carrero, E., Rodríguez-Barro, P., Herrero Rodríguez, R., Sastre de Vicente, M. 2006. Biosorption of phenolic compounds by the brown alga Sargassum muticum. *Journal of Chemical Technology and Biotechnology* 81(7): 1093–1099.

Sáenz-Marta, C. I., de Lourdes Ballinas-Casarrubias, M., Rivera-Chavira, B. E., Nevárez-Moorillón, G. V. 2015. Biosurfactants as useful tools in bioremediation. In *Advances in Bioremediation of Wastewater and Polluted Soil*, ed. N. Shiomi, 93–109. Crotia: InTech.

Sakaguchi, T., Horikoshi, T., Nakajima, A. 1978. Uptake of uranium from sea water by microalgae. *Journal of Fermentation Technology* 56(6): 561–565.

Samantaray, D., Mohapatra, S., Mishra, B. B. 2014. Microbial bioremediation of industrial effluents. In *Microbial Biodegradation and Bioremediation*, ed. S. Das, 325–339. Amsterdam, Netherlands: Elsevier.

Sandau, E., Sandau, P., Pulz, O., Zimmermann, M. 1996. Heavy metal sorption by marine algae and algal by-products. *Acta Biotechnologica* 16(2-3): 103–119.

Sanscartier, D., Zeeb, B., Koch, I., Reimer, K. 2009. Bioremediation of diesel-contaminated soil by heated and humidified biopile system in cold climates. *Cold Regions Science and Technology* 55(1): 167–173.

Saranraj, P., Stella, D. 2014. Impact of sugar mill effluent to environment and bioremediation: A review. *World Applied Sciences Journal* 30(3): 299–316.

Saritha, V., Maruthit, Y. A., Mukkanti, K. 2010. Potential fungi for bioremediation of industrial effluents. *BioResources* 5(1): 8–22.

Saunders, R. J., Paul, N. A., Hu, Y., de Nys, R. 2012. Sustainable sources of biomass for bioremediation of heavy metals in waste water derived from coal-fired power generation. *PloS One* 7(5): e36470.

Seeger, M., Hernández, M., Méndez, V., Ponce, B., Córdova, M., González, M. 2010. Bacterial degradation and bioremediation of chlorinated herbicides and biphenyls. *Journal of Soil Science and Plant Nutrition* 10(3): 320–332.

See-Too, W. S., Ee, R., Lim, Y. L., Convey, P., Pearce, D. A., Mohidin, T. B. M., Yin, W-F., Chan, K. G. 2017. Complete genome of Arthrobacter alpinus strain R3. 8, bioremediation potential unraveled with genomic analysis. *Standards in Genomic Sciences* 12(1): 2–7.

Shah Maulin, P. 2017. Environmental bioremediation of industrial effluent. *Journal of Molecular Biology and Biotechnology* 2(1–2): 1–3.

Shalini, K., Anwer, Z., Sharma, P. K., Garg, V. K., Kumar, N. 2010. A review on pharma pollution. *International Journal of PharmTech Research* 2(4): 2265–2270.

Shanab, S., Essa, A., Shalaby, E. 2012. Bioremoval capacity of three heavy metals by some microalgae species (Egyptian Isolates). *Plant Signaling & Behavior* 7(3): 392–399.

Shanatan, P. 2004. Bioremediation. Waste containment and remediation technology, Massachusetts Institute of Technology, MIT Open Courseware.

Shanthi, J., Krubakaran, C. T. B., Balagurunathan, R. 2012. Characterization and isolation of Paper mill effluent degrading microorganisms. *Journal of Chemical and Pharmaceutical Research* 4(10): 4436–4439.

Shi, W., Shao, H., Li, H., Shao, M., Du, S. 2009. Progress in the remediation of hazardous heavy metal-polluted soils by natural zeolite. *Journal of Hazardous Materials* 170: 1–6.

Shobha, G., Moses, V., Ananda, S. 2014. Biological synthesis of copper nanoparticles and its impact. *International Journal of Pharmaceutical Science Invention* 3(8): 6–28.

Sikkema, J., de Bont, J. A., Poolman, B. 1995. Mechanisms of membrane toxicity of hydrocarbons. *Microbiological Reviews* 59: 201–222.

Siñeriz, L. M., Benito, J. M., Albarracín, V. H., Lebeau, T., Amoroso, M. J. Abate, C. M. 2005. Heavy-metal resistant actinomycetes. In *Environmental Chemistry: Green Chemistry and Pollutants in Ecosystems,* eds. E. Lichtfouse, J. Schwarzbauer, and D. Robert, 757–776. Berlin, Germany: Springer.

Singh, D., Tiwari, A., Gupta, R. 2012. Phytoremediation of lead from wastewater using aquatic plants. *Journal of Agricultural Technology* 8(1): 1–11.

Singh, D. K. 2008. Biodegradation and bioremediation of pesticide in soil: Concept, method and recent developments. *Indian Journal of Microbiology* 48(1): 35–40.

Singh, J., Kaushik, N., Biswas, S. 2014. Bioreactors – Technology and design analysis. *The Scitech Journal* 01(6): 28–36.

Singhal, A., Thakur, I. S. 2009. Decolourization and detoxification of pulp and paper mill effluent by Cryptococcus sp. *Biochemical Engineering Journal* 46: 21–27.

Sivagamasundari, T., Jayakumar, N. 2017. Optimization of diesel oil degrading bacterial strains at various culture parameters. *International Journal of Research and Development in Pharmacy & Life Sciences* 6(6): 2840–2844.

Stackelberg, P. E., Furlong, E. T., Meyer, M. T., Zaugg, S. D., Henderson, A. K., Reissman, D. B. 2004. Persistence of pharmaceutical compounds and other organic wastewater contaminants in a conventional drinking-water treatment plant. *Science of the Total Environment* 329: 99–113.

Stirk, W. A., Van Staden, J. 2000. Removal of heavy metals from solution using dried brown seaweed material. *Botanica Marina* 43(5): 467–473.

Suja, F., Rahim, F., Taha, M. R., Hambali, N., Razali, M. R., Khalid, A., Hamzah, A. 2014. Effects of local microbial bioaugmentation and biostimulation on the bioremediation of total petroleum hydrocarbons (TPH) in crude oil contaminated soil based on laboratory and field observations. *International Biodeterioration & Biodegradation* 90: 115–122.

Surti, H. 2016. Physico-chemical and microbial analysis of waste water from different industry and cod reduction treatment of industrial waste water by using selective microorganisms. *International Journal of Current Microbiology and Applied Sciences* 5(6): 707–717.

Tayabali, A. F., Coleman, G., Crosthwait, J., Nguyen, K. C., Zhang, Y., Shwed, P. 2017. Composition and pathogenic potential of a microbial bioremediation product used for crude oil degradation. *PloS One* 12(2): 1–21.

Tayade, S., Patel, Z. P., Mutkule, D. S., Kakde, A. M. 2013. Pesticide contamination in food: A review. *IOSR Journal of Agriculture and Veterinary Science* 6: 7–11.

Tchobanoglous, G., Theisen, H., Vigil, S. A. 1993. *Integrated Solid Waste Management: Engineering Principles and Management Issues.* Boston USA: McGraw-Hill.

Thassitou, P. K., Arvanitoyannis, I. S. 2001. Bioremediation: A novel approach to food waste management. *Trends in Food Science & Technology* 12(5–6): 185–196.

Torres, E. M. 2016. Microalgae sorption of ten individual heavy metals and their effects on growth and lipid accumulation, Master Thesis, Utah State University, USA.

Tripathi, A. K., Harsh, N. S. K., Gupta, N. 2007. Fungal treatment of industrial effluents: A mini-review. *Life Science Journal* 4(2): 78–81.

Türker, M. 2014. Yeast biotechnology: Diversity and applications. Paper presented at 27th VH Yeast Conference, Advances in Science and Industrial Production of Baker's Yeast, Istanbul, 1–26.

Tzirita, M., Papanikolaou, S., Quilty, B. 2018. Degradation of fat by a bioaugmentation product comprising of Bacillus spp. before and after the addition of a Pseudomonas sp. *European Journal of Lipid Science and Technology* 120(2): 1–9.

U.S. Environmental Protection Agency (US EPA). Land farming definition | ecology dictionary, 2013. www.ecologydictionary.org/Land_Farming (accessed March 12, 2018).

Uqab, B., Mudasir, S., Nazir, R. 2016. Review on bioremediation of pesticides. *Journal of Bioremediation & Biodegradation* 7(3): 1–5.

Verma, A., Singh, H., Anwar, M. S., Kumar, S., Ansari, M. W., Agrawal, S. 2016. Production of thermostable organic solvent tolerant keratinolytic protease from Thermoactinomyces sp. RM4: IAA production and plant growth promotion. *Frontiers in Microbiology* 7: 1189.

Verma, N., Sharma, R. 2017. Bioremediation of toxic heavy metals: A patent review. *Recent Patents on Biotechnology* 11(3): 171–187.

Vidali, M. 2001. Bioremediation. an overview. *Pure and Applied Chemistry* 73(7): 1163–1172.

Volesky, B. 1990. *Biosorption of Heavy Metals.* Boca Raton, FL: CRC Press.

Wang, Q., Zhang, S., Li, Y., Klassen, W. 2011. Potential approaches to improving biodegradation of hydrocarbons for bioremediation of crude oil pollution. *Environ Protection Journal* 2: 47–55.

Wang, S., Hu, Y., Wang, J. 2018. Biodegradation of typical pharmaceutical compounds by a novel strain Acinetobacter sp. *Journal of Environmental Management* 217: 240–246.

Wang, Y., Wang, C., Li, A., Gao, J. 2015. Biodegradation of pentachloronitrobenzene by arthrobacter nicotianae DH19. *Letters in Applied Microbiology* 61: 403–410.

Wase, J., Forster, C. F. 1997. *Biosorbents for Metals Ions.* New York USA: Taylor& Francis.

Wei, M.-S., Huang, K.-H. 2001. Recycling and reuse of industrial wastes in Taiwan. *Waste Management* 21: 93–97.

Wen, X. 2009. Industrial pollution. Encyclopedia of Life Support Systems (EOLSS). Point Sources of Pollution: Local Effects and It's Control - Vol I.

Williams, R. T., Myler, C. A. 1990. Bioremediation using composting. *Biocycle* 31(11): 78–82.

Xue, L., Famous, E., Jiang, J., Shang, H., Ma, P. 2016. Experimental survey on microbial bioremediation of food wastewaters. *International Journal of Scientific and Research Publications* 6(9): 110–118.

Yadav, A. N., Verma, P., Kumar, V., Sangwan, P., Mishra, S., Panjiar, N., Gupta, V. K., Saxena, A. K. 2017. Biodiversity of the genus Penicillium in different habitats. In *New and Future Developments in Microbial Biotechnology and Bioengineering*, eds. V. K. Gupta, and S. Rodriguez-Couto, 3–18. Amsterdam, Netherlands: Elsevier.

Yang, Q., Angly, F. E., Wang, Z., Zhang, H. 2011. Wastewater treatment systems harbor specific and diverse yeast communities. *Biochemical Engineering Journal* 58: 168–176.

Zhang, L., Hu, J., Zhu, R., Zhou, Q., Chen, J. 2013. Degradation of paracetamol by pure bacterial cultures and their microbial consortium. *Applied Microbiology and Biotechnology* 97(8): 3687–3698.

Zulaikha, S., Lau, W. J., Ismail, A. F., Jaafar, J. 2014. Treatment of restaurant wastewater using ultrafiltration and nanofiltration membranes. *Journal of Water Processing Engineering* 2: 58–62.

9 Biosorption
A Promising Technique against Dye Removal

G. Ersöz and S. Atalay
Department of Chemical Engineering, Ege University, Izmir, Turkey

9.1 INTRODUCTION

Dyes are colored compounds that are widely used in many industries such as textiles, cosmetics, printing, rubber, plastics and leather for using in coloring their products. As a result of these industrial activities, generally considerable amount of colored wastewater is generated.

The discharge of huge amounts of dye-bearing wastewater brings about severe environmental problems. The main characteristics of dye containing wastewater are that they have high in organic content, high salt content. Most of the dyes have complicated, stable aromatic structures having a tendency to accumulate in environment (Fu and Viraraghavan, 2001; Kadirvelu et al., 2003). Consequently, the treatment of the dye containing effluents is very important.

The conventional treatment methods including coagulation (Gao et al., 2007; Tan et al., 2000), flocculation (Kono and Kusumoto, 2015; Melo et al., 2018), ion exchange (Hassan and Carr, 2018), precipitation (Zhu et al., 2007), ozonation (Ghuge and Saroha, 2018), adsorption (Lipatova et al., 2018; Yang et al., 2018) or a various combinations of these processes have been used for the dye treatment.

Unfortunately, application of these methods have some drawbacks such as operating costs, by-product or undesired product generation, high-energy requirement (Padmesh et al., 2005) and limited application to a wide range of effluents (Fu and Viraraghavan, 2001).

The application of biological processes that use biosubstances such as microorganisms and plants can be an alternative to recently applied chemical and physical methods. Biosorption is an emerging and attractive technology by which the desired substances are removed from aqueous solution by biological substances. These substances can be organic and inorganic and are can be either in soluble or insoluble forms. It involves the interaction between pollutant and biomass (Gadd, 2009; Michalak et al., 2013). Recently, biosorption processes have gained great importance from an environmental point of view that can degrade toxic compound in industrial waste water efficiently.

This chapter aims to provide a better view of the principles of the biosorption process by comparing with other conventional methods. The various practical aspects related to new applications of biosorption are discussed. Through examples and case studies, adsorption, kinetics, equilibrium and thermodynamic researches on the treatment of dyes are also presented.

9.2 DYE REMOVAL: AN OVERVIEW

Dye containing effluents discharged from various industries are dangerous toward the nature and human health. Hence, the existence of dye effluent in water sources is becoming a great concern for researchers.

There are various treatment techniques for the treatment of dyes from effluents and these methods can be investigated in three different categories. These technologies include physical techniques (membrane filtration processes and sorption techniques), chemical techniques (coagulation or flocculation and some simple oxidation processes) and advanced chemical techniques and biological techniques (microbial and enzymatic degradation) (Adegoke and Bello, 2015; Nidheesh et al., 2018). However, all of them have some advantages and drawbacks as well.

9.2.1 Physical Methods

Physical methods are usually straightforward techniques generally realized by the mass transfer mechanism. Traditional physical methods for dye removal are membrane filtration processes such as nanofiltration, electrodialysis, reverse osmosis and adsorption techniques (Katheresan et al., 2018).

The main drawback of using membrane filtration processes is that membranes have a limited lifetime and fouling problems. The periodic replacement of the membrane is required and hence the cost of this replacement must be included in the economic analysis of the process. Adsorption is an effective and one of the best equilibrium processes for the treatment of contaminants from the wastewater. Recently, adsorption methods have gained great concern due to their success in the treatment of pollutants.

9.2.2 Chemical Methods

Dye removal by chemical processes includes methods using chemistry or chemistry theories in accomplishing dye removal. Traditional methods include a various processes such as coagulation or flocculation electro-coagulation, irradiation, electro-flotation or electrochemical processes. These methods are successful in the removal of dyes but are rather expensive. In addition, they have the disadvantage of disposal problem of the sludge generated during the process (Kharub, 2012).

9.2.3 Biological Methods

Biological method is removal of pollutants with the biological phenomenon such as bioremediation which is a green technique to remove or degrade dye from aqueous solution. The bioremediation is economically feasible, environmental-friendly process. If the biological methods are compared with physical and chemical treatment processes, the biological treatment process is known to be one of the most economical techniques. By biological degradation, the pollutants, by breaking of bonds, turn into comparatively less hazardous compounds. For example, in biological oxidation of a dye containing water, the chromophoric group breaks down and this helps in removal of the dye color (Bhatia et al., 2017).

9.2.4 Advanced Chemical Methods

Advanced chemical methods are various chemical treatment processes designed to remove the toxic, hazardous organic compounds in wastewater by various oxidation reactions. The most used advanced oxidation processes are ozonation, photocatalytic degradation, Fenton's reagent, photo-Fenton and wet air oxidation. They have proven to be successful for dye containing aqueous solutions.

Nowadays, researchers have been mainly interested in the bioaccumulation and the biosorption processes. They have been shown as candidates to replace traditional methods for the treatment of aqueous solutions. By the help of inactive and dead biomass, biosorption has found to be an efficient technology to remove the dye molecules from dilute effluents (Malik, 2004; Vijayaraghavan and Yun, 2008; Volesky and Holan, 1995).

9.3 BIOSORPTION

9.3.1 An Overview

Biosorption is a treatment process that is applied for the treatment of pollutants that are not easily biodegradable such as metals and dyes in waters.

The concept biosorption can be investigated in two terms. The "bio" term stands for the involvement of a biological that the sorbent has a biological origin and the "sorption" term denotes physical and chemical process by which the attachment of substances are realized. In general, it is a term used for both absorption and adsorption. In this case, biosorption is a subclass of adsorption with biological matrix, sorbent (Chojnacka, 2009; Michalak et al., 2013).

Consequently, biosorption can be defined as a process that occurs naturally in biomass and allows it to concentrate and binds pollutants onto its structure by an equilibrium process (Chojnacka, 2009; Sameera et al., 2011).

The biosorption process involves a solid phase (biosorbent) and a liquid phase as a solvent (normally water; Fomina and Gadd, 2014; Mosbah and Sahmoune, 2013).

Previously, biosorption was categorized according to the cell metabolism dependency: metabolism dependent or metabolism independent. But, nowadays, metabolism-dependent processes are named as bioaccumulation, and metabolic-independent methods are named as biosorption (Bilal et al., 2018; Kushwah et al., 2015).

Bioaccumulation consists of two processes. The first process is similar to biosorption which involves the attachment of toxic elements to the surface and in the second step metal ions transports into cells. Bioaccumulation is a nonequilibrium process and more complex than biosorption itself (Flouty and Estephane, 2012).

Biosorption is a spontaneous technique that has a faster rate bioaccumulation and is realized by a biosubstance (dead, inactive or biologically derived substances). This process has the advantage of using dead materials because there is no need of growing (keeping biomass alive requires extra care and extra cost), and these materials are present as wastes or by-products. In contrast, the bioaccumulation is metabolically active process realized by living substances (Chojnacka, 2009; Farooq et al., 2010).

9.3.2 Advantages and Disadvantages of Biosorption Process

A biosorption-based process has a number of advantages when compared with the traditional methods used. The major advantages of biosorption are (Ahalya et al., 2003; Sameera et al., 2011) as follows:

- Low cost (both capital and operation)
- High efficiency
- Sludge minimization
- No additional nutrient requirement
- Metal recovery possibility – selective removal of metals
- Very fast (dead material acts as an ion-exchange resin)

However, there are some disadvantages as well. The disadvantages of biosorption are (Farooq et al., 2010; Kushwah et al., 2015) as follows:

- process reaches saturation early
- The potential for biological process improvement
- The properties of the biosorbents cannot be biologically controlled

9.3.3 Biosorption Mechanisms

Biosorption is a physiochemical process. It takes place naturally in certain biosubstance which allows it to passively concentrate and binds contaminants onto its cellular structure (Sameera et al., 2011).

Biosorption is a physicochemical and metabolically independent process based on a variety of mechanism which is rather complex. Fig. 9.1 shows different types of mechanisms involved in biosorption process. These mechanisms are mainly based on sorbate/sorbent or solute/solvent interactions. In other words, the process consists of the binding of sorbate onto the biosorbent (Farooq et al., 2010; Fomina and Gadd, 2014; Gadd, 2009; Mosbah and Sahmoune, 2013).

Microprecipitation is a type of precipitation which occurs locally on or within the pores of the biosorbent depending on the conditions (Robalds

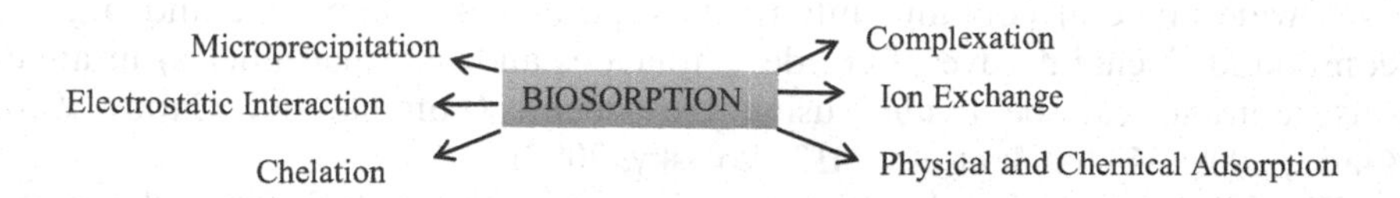

FIGURE 9.1 The mechanisms involved in biosorption process

et al., 2016). Precipitation may be either dependent on the cellular metabolism or not.

The removal of the pollutant from aqueous media may also be realized by the formation of the complex on the cell surface after the interaction between the metal and the active groups (Kushwah et al., 2015). Complex formation of ions with organic molecules involves ligand centers in the organic and this may be electrostatic or covalent (Tsezos et al., 2006).

Physical adsorption takes place by van der Waals' forces (Kushwah et al., 2015) Organic molecules which have functional group more than one with donor electron pairs can simultaneously give these to a metal. This ends in the formation of a ring structure called chelation. The ring structure which has already been formed prevents the mineral from entering into undesired reactions. Chopra and Pathak concluded that cheated compounds are more stable than complexes involving monodentate ligands since the chelating agent bonds to a metal ion in more than one place simultaneously (Chopra and Pathak, 2010; Tsezos et al., 2006).

The metal ions in the dilute solutions are exchanged with ions which are staying on the exchange resin by electrostatic forces. Ion exchange is a reversible chemical reaction. In this reaction, an ion in a solution is exchanged with a similarly charged ion attached to an immobile solid particle (Chopra and Pathak, 2010). As mentioned previously, various processes are known to contribute to biosorption: surface complexation and precipitation, physical adsorption or ion exchange. Among them the ion exchange is accepted to be the principle mechanism of biosorption (Farooq et al., 2010). Since the mechanism depends on ion exchange, protons compete with metal cations for the binding sites and hence pH is mentioned to be one of the main operating parameters, which influences the process (Chojnacka, 2009; Farooq et al., 2010; Schiewer and Volesky, 2000).

The key points controlling and affecting the efficiency of these mechanisms are (Kanamarlapudi et al., 2018; Park et al., 2010) as follows:

- The chemical and coordination characteristics of pollutant
- Properties of the biosorbent
- Type and availability of the binding site
- The process parameters (pH and temperature)

9.3.4 Biosorbents

Biosorbents are biological materials that are used as chelating and complexing sorbents for the treatment of pollutants.

A wide range of contaminants (metals, particulates, inorganic and organic compounds including dyes, fluoride, phthalates and pharmaceuticals) in aqueous solutions can be treated using biosorbents (Fomina and Gadd, 2014; Gadd, 2009; Michalak et al., 2013; Volesky, 2007).

The biosorbents and their derivatives contain different functional groups. They involve the binding of ions by physically or chemically. Electrostatic interaction or van der Waals forces are considered in physical binding and in chemical binding the displacement of cations (ion exchange) or protons are considered. After binding, the biosorbents involve in chelation, reduction, precipitation and complexation. The biosorbents are usually much more selective than conventional ion-exchange resins and especially it is known that the commercial activated carbons can treat dye containing wastewaters very efficiently.

In general, an ideal biosorbent should possess properties such as availability, nonhazardous, good metal binding capacity, large-scale usability, sustainable and reusable (Bilal et al., 2018; Wang and Chen, 2009).

There is a wide range of natural materials that has been used as biosorbents for the desired pollutant treatment.

Biosorbents for the removal of pollutants may be investigated under the following categories (Park et al., 2010):

Bacteria (Gram-positive, Gram-negative bacteria and cyanobacteria)

Fungi (mold mushrooms and yeast)

Algae (microalgae and macroalgae)

Wastes of industry (fermentation wastes, food/beverage wastes, activated sludges and anaerobic sludges)

Wastes of agricultural industry (fruit/vegetable wastes, rice straws, wheat bran and soybean hulls)

Natural residues (plant residues, sawdust, tree barks and weeds)

Others

9.3.5 Factors Affecting Biosorption

The mechanism of biosorption is affected by the biomass itself, the characteristics of its surface, and the process parameters such as pH, temperature, initial concentration of pollutant, contact time and biosorbent amount (Bilal et al., 2018; Fan, 2013).

9.3.5.1 pH

Solution pH is the main parameter in the biosorption process. The solution chemistry of the pollutants is directly affected by pH affects. Also, the activity of functional groups in the biosorbents and competition with the present ions in solution are influenced by the solution pH (Vijayaraghavan and Yun, 2008).

It determines the speciation and solubility of ions and the charges on the sorption sites on biosorbent surface. The solution pH strongly affects the uptake capacity of biomass. Optimum pH takes different values for different biosorption systems.

At low pH, in the medium excess H^+ and H_3O^+ ions are present and hence carboxylic groups protonate, and the repulsive forces of these protonated groups with positively charged heavy metal ions are responsible for the lower biosorption capacity at low pH. If pH is increased, the functional groups (hydroxyl groups, amine and carboxyl) are deprotonated by raising the electrostatic attraction with ions due to a negative charge. If the increase in pH is high, hydroxide anionic complexes are formed and precipitation may occur leading a low biosorption capacity (Bilal et al., 2018).

In other words, it can be concluded that increasing pH increases the removal of cationic metals but reduces that of anionic metals however; too much increase in pH can result in precipitation, which should be better prevented (Fomina and Gadd, 2014).

9.3.5.2 Temperature

Temperature variations affect the thermodynamic parameters and this changes the sorption capacity. In sorption processes which take place endothermically, if the temperature is increased, an increase in biosorption efficiency is observed; in contrast to this, an increase in temperature lowers the biosorption efficiency in the case of exothermic sorption processes. A very high temperature can also denature the biomass structure (Bilal et al., 2018). In general, biosorption is an exothermic process, and hence the biosorption efficiency usually increases with decreasing temperature.

9.3.5.3 Pollutant Concentration

When pollutant concentration is increased, the amount of biosorbed pollutant per unit weight of biosorbent increases. However, this possibly decreases the removal efficiency (Fomina and Gadd, 2014).

9.3.5.4 Biosorbent Amount

Increasing the biosorbent amount increases the available adsorption sites and hence the biosorption capacity increases. However, using excess amount of biomass can lead to interference between binding sites and consequently, the removal efficiency may be affected.

9.3.5.5 Contact Time

There is an optimum contact time for the biosorption processes. If the contact time is increased up to this time, biosorption efficiency will increase; afterward the removal may stay constant (Bilal et al., 2018).

9.3.6 Modeling of Biosorption: Isotherm and Kinetic Models

9.3.6.1 Biosorption Isotherms

Equilibrium isotherm equations are used to model the experimental biosorption data and investigate the parameters gained from the different models.

Adsorption isotherms are very important for the optimization of the biosorbent usage because they describe the relation between sorbate and biosorbent. Thus, analysis of equilibrium data is important for practical design and operation of biosorption–adsorption systems.

Experimental data are fitted by applying well known and most commonly used isotherm models, such as Langmuir, Freundlich, Temkin, Sips and Redlich–Peterson (Haddad et al., 2014; Kılıç et al., 2014).

9.3.6.1.1 Langmuir Isotherm Model

Langmuir model assumes that biosorption occurs as monolayer on homogeneous surface having a finite number of vacant sites energetically equivalent to each other (Ahsan et al., 2018; Langmuir, 1918).

Langmuir isotherm defines the surface coverage by dynamic equilibrium, balancing the relative rates of adsorption and desorption.

It is based on the following four assumptions (Abbas et al., 2014; Weber and Chakravorti, 1974):

- Adsorbent surface is uniform.
- There is no interaction between molecules adsorbed on neighboring sites.
- Adsorption takes place through the same mechanism.
- Adsorption of the molecules are at the define sites on the surface.

Basically, the Langmuir isotherm equation has a hyperbolic form:

$$q_e = q_{th}^{e} \frac{K_{\text{eq}} C_e}{1 + K_{\text{eq}} C_e}$$

where

q_e: the adsorption capacity by weight at equilibrium.

q_{th}^{e}: the theoretical maximum adsorption capacity by weight.

K_{eq}: the equilibrium constant of adsorption reaction.

C_{e}: concentration of adsorbate at equilibrium.

This isotherm model has been generally applied in equilibrium study of biosorption; however, Liu and Liu mentioned that the Langmuir isotherm gives no detail aspects of biosorption mechanism (Liu and Liu, 2008).

9.3.6.1.2 The Freundlich Isotherm Model

The Freundlich isotherm model is generally used for multilayer biosorption and non-ideal biosorption on heterogeneous surfaces. This isotherm tells about the surface heterogeneity and the exponential distribution of active sites and their energies. It is defined by the following equation (Ahsan et al., 2018; Ayawei et al., 2015,).

$$q_{e=} k_{\text{F}} C_e^{\frac{1}{n_{\text{F}}}}$$

where k_{F} and n_{F} are Freundlich constants.

Because the Freundlich isotherm equation is exponential, Liu and Liu suggested that this model can only be applied in the concentration ranges of low to intermediate (Liu and Liu, 2008).

9.3.6.1.3 Sips Isotherm Model

The Sips model is an additional empirical model, which holds the characteristics of both Langmuir and Freundlich isotherm models (Ebrahimian et al., 2014).

$$q_e = q_{th}^{e} \frac{K_{eq} C_e^{n_s}}{1 + K_{eq} C_e^{n_s}}$$

where n_s is the Sips constant.

When n_s equals unity, the Sips isotherm tends toward the Langmuir isotherm, indicating the homogeneous adsorption. If the value of n_s is different from the unity, it will indicate a heterogeneous surface (Nethaji et al., 2013).

9.3.6.1.4 Redlich–Peterson Isotherm

The Redlich–Peterson isotherm has the properties of both Langmuir and Freundlich isotherms. It can be represented as (Repo et al., 2011)

$$q_e = \frac{q_m (K_{RP} C_e)}{1 + (K_{RP} C_e)^{n_{RP}}}$$

K_{RP} (L/mg) and n_{RP} are Redlich–Peterson constants.

Nethaji et al. mentioned that Redlich–Peterson isotherm parameters are similar to that of Sips isotherm parameters. (Nethaji et al., 2013).

9.3.6.1.5 Temkin Isotherm Model

Temkin isotherm model assumes that the biosorption heat drops linearly with the increase of the biosorbent amount or coverage of the adsorbent. The model comprises a factor that shows the interactions between the biosorbent and the adsorbed particles (Ahsan et al., 2018; Liu and Liu, 2008).

The model can be represented as

$$q_e = \frac{RT}{b_t} \ln(a_t C_e)$$

R: gas constant.
T: absolute temperature in Kelvin.
b_t: constant related to the heat of adsorption.
a_t: Temkin isotherm constant.

9.3.6.2 Biosorption Kinetics

The rate and mechanism of biosorption is controlled by various factors like physical and/or chemical features of biosorbents and also mass transfer process. Various kinetic models have been investigated to explain the mechanism of biosorption processes in the literature. These proposed kinetic models are helpful designing and optimizing the treatment processes.

Biosorption is a multi-step process. It is realized by four consecutive elementary steps (Abbas et al., 2014; Khraisheh et al., 2002).

1. Transfer of solute from the bulk of solution to the liquid film.
2. Transport of the solute from the boundary liquid film to the surface of the particles (external diffusion).
3. Transfer of solute from the surface to the internal active binding sites (internal diffusion).
4. Interaction of the solute with the active binding sites.

Generally, the time-dependent experimental batch adsorption data are used for the kinetic modeling. To determine the best fitting kinetic rate generally, linear regression analysis is to be used.

The most common kinetic models used to describe the mechanism of the biosorption process are summarized below (Samuel et al., 2015):

9.3.6.2.1 Reaction-Based Models

Pseudo-First Order

The pseudo-first-order rate expression based the most common used rate equation and is known as the Lagergren rate equation. The pseudo-first-order rate expression is related to the capacity of solid.

It is represented as Lagergren and Svenska (1898):

$$\frac{dq}{dt} = k_1(q_e - q)$$

where q_e (mg/g) and q (mg/g) are the adsorption capacity at equilibrium and at time t, respectively.

k_1 (min^{-1}) is the rate constant of the pseudo-first-order adsorption reaction.

Pseudo-Second Order

The pseudo-second-order kinetic expression is also focused on the solid-phase sorption capacity. It is assumed that there is a monolayer of adsorbate on the surface of adsorbent. The adsorption process takes place only on specific sites. It involves no interactions between the adsorbed pollutants (Abbas et al., 2014; Ho et al., 2000).

The rate of adsorption is nearly negligible when compared with the initial rate of adsorption.

It is represented as

$$\frac{dq}{dt} = k_2(q_e - q)^2$$

where q_e (mg/g) and q (mg/g) are the adsorption capacity at equilibrium at time t, respectively, and k_2 ((g/mg)min^{-1}) is the rate constant of the pseudo-second-order adsorption reaction.

Elovich

Elovich model equation is another model to describe activated chemical adsorption. It can be expressed as

$$\frac{dq}{dt} = a\exp(-bq)$$

where a and b are constants. Wu et al. mentioned that the constant a can be taken as the initial rate because the ratio: dq/dt approaches a when q approaches zero (Wu et al., 2009).

9.3.6.2.2 Diffusion-Based Models

Boyd

This kinetic model is related to the diffusion through the boundary liquid film. In this model, adsorption kinetics is considered as a chemical phenomenon. The simplified form of the rate expression can be written as

$$\frac{q}{q_e} = 1 - \frac{6}{\pi^2}\exp(-b_B t)$$

where b_B is a constant.

Intraparticle Diffusion

Intraparticle diffusion model is the most commonly used model for determination of the mechanism involved in the adsorption process (Weber and Morris, 1963). The rate equation can be expressed as

$$q = K_d t^{1/2} + I$$

where K_d (mg/g min$^{-1/2}$) is a measure of diffusion coefficient the intraparticle diffusion rate constant and I (mg g^{-1}) is intraparticle diffusion constant.

If there is intraparticle diffusion, then q versus $t^{1/2}$ plot will be linear passing through the origin. In such case, the intraparticle diffusion will be the only rate limiting parameter controlling the process. In other words, the intraparticle diffusion or pore diffusion is not the sole rate limiting step, adsorption kinetics may be governed by film diffusion and intraparticle diffusion at the same time.

Abdelwahab mentioned that the values of I (intercept) may give an idea for the boundary layer thickness. It is mentioned that the larger the value of I, the greater will be the boundary layer effect (Abdelwahab, 2007).

9.3.6.3 Estimation of Thermodynamic Parameters

For thermodynamic studies, especially Gibb's free energy, enthalpy and entropy are the main parameters that have a significant role to predict the spontaneousness and heat change involved in the adsorption process.

The changes in free energy ($\Delta G°$), enthalpy ($\Delta H°$) and entropy ($\Delta S°$) of the processes can be determined by using the following equations:

$$\Delta G^o = -RT \ln K_c$$

$$\Delta G^o = \Delta H^o - T\Delta S^o$$

$$K_c = \frac{q_e}{C_e}$$

where

$\Delta G°$: standard change Gibbs free energy (kJ mol^{-1})

$\Delta H°$: standard change enthalpy (J mol^{-1})

$\Delta S°$: standard change entropy (J $mol^{-1}K^{-1}$)

R: universal gas constant (J $mol^{-1}K^{-1}$)

T: the absolute temperature (K)

K_c: the equilibrium constant.

The negative and positive values of $\Delta H°$ show the adsorption processes if exothermic or endothermic, respectively. The negative $\Delta G°$ values indicate that the biosorption is spontaneous. The higher negative $\Delta G°$ values indicate a more energetically, highly favorable biosorption processes.

9.4 BIOSORPTION OF DYES IN LITERATURE

Using biomass in treatment processes has been in practice for a while, scientists and engineers are hoping this phenomenon will allow an economical alternative for removing pollutants especially dyes from wastewater.

In literature, there are many studies that are available discussing the use of biosorption of water and wastewater containing dyes. Below some studies in literature covering the biosorption of dye containing wastewater with different types of biosorbents are summarized:

9.4.1 Bacteria as Biosorbents

Biosorption of solutions that mainly contain dyes has been reported using various bacterial strains, such as *Acidithiobacillus thiooxidans, Bacillus catenulatus JB-022* and *Nostoc linckia.*

The efficiency of the bacterial strain *Acidithiobacillus thiooxidans* to treat sulfur blue 15 (SB15) dye from aqueous solutions was investigated by Nguyen et al. They found out that this type of bacterium can be an effective biosorbent. Langmuir, Freundlich and Dubinin-Radushkevich models were tested to investigate the biosorption mechanisms. The biosorption isotherms were described by the Langmuir equation better than by the Freundlich or Dubinin-Radushkevich models. The process was determined as following the pseudo-second-order kinetics. The dye removal and decolorization were 87.5% and 91.4%, respectively, by the biosorption process. Biodegradation was suggested as a subsequent process for the dye which is remained. Under the optimum conditions determined approximately 50% of SB15 was removed after 4 days of biodegradation (Nguyen et al., 2016).

Biosorption of cationic basic dye and cadmium by the biosorbent *Bacillus catenulatus JB-022* strain was investigated by Kim et al. They aimed to follow the bacterial strains isolated from soils and polluted pond for biosorption of both cationic dye and cadmium. *Bacillus catenulatus JB-022* strain removed 58% and 66% of cationic basic blue 3 (BB3) and cadmium (Cd(II)), respectively. The biosorption equilibrium data were well explained by the Langmuir adsorption isotherm, and the kinetic studies showed that the biosorption process followed the second-order model. They concluded that *Bacillus catenulatus JB-022* can be proposed as an perfect biosorbent with potentially important applications in treatment of cationic pollutants in aqueous solutions (Kim et al., 2015).

Biosorption of reactive dye by waste biomass of *Nostoc linckia* was realized by Mona and his group. They used a laboratory sized hydrogen bioreactor after its immobilization as a biosorbent for treatment of the reactive dye RR 198. Kinetic behavior of the removal process was examined in batch mode. For optimization of the interacting parameters such as pH, temperature and initial dye concentration, response surface methodology was used. FT-IR analysis was performed to analyze the surface features of the biosorbent involved in the biosorption process. The maximum capacity of adsorption for the immobilized biomass was 3.5 mg/g at pH 2.0, initial concentration of 100 mg/L and 35°C, when 94% of the dye was removed (Mona et al., 2011).

9.4.2 Fungi as Biosorbents

Fungal biomass is nonhazardous to people and animals and it may be produced inexpensively by using simple fermentation methods. Fungal biomass is widely recommended and used in the biosorption of dyes (Aksu and Karabayır, 2008).

Almeida and Corso evaluated the biodegradation and biosorption of the azo dye Procion RedMX-5B in solutions with the filamentous fungi *Aspergillus niger* and *Aspergillus terreus*. The decolorization and toxicity tests were realized. They concluded that the seeds from *L. sativa* and the larvae of *A. salina* are very good indicators of toxicity. They were also found to be sensitive to changes in the toxicity of the solutions after the treatments. The fungal biomass showed a considerable affinity for Procion Red MX-5B, having ~100% decolorization (Almeida and Corso, 2014).

Aksu and Karabayır compared the biosorption properties of different kinds of fungi for the treatment of Gryfalan Black RL metal-complex dye. Three kinds of filamentous fungi (*Rhizopus arrhizus, Trametes versicolor* and

Aspergillus niger) were tested to determine the activities in adsorption of metal-complex dye as a function of pH, temperature and dye concentration. Among the three of them, they mentioned that *R. arrhizus* was the most effective biosorbent. The equilibrium data were described by Langmuir model for the each fungus system. Kinetic studies showed that as well as adsorption kinetics and internal diffusion had also an important role on controlling the overall adsorption rate for all cases. They also performed thermodynamic analysis and this analysis indicated that biosorption by fungi *A. niger* was endothermic while the other two were exothermic (Aksu and Karabayır, 2008).

9.4.3 Algae as Biosorbents

The use of dead algal biomass in the form of microstructure and nanostructured particles for dye removal is also studied in the literature.

S. platensis is a type of blue-green algae. It is known to be an alternative source of protein for food. Dotto et al. investigated the potential of micro- and nanoparticles of *S. platensis* in the treatment of two synthetic dyes by biosorption. They characterized particles by scanning electron microscopy, specific surface area, size distribution, pore volume and average pore radius, infrared analysis and X-ray diffraction. The effects of particle size and biosorbent amount on biosorption capacity and percentage dye removal were investigated. The kinetic models pseudo-first- and second-order and Elovich were used to describe the process kinetics. They concluded that the pseudo-first-order model fit well in the biosorption of both dyes, and Elovich model was more appropriate to the biosorption onto nanoparticles (Dotto et al., 2012).

The usage of microalgae has shown good results for dye removal by Rosa et al. too. They used microalgal species *Chlorella pyrenoidosa* as a biosorbent for the treatment of rhodamine B. Biosorption kinetics was studied using pseudo-first-order, pseudo-second order and Elovich models. Equilibrium isotherms were analyzed by the Langmuir, Freundlich, Sips and Temkin models. By the experimental results, they concluded that when the initial dye concentration was 100 mg/L, the microalgae biomass presented the highest biosorption capacity, at pH value of 8.0 and at a temperature of 25°C. The pseudo-second-order kinetic model better fitted the experimental data. Among the isotherm models they tested, they mentioned that the Sips model was the best fit to the experimental results (Rosa et al., 2018).

The biosorption efficiency and capacity of methylene blue (MB) dye from aqueous solution by untreated and pretreated (physical and chemical) brown (*Nizamuddinia zanardinii*), red (*Gracilaria parvispora*) and green (*Ulva fasciata*) algae were investigated by Daneshvar et al. The effect of different physical treatments such as particle size, wet or dry heating, boiling water and chemical treatments which are basic, acidic, salts, alcohols and organic agents were investigated to have the modified forms of different macroalgae (brown, red and green). The biosorption features of the treated biomasses were examined as a function of initial dye concentration and reaction time. Pretreatment with sodium chloride enhanced the dye removal by macroalgae. Equilibrium data were best fitted by Langmuir isotherm model. The biosorption kinetics was described well by the pseudo–second-order kinetic model (Daneshvar et al., 2017).

9.4.4 Agricultural Wastes as Biosorbents

Dye containing solutions are toxic in nature change the physicochemical properties of receiving bodies. Using agricultural wastes as alternatives for activated carbons has gain considerable interest. As a result, cheap and environment friendly adsorbents from agricultural wastes are investigated to treat dye containing wastewater.

Among the organic wastes that are generated by industrial activities, the waste formed in the production of wine can be a promising biosorbent. Oliveira et al. studied on the potential use of grape pomace in the treatment of the Brown KROM KGT dye in an aqueous solution. The characterization of biosorbent was realized. They also determined the optimum operating conditions. In addition to these, a kinetic, thermodynamic and equilibrium studies were performed. The kinetic model of the pseudo-first-order best represented the experimental data. The adsorption equilibrium data showed a process of monolayers, according to the Langmuir model. The thermodynamic data resulted in a thermodynamically favorable and exothermic process. If the obtained results are investigated, it can be concluded that the grape pomace is a good candidate to be used as biosorbent in the treatment wastewaters containing dyes (Oliveira et al., 2018).

It is well known that the wastes from different sources may be used as effective materials in bioremediation of wastewater. Deniz and Kepekçi studied the possibility of using pistachio shell for the treatment of a reactive-azo present in water. They carried out biosorption studies batch wise and they found out that the biosorption process was fast and the yield of biosorption decreased when the dye concentration was increased. Freundlich, Langmuir, Sips and Dubinin-Radushkevich isotherm models were investigated, and the results showed that the biosorption data were best represented by the pseudo-second-order and Sips models. They calculated the standard Gibbs free energy change as −5.184 kJ mol^{-1}, indicating that the physical forces were involved in the spontaneous biosorption of dye onto the biosorbent. They concluded that the pistachio can be a good and cheap material for the biosorption of such hazardous dyes from water (Deniz and Kepekci, 2016).

Temesgen and his friends investigated the biosorption of reactive red dye (RRD) from wastewater using banana and orange peels as a cheap bioadsorbent. Biosorbents were characterized by surface area and point zero charge. The effect of operating conditions such as pH, temperature, contact time, adsorbent amount and initial dye concentration was studied in a batch adsorption experiment. The result showed maximum removal efficiency of 89.41% and 70.25% at pH of 4, initial dye concentration of 25 mg/L, adsorbent dosage 1 g/100 mL and temperature of 30°C on the activate surface of orange and banana peels, respectively. The adsorption isotherm, adsorption kinetics and adsorption thermodynamics of the activated banana and orange peels were also studied. The gained results were well presented with Langmuir and Freundlich isotherm models and adsorption process fitted well the pseudo–second-order model for both adsorbents. Overall both peels have good potential to treat the dye in wastewater. They recommended these biosorbents to be applied in industrial scale to optimize the treatment cost (Temesgen et al., 2018).

Deniz studied the methyl orange dye treatment by almond shell wastes and tested the biosorption efficiency. The pseudo-second-order kinetic model described the dye biosorption process. The equilibrium data fit well with the Langmuir isotherm model by concluding the biosorption was the monolayer coverage of dye on the biosorbent and the homogeneity of active sites. Besides, to determine the nature of biosorption process the standard, Gibbs free energy change was also calculated. These results showed that using almond shell residues as dye biosorbent could be a good choice for both economical and environmental point of view (Deniz 2013).

9.5 CONCLUDING REMARKS

The effluents containing dyes are one of the major threats to human and environment health. Some of the traditional methods that are presently applied are not sufficient for the total removal of these persistent organic pollutants. Recently, much care has been given for the treatment of various organic pollutants by biosorption.

Biosorption is found to be a potential alternative technique for treatment of dye containing effluents because it has considerable advantages, especially from economical and environmental aspects.

The process has the potential to find environmental applications but also in the manufacture of new high-value products. The use of low-cost biosorbents for the treatment of aqueous effluents to remove dyes and various organics has been recognized as the main innovative development in the area of biosorption for the last few years.

However, it can be concluded that the improvement of biosorption process requires further investigation in the direction of modeling, immobilization of biosorbents and regeneration and of treating the industrial effluents.

REFERENCES

Abbas, S. H., Ismail, I. M., Mostafa, T. M., Sulaymon, A. H. 2014. Biosorption of heavy metals: A review. Journal of Chemical Science and Technology, 3(4), 74–102.

Abdelwahab, O. 2007. Kinetic and isotherm studies of copper (II) removal from wastewater using various adsorbents. 33, 125–143.

Adegoke, K. A., Bello, O. S. 2015. Dye sequestration using agricultural wastes as adsorbents. Water Resources and Industry, 12, 8–24.

Ahalya, N., Ramachandra, T. V., Kanamadi, R. D. 2003. Biosorption of heavy metals. Journal of Chemistry and Environment, 7, 71–79.

Ahsan, M. A., Islam, M. T., Imam, M. A., Hyder, A. G., Jabbari, V., Dominguez, N., Noveron, J. C. 2018. Biosorption of bisphenol A and sulfamethoxazole from water using sulfonated coffee waste: Isotherm, kinetic and thermodynamic studies. Journal of Environmental Chemical Engineering, 6, 6602–6611.

Aksu, Z., Karabayır, G. 2008. Comparison of biosorption properties of different kinds of fungi for the removal of Gryfalan Black RL metal-complex dye. Bioresource Technology, 99(16), 7730–7741.

Almeida, E. J. R., Corso, C. R. 2014. Comparative study of toxicity of azo dye Procion Red MX-5B following biosorption and biodegradation treatments with the fungi Aspergillus niger and Aspergillus terreus. Chemosphere, 112, 317–322.

Bhatia, D., Sharma, N. R., Singh, J., Kanwar, R. S. 2017. Biological methods for textile dye removal from wastewater: A review. Critical Reviews in Environmental Science and Technology, 47(19), 1836–1876.

Bilal, M., Rasheed, T., Sosa-Hernández, J., Raza, A., Nabeel, F., Iqbal, H. 2018. Biosorption: An interplay between marine algae and potentially toxic elements—A review. Marine Drugs, 16(2), 65.

Chojnacka, A. 2009. Biosorption and bioaccumulation in practice. Nova Science Publishers, New York

Chopra, A. K., Pathak, C. 2010. Biosorption technology for removal of metallic pollutants—An overview. Journal of Applied and Natural Science, 2(2), 318–329.

Daneshvar, E., Vazirzadeh, A., Niazi, A., Sillanpää, M., Bhatnagar, A. 2017. A comparative study of methylene blue biosorption using different modified brown, red and green macroalgae—Effect of pretreatment. Chemical Engineering Journal, 307, 435–446.

Deniz, F. 2013. Dye removal by almond shell residues: Studies on biosorption performance and process design 2013. Materials Science and Engineering: C, 33(5), 2821–2826.

Deniz, F., Kepekci, R. A. 2016. Dye biosorption onto pistachio by-product: A green environmental engineering approach. Journal of Molecular Liquids, 219, 194–200.

Dotto, G. L., Cadaval, T. R. S., Pinto, L. A. A. 2012. Use of Spirulina platensis micro and nanoparticles for the removal synthetic dyes from aqueous solutions by biosorption. Process Biochemistry, 47(9), 1335–1343.

Ebrahimian, A., Saberikhah, E., Emami, M. S., Sotudeh, M. 2014. Study of biosorption parameters: Isotherm, kinetics and thermodynamics of basic blue 9 biosorption onto Foumanat tea waste. Cellulose Chemistry Technology, 48, 735–743.

Fan, J., 2013. Application of Cupriavidus metallidurans and ochrobactrum intermedium for copper and chromium biosorption. Doctoral dissertation, Master of Science Thesis, Faculty of the Department of Environmental Engineering University of Houston.

Farooq, U., Kozinski, J. A., Khan, M. A., Athar, M. 2010. Biosorption of heavy metal ions using wheat based biosorbents—A review of the recent literature. Bioresource Technology, 101(14), 5043–5053.

Flouty, R., Estephane, G. 2012. Bioaccumulation and biosorption of copper and lead by a unicellular algae Chlamydomonas reinhardtii in single and binary metal systems: A comparative study. Journal of Environmental Management, 111, 106–114.

Fomina, M., Gadd, G. M. 2014. Biosorption: Current perspectives on concept, definition and application. Bioresource Technology, 160, 3–14.

Fu, Y., Viraraghavan, T. 2001. Fungal decolorization of dye wastewaters: A review. Bioresource Technology, 79, 251–262.

Gadd, G. M. 2009. Biosorption: Critical review of scientific rationale, environmental importance and significance for pollution treatment. Journal of Chemical Technology & Biotechnology: International Research in Process, Environmental & Clean Technology, 84(1), 13–28.

Gao, B. Y., Yue, Q. Y., Wang, Y., Zhou, W. Z. 2007. Color removal from dye-containing wastewater by magnesium chloride. Journal of Environmental Management, 82(2), 167–172.

Ghuge, S. P., Saroha, A. K. 2018. Catalytic ozonation of dye industry effluent using mesoporous bimetallic Ru-Cu/SBA-15 catalyst. Process Safety and Environmental Protection, 118, 125–132.

Haddad, E. M., Regti, A., Slimani, R., Lazar, S. 2014. Assessment of the biosorption kinetic and thermodynamic for the removal of safranin dye from aqueous solutions using calcined mussel shells. Journal of Industrial and Engineering Chemistry, 20(2), 717–724.

Hassan, M. M., Carr, C. M. 2018. A critical review on recent advancements of the removal of reactive dyes from dyehouse effluent by ion-exchange adsorbents. Chemosphere, 209, 201–219.

Ho, Y. S., Ng, J. C. Y., McKay, G. 2000. Kinetics of pollutant sorption by biosorbents. Separation and Purification Methods, 29(2), 189–232.

Kadirvelu, K., Kavipriya, M., Karthika, C., Radhika, M., Vennilamani, N., Pattabhi, S. 2003. Utilization of various agricultural wastes for activated carbon preparation and application for the removal of dyes and metal ions from aqueous solutions. Bioresource Technology, 87(1), 129–132.

Kanamarlapudi, S. L. R. K., Chintalpudi, V. K., Muddada, S. 2018. Application of Biosorption for Removal of Heavy Metals from Wastewater. In Biosorption. IntechOpen.

Katheresan, V., Kansedo, J., Lau, S. Y. 2018. Efficiency of various recent wastewater dye removal methods: A review. Journal of Environmental Chemical Engineering, 6, 4676–4697.

Kharub, M. 2012. Use of various, technologies, methods and adsorbents for the removal of dye. Journal of Environmental Research and Development, 6(3A).

Khraisheh, M. M., Al-Degs, Y. S., Allen, S. J., Ahmad, M. N. 2002. Elucidation of controlling steps of reactive dye adsorption on activated carbon. Industrial & Engineering Chemistry Research, 41(6), 1651–1657.

Kılıç, Z., Atakol, O., Aras, S., Cansaran-Duman, D., Çelikkol, P., Emregul, E. 2014. Evaluation of different isotherm models, kinetic, thermodynamic, and copper biosorption efficiency of Lobaria pulmonaria (L.) Hoffm. Journal of the Air & Waste Management Association, 64(1), 115–123.

Kim, S. Y., Jin, M. R., Chung, C. H., Yun, Y. S., Jahng, K. Y., Yu, K. Y. 2015. Biosorption of cationic basic dye and cadmium by the novel biosorbent Bacillus catenulatus JB-022 strain. Journal of Bioscience and Bioengineering, 119(4), 433–439.

Kono, H., Kusumoto, R. 2015. Removal of anionic dyes in aqueous solution by flocculation with cellulose ampholytes. Journal of Water Process Engineering, 7, 83–93.

Kushwah, A., Srivastav, J. K., Palsania, J. 2015. Biosorption of heavy metal. A review. European Journal of Biotechnology and Bioscience, 3, 51–55.

Lagergren, S., Svenska, B. K. 1898. On the theory of so-called adsorption of materials. Royal Sweden Academy Science Document, Band, 24, 1–13.

Langmuir, I. 1918. The adsorption of gases on plane surfaces of glass, mica and platinum. Journal of the American Chemical Society, 40(9), 1361–1403.

Lipatova, I. M., Makarova, L. I., Yusova, A. A. 2018. Adsorption removal of anionic dyes from aqueous solutions by chitosan nanoparticles deposited on the fibrous carrier. Chemosphere, 212, 1155–1162.

Liu, Y., Liu, Y. J. 2008. Biosorption isotherms, kinetics and thermodynamics. Separation and Purification Technology, 61(3), 229–242.

Malik, A. 2004. Metal bioremediation through growing cells. Environment International, 30(2), 261–278.

Melo, R. P. F., Neto, E. B., Nunes, S. K. S., Dantas, T. C., Neto, A. D. 2018. Removal of reactive Blue 14 dye using micellar solubilization followed by ionic flocculation of surfactants. Separation and Purification Technology, 191, 161–166.

Michalak, I., Chojnacka, K., Witek-Krowiak, A. 2013. State of the art for the biosorption process—A review. Applied Biochemistry and Biotechnology, 170(6), 1389–1416.

Mona, S., Kaushik, A., Kaushik, C. P. 2011. Biosorption of reactive dye by waste biomass of Nostoc linckia. Ecological Engineering, 37(10), 1589–1594.

Mosbah, R., Sahmoune, M. N. 2013. Biosorption of heavy metals by Streptomyces species—An overview. Central European Journal of Chemistry, 11(9), 1412–1422.

Nethaji, S., Sivasamy, A., Mandal, A. B. 2013. Adsorption isotherms, kinetics and mechanism for the adsorption of cationic and anionic dyes onto carbonaceous particles prepared from Juglans regia shell biomass. International Journal of Environmental Science and Technology, 10(2), 231–242.

Nguyen, T. A., Fu, C. C., Juang, R. S. 2016. Biosorption and biodegradation of a sulfur dye in high-strength dyeing wastewater by Acidithiobacillus thiooxidans. Journal of Environmental Management, 182, 265–271.

Nidheesh, P. V., Zhou, M., Oturan, M. A. 2018. An overview on the removal of synthetic dyes from water by electrochemical advanced oxidation processes. Chemosphere, 197, 210–227.

Oliveira, A. P., Módenes, A. N., Bragião, M. E., Hinterholz, C. L., Trigueros, D. E., Isabella, G. D. O. 2018. Use of grape pomace as a biosorbent for the removal of the Brown KROM KGT dye. Bioresource Technology Reports, 2, 92–99.

Park, D., Yun, Y. S., Park, J. M. 2010. The past, present, and future trends of biosorption. Biotechnology and Bioprocess Engineering, 15(1), 86–102.

Repo, E., Malinen, L., Koivula, R., Harjula, R., Sillanpää, M. 2011. Capture of Co (II) from its aqueous EDTA-chelate by DTPA-modified silica gel and chitosan. Journal of Hazardous Materials, 187(1-3), 122–132.

Robalds, A., Naja, G. M., Klavins, M. 2016. Highlighting inconsistencies regarding metal biosorption. Journal of Hazardous Materials, 304, 553–556.

Rosa, A. L. D., Carissimi, E., Dotto, G. L., Sander, H., Feris, L. A. 2018. Biosorption of rhodamine B dye from dyeing stones effluents using the green microalgae Chlorella pyrenoidosa. Journal of Cleaner Production, 198, 1302–1310.

Sameera, V., Naga Deepthi, C. H., SrinuBabu, G., Ravi Teja, Y. 2011. Role of biosorption in environmental clean-up. Microbial Biochem Technology, Review-1.

Samuel, M. S., Abigail, M. E. A., Chidambaram, R. 2015. Isotherm modelling, kinetic study and optimization of batch parameters using response surface methodology for effective removal of Cr(VI) using fungal biomass. PLoS One, 10(3).

Tan, B. H., Teng, T. T., Omar, A. M. 2000. Removal of dyes and industrial dye wastes by magnesium chloride. Water Research, 34(2), 597–601.

Temesgen, F., Gabbiye, N., Sahu, O. 2018. Biosorption of reactive red dye (RRD) on activated surface of banana and orange peels: Economical alternative for textile effluent. Surfaces and Interfaces, 12, 151–159.

Tsezos, M., Remoundaki, E., Hatzikioseyian, A. 2006. Biosorption-principles and applications for metal immobilization from waste-water streams. In Proceedings of EU-Asia workshop on clean production and nanotechnologies, Seoul, 23–33.

Vijayaraghavan, K., Yun, Y.-S. 2008. Bacterial biosorbents and biosorption. Biotechnology Advances, 26, 266–291.

Volesky, B., Holan, Z. R. 1995. Biosorption of heavy metals. Biotechnology Progress, 11 (3), 235–250.

Wang, J., Chen, C. 2009. Biosorbents for heavy metals removal and their future. Biotechnology Advances, 27, 195–226.

Weber, T. W., Chakravorti, R. K. 1974. Pore and solid diffusion models for fixed-bed adsorbers. AIChE Journal, 20(2), 228–238.
Weber, W. J., Morris, J. C. 1963. Kinetics of adsorption on carbon from solution. Journal of the Sanitary Engineering Division, 89(2), 31–60.
Wu, F. C., Tseng, R. L., Juang, R. S. 2009. Characteristics of Elovich equation used for the analysis of adsorption kinetics in dye-chitosan systems. Chemical Engineering Journal, 150(2–3), 366–373.
Yang, R., Li, D., Li, A., Yang, H. 2018. Adsorption properties and mechanisms of palygorskite for removal of various ionic dyes from water. Applied Clay Science, 151, 20–28.
Zhu, M. X., Lee, L., Wang, H. H., Wang, Z. 2007. Removal of an anionic dye by adsorption/precipitation processes using alkaline white mud. Journal of Hazardous Materials, 149(3), 735–741.

10 Green Synthesis of Carbonaceous Adsorbents and Their Application for Removal of Polyaromatic Hydrocarbons (PAHs) from Water

S. R. Barman and A. Mukhopadhyay
Department of Environmental Science, University of Calcutta, Kolkata, India
P. Das
Department of Chemical Engineering, Jadavpur University, Kolkata, India

10.1 INTRODUCTION

The biosphere of earth is dependent on few finite resources of which water is an important component. Though the earth surface is composed of approximately 71% water, very little of this is accessible for use. This small amount of water supports all the life form in our planet. However, due to rapid industrialization and growth of civilization, the problem of water pollution has come up as a burning issue. In the last decade, a substantial understanding of water quality and its links to human health has been established. With increasing understanding of the vital health effects associated with contaminated water, the maximum permissible limits (MPL) of various contaminants have been decreased with time. There is no access to safe drinking water in many parts of the world (Ali and Gupta, 2006). It has been earlier established that more than 1.1 billion people do not have adequate supply of drinking water (WHO, 2015) and millions die yearly (approximately 3,900 children per day) due to waterborne diseases through consumption of contaminated water (Vasudevan and Oturan, 2014). Because of mismanagement of water resources, potable water is getting increasingly scarce and costly (Adeleye et al., 2016). Pollution of water can also lead to huge losses incurred by people whose livelihood chiefly depends on fishing due to loss of habitat and contamination. As per international standards, a country is classified as water-stressed or water-scarce if the per capita availability of water is less than 1700 m^3 and 1,000 m^3 per year, respectively (Chakraborti et al., 2009). It is suspected that lot of developing countries, such as India and China, as well as highly developed countries of Europe will face water scarcity by 2025, if proper measures are not taken (Vasudevan and Oturan, 2014). Presently, developing countries have low consumption of water; however, overpopulation and unchecked pollution will soon lead to scarcity of water in these regions as well.

Industrial effluents including dyes, heavy metals, polyaromatic hydrocarbons (PAHs), phenols, pharmaceuticals, organic waste, pesticides and microbial colonies has been reportedly released into water (Manoli and Samara, 1999). Some of these pollutants are toxic to various life forms and can even alter the pH, salinity and other properties of water. This in turn affects terrestrial as well as aquatic life. In the last few decades, micropollutants in water have emerged complicating the already existing problem of water pollution. Micropollutants are a group of pollutants that are generally present in trace amounts (in the range of ng/L to μg/L) (Luo et al., 2014). They are constituted of both natural and synthetic products which include hormones, pharmaceuticals, cosmetic products, chlorinated pesticides, polycyclic aromatic hydrocarbons (PAHs) and polychlorinated biphenyls (PCBs) (Pavoni et al., 2003). The huge diversity of these pollutants and simultaneous occurrence in low concentration makes them difficult to detect and remove from water.

The existing wastewater treatment technologies are not adequately equipped to remove these emerging pollutants resulting in their occurrence and accumulation in water (Bolong et al., 2009; Luo et al., 2014; Priac et al., 2017). Organic micropollutants such as PAHs are hydrophobic, making them difficult to degrade (Barman et al., 2017). These may thus persist in the environment for a long time and bio-magnify in the food chain.

10.1.1 Polyaromatic Hydrocarbons (PAHs) in Water

Amongst the various micropollutants present in water, PAHs have caused increasing concern. Sixteen PAHs have been listed as primary pollutants by the US Environmental Protection Agency (Liu et al., 2001). The 16 priority PAHs (US EPA, 1984; Yan et al., 2004) have been shown in Figure 10.1.

PAHs are organic compounds comprising of multi-rings of benzene. There are around 10,000 known PAHs of which some are found in substantial amount in the environment and some even in food items (ATSDR, 1995). Depending on the structure of PAHs, i.e., the number of fused aromatic rings, these compounds exhibit varying chemical properties (Mrozik et al., 2003). According to their structure, PAHs have been divided into two groups, namely light and heavy PAHs. Light PAHs consist of compounds with four or less aromatic rings, whereas more than four-ringed compounds are considered as

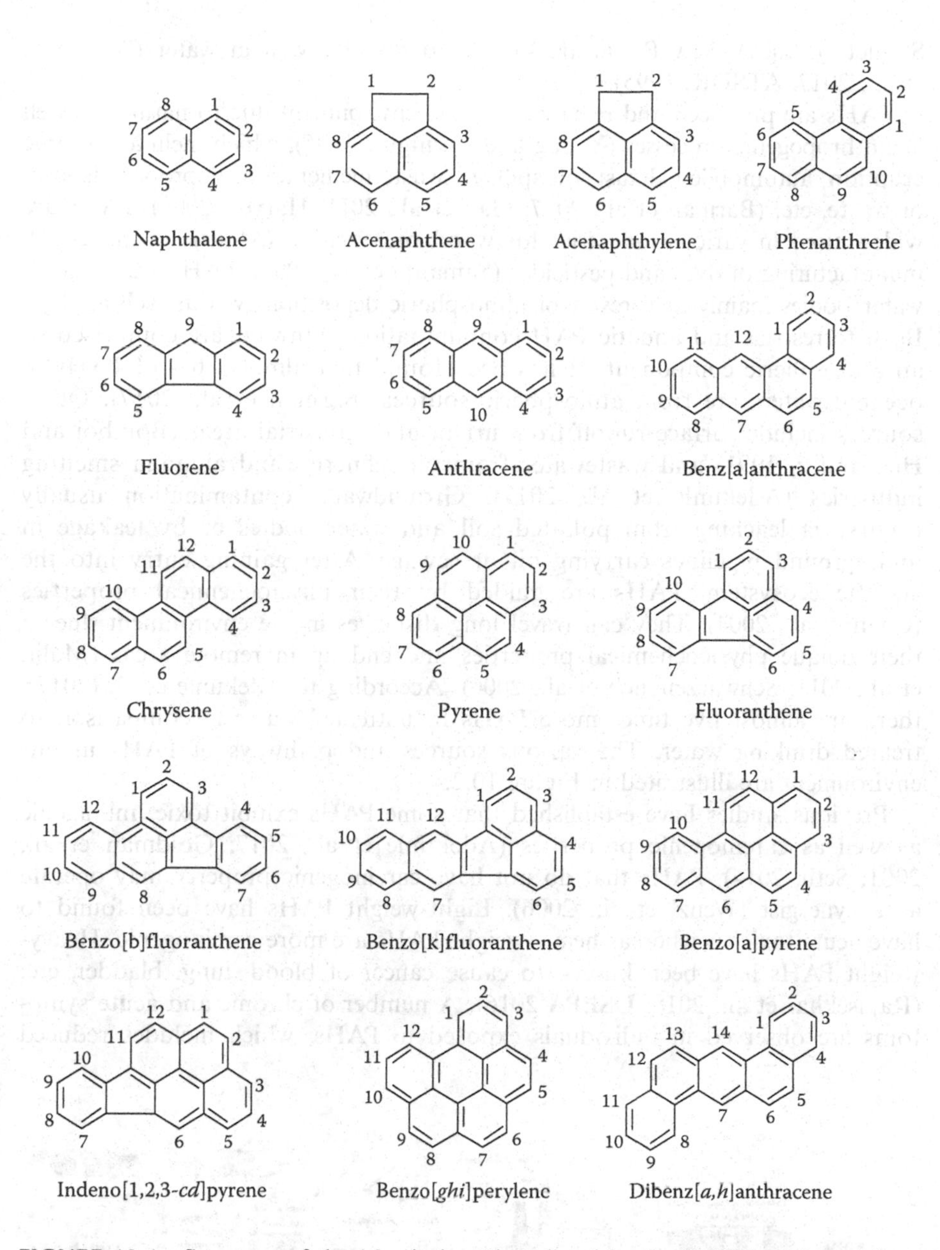

FIGURE 10.1 Structure of the 16 priority PAHs listed by the USEPA (Adapted from Yan et al., 2004).

heavy PAHs. It has been studied that heavy PAHs are more toxic as well as stable compounds as compared to light PAHs (ATSDR, 1995). Solubility of PAHs in water is highly dependent on their structure, however, PAHs are lipid soluble and thus can get adsorbed in the body if ingested (Bamforth and

Singleton, 2005). Few PAHs are known to dissolve well in water (Adekunle et al., 2017; ATSDR, 1995).

PAHs are produced and released into the environment due to natural as well as anthropogenic processes (Young and Cerniglia, 1995), which include volcanic eruption, automobile exhaust, oil spillage, waste incineration, improper disposal of waste, etc. (Barman et al., 2017; Gao et al., 2011; Harvey, 2000). PAHs are widely used in various industries for wood preservation (Maletić et al., 2011), manufacturing of dyes and pesticides (Samanta et al., 2002). PAHs are found in water bodies mainly as a result of atmospheric deposition (wet as well as dry). Both terrestrial and aquatic PAH contamination pathways are comprised of an atmospheric component. It has been found that almost 80% of PAHs in oceanic waters are from atmospheric sources (Barman et al., 2017). Other sources include surface runoff from urban and industrial areas (Bomboi and Hernandez, 1991) and wastewater from oil refineries and alumina smelting industries (Adekunle et al., 2017). Groundwater contamination usually occurs via leaching from polluted soil and water bodies or by leakage in underground pipelines carrying oil or sewage. After gaining entry into the aquatic ecosystem, PAHs are guided by their physiochemical properties (Chen et al., 2004). They can travel long distances in the environment due to their unique physicochemical properties and end up in remote areas (Malik et al., 2011; Schwarzenbach et al., 2006). According to Adekunle et al. (2017), there are almost five times more PAHs in untreated water in comparison to treated drinking water. The various sources and pathways of PAHs in our environment are illustrated in Figure 10.2.

Previous studies have established that some PAHs exhibit toxic, mutagenic as well as carcinogenic properties (Adekunle et al., 2017; Goldman et al., 2001; Seth, 2014). PAHs that do not have carcinogenic property may operate as a synergist (Wenzl et al., 2006). Light-weight PAHs have been found to have acute toxicity, whereas heavy-weight PAHs are more carcinogenic. Heavy-weight PAHs have been known to cause cancer of blood, lung, bladder, etc. (Rajasekhar et al., 2018; USEPA 2016). A number of chronic and acute symptoms are observed in individuals exposed to PAHs, which includes reduced

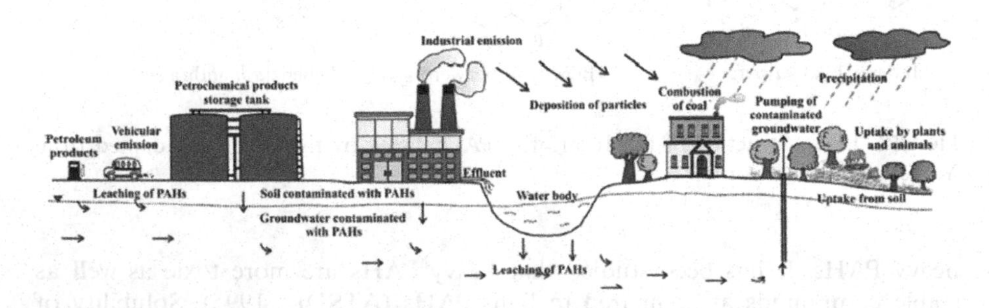

FIGURE 10.2 Various sources of PAHs in the environment (Adapted from Brindha and Elango, 2014).

immunity, kidney and liver damage, asthma, eye inflammation, nausea and confusion (Seth, 2014). It has been reported that the metabolite formed during the breakdown of PAHs are electrophilic and can easily bind with DNA (Pitarch et al., 2007). Because of their detrimental properties, PAHs are raising considerable concern worldwide.

In India there are very few reports of levels of PAH in water. Very often PAHs go unchecked in aqueous system, affecting both flora and fauna. According to the World Health Organization (WHO) guidelines (WHO, 2017), the maximum permissible limit of PAHs in drinking water is 11μg/L. In a study conducted by Rajasekhar et al. (2018), it was found that in a metropolitan city of India there was a 98% cancer risk in adults and 89% cancer risk in children if groundwater from wells were used for bathing. Five predominant PAHs were detected in these wells which were earlier reported by Brindha and Elango (2014), which included acenaphthene, fluorene, phenanthrene, anthracene and fluoranthene. In a study by Malik et al. (2011), it was reported that the water of Gomti River mostly contained two- and three-ringed PAHs among which naphthalene and phenanthrene were the most detectable compounds. It was also established that river water near cities had higher concentrations of PAHs as compared to the other parts. In another study by Amaraneni (2006), it was reported that Kolleru lake wetlands in India, which were used for prawn cultivation, had PAH contamination higher than permissible limits. The high level of PAHs in the sediments and water of these wetlands resulted in contamination of the prawns, which ultimately lead to human health hazard. Reddy et al. (2005) conducted a study in Alang–Sosiya situated in the Gulf of Cambay. It was reported that the total PAH concentrations ranged from 78 μg/L to 1565 μg/L, which were much higher than other reference stations.

Studies conducted in other parts of the world too exhibit similar scenario. In the study by Adekunle et al. (2017) in Osun state, Nigeria, it was found that the levels of PAHs in groundwater in the area were higher than the safe limit prescribed by WHO. In another study conducted by Gasperi et al. (2009), it was found that PAH levels in the settleable particles in the Seine River of France was high and was potentially hazardous to the aquatic ecosystem. Research carried out by Pavoni et al. (2003) on seven types of sea weeds from the lagoon of Venice revealed that this bioindicator species was mainly contaminated with PAHs with concentrations up to 56 ng/g of dry weight of the weed. Agah et al. (2017) conducted a study on the concentrations of the 16 priority PAH in the water and sediments of Chabahar Bay, Oman Sea. It was found that the post-monsoon levels of PAHs in water and sediments, 59.6 ng/L and 92.8 ng/L, respectively, were higher than pre-monsoon levels of PAHs. Vilanova et al. (2001) established that low as well as high molecular weight PAHs can be found in remote mountain lakes of Europe despite their geographical location. The PAH concentrations in these lakes were higher than that of North Atlantic marine zone and one or two times lower than urban polluted areas.

Thus, with increasing industrialization and population, there is an immediate requirement of proper detection and removal of PAHs from water. Various methods have been used to remediate PAHs from wastewater in the past (Bhatnagar et al., 2015). These included costly techniques like electrodialysis, reverse osmosis, ion exchange and electrolysis, which also have high energy requirements making them less cost-effective. Other than high costs of operation, there is also a problem of toxic sludge generation, which may further complicate the treatment process (Bhatnagar et al., 2015). The other treatment procedures include biodegradation and adsorption. Adsorption is a quick and cost-effective means for wastewater treatment. A lot of interest has developed in utilizing the adsorption phenomenon in the removal of toxic, non-biodegradable pollutants from wastewater due to its high efficiency. The present times demands formulating of new cheap and efficient adsorbents for the removal of PAHs.

10.2 ADSORPTION FOR WASTEWATER TREATMENT

The phenomenon of adhesion of different types of molecules to a surface is called "adsorption." A film of adsorbate is formed on the adsorbent in this process. The mass transfer takes place in the interface region where molecules of the adsorbent accumulate. Adsorption components are chiefly responsible for particulate pollution removal from wastewater (Khattri and Singh, 2009).

Based on the type of attraction between the adsorbent and adsorbate molecule, the process can be classified into physical adsorption (physisorption) and chemical adsorption (chemisorption). When weak forces exist between the adsorbate and adsorbent molecules and the adsorption process can be easily reversed, then the adsorption is known as "physisoprtion," and when the forces are strong like chemical bond then the adsorption is called "chemisorptions." It has been found that under certain conditions both processes can simultaneously take place. It has been seen that physisorption are usually exothermic in nature, which results in a decline in entropy (S) and Gibbs free energy (ΔG) (Cooney, 1998). Figure 10.3 graphically represents the two types of adsorption. In case of adsorption of liquid pollutants on solid surface, accumulation of liquid particles on the solid–liquid interface occurs, resulting in a dynamic equilibrium between the solute adsorbed and the residual solute in the solution.

There are various advantages of adsorption as compared to other conventional treatment techniques. Its wide acceptability lies on its ability to provide a continuous treatment for huge quantities of wastewater, simplicity and convenience (Banerjee et al., 2017; De Gisi et al., 2016; Saad et al., 2014). In addition, a unique high selectivity of adsorbent is also attainable. Adsorbents are usually unaffected by the toxicity or salinity of the wastewater, thus making them cost-effective and reusable for multiple cycles of waste treatment.

In industrial set up, adsorption is used as a tertiary treatment of dissolved pollutants that elude the biological as well as the chemical oxidation phases (De Gisi et al., 2016).

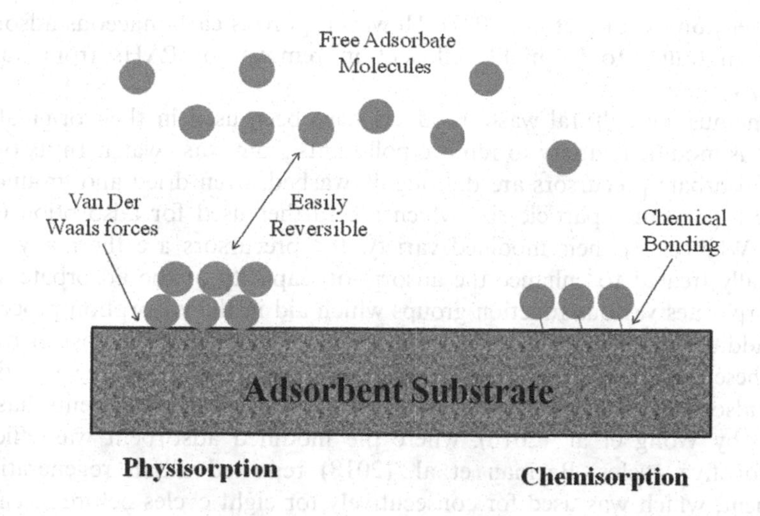

FIGURE 10.3 Graphical representation of chemical and physical adsorption.

It has been established by previous studies that the adsorption of PAHs onto porous carbon is relatively better than onto sediments, suspended organic particles or soils (Yuan et al., 2010). Because of low solubility in water (Barman et al., 2017) and strong attraction towards solid substrate, adsorption of PAHs has become an indisputable technique (Hu et al., 2014). The simple design and low cost of maintenance and operation makes adsorption a more feasible option for PAH removal. Further non-toxic byproducts are produced during the process (Tong et al., 2010) making it an environmentally friendly method. A huge number of pollutants have been removed from aqueous solutions using carbonaceous adsorbents. The proper adsorbent selection and optimization of the factors influencing the process can help in obtaining maximum removal of PAHs from various medium (Wang et al., 2015a).

10.3 GREEN CARBONACEOUS ADSORBENTS

Carbon-based adsorbents are usually porous with the presence of various hetero atoms including oxygen and hydrogen which aid in adsorption process. They have superior kinetic properties and can be easily regenerated. Furthermore, they have been found to be excellent adsorbents for non-polar organic compounds from their aqueous solutions (Hu and Vansant, 1995). A lot of industrial and agricultural byproducts such as rice husk (Masoud et al., 2016), chemical-treated sewage sludge (Martín et al., 2004), tree bark (Şen et al., 2015), vegetable peels, plant leaves (Sulyman et al., 2017), spent tea leaves (Attia, 2012) have been utilized as adsorbents for removal of a range of pollutants from wastewater. Removal of PAHs from water has been performed using cheap products, such as wood fiber and leonardite (Boving and Zhang,

2004; Zeledon-Toruno et al., 2007). However, porous carbonaceous adsorbents have been found to be highly efficient in removal of PAHs from aqueous solutions.

Numerous agricultural waste products have been used in their original form as well as modified variety to remove pollutants from wastewater. In its original form, adsorbate precursors are thoroughly washed, oven dried and grounded to achieve a particular particle size which are further used for adsorption experiments. Whereas in their modified variety, the precursors are thermally and/or chemically treated to enhance the adsorption capacity of the adsorbate as well as incorporates various function groups which aid in the adsorption process.

In addition to being produced from waste and reducing the cost of production, these adsorbents can also be reused several times after regeneration of spent adsorbent. Thermal regeneration of carbon-based adsorbents has been studied by Wong et al. (2016), where the modified adsorbent was efficiently used for five cycles. Barman et al. (2018) reported alkali regeneration of adsorbent, which was used for consecutively for eight cycles before discarding. Acid treatment has also been done for adsorbent regeneration by Kwak et al. (2018), and the adsorbent was usable for five cycles. Mohanta and Ahmaruzzaman (2018) used both alkali and thermal treatment for adsorbent regeneration. The spent adsorbent which loses its efficiency may be utilized as fuel for boiler to recover energy. They can also be used in landfills, brick making, and other construction raw materials (Mohanta and Ahmaruzzaman, 2018).

10.3.1 Biochars

Biochar, a product of pyrolysis of a range of organic materials, is a carbonaceous adsorbent which has been gaining popularity. Biochars have porous structure, large surface area and good affinity for non-polar compounds, which makes it a good adsorbent for organic pollutants from water. It is also used for soil amendment as it has high carbon content which can improve soil fertility (Qambrani et al., 2017). The fabrication of biochar is simple and is comparable to the most ancient technique of charcoal production. They have been produced from various agricultural and industrial byproducts which have been represented in Table 10.1. The production of biochar can be accompanied with production of syngas as well as usable biofuel which is an added advantage (Tan et al., 2016).

Production of biochar from waste products thus provides an excellent closed-loop approach to waste and environmental management in an eco-friendly manner. The availability of low-cost raw materials, large surface area for adsorption and cost effectiveness has made biochars a potent contestant in the field of wastewater treatment. Biochars have the ability to remove organic contaminants such as PAHs from aqueous solutions. They have been used to remove selective PAHs from sewage (Oleszczuk et al., 2012) and other aqueous solutions (Chen et al., 2008; Wang et al., 2006). Biochars have been reported to contain various functional groups such as lipids, cellulose, and proteins, which upon activation via pyrolysis result in better adsorptive capacity (Inyang et al., 2011). The pollutant removal

TABLE 10.1

Significant works in the field of biochar formulation from waste materials undertaken in the last 10 years

Raw material used	Pyrolysis temperature and time	Atmosphere	Reference
Pine needle	373–973 K for 6 h	Limited O_2	Chen et al., 2008
Orange peel	473 K for 6 h	Limited O_2	Chen and Chen., 2009
Dairy manure	473–623 K for 4 h	Limited O_2	Cao et al., 2009
Plant biomass	373–973 K for 1 h	Limited O_2	Keiluweit et al., 2010
Swine manure	893 K for 2 h	Limited O_2	Cao et al., 2010
Paper mill waste	823 K for 2 h	Limited O_2	Van Zwieten et al., 2010
Crop residues	573–973 K for 4 h	Limited O_2	Yuan et al., 2011
Wastewater sludge	573–973 K at a rate of 10°C min^{-1}	N_2 atmosphere	Hossain et al., 2011
Soybean stover and peanut shells	573 and 973 K for 3 h	Limited O_2	Ahmad et al., 2012
Pitch pine wood chip	573–773 K for 2 s	O_2 free	Kim et al., 2012
Conocarpus wastes	473–1073 K for 4 h	Limited O_2	Al-Wabel et al., 2013
Waste rubber-wood-sawdust	723–1123 K for 1 h	N_2 atmosphere	Ghani et al., 2013
Biomass waste solids of eucalyptus and palm bark	673 K for 30 min	Limited O_2	Sun et al., 2013
Willow wood chips	573–873 K at a rate of ~25°C min^{-1}	Limited O_2	Wang et al., 2014
Goat manure	673–1073 K for 30 min	N_2 atmosphere	Touray et al., 2014
Human manure	573–1073 K for 40 min	N_2 atmosphere	Liu et al., 2014
Bamboo	673 K for 2 h	Limited O_2	Chen et al., 2015
Date palm waste	573–1073K for 4 h	Limited O_2	Usman et al., 2015
Pine sawdust, paunch grass, broiler litter, sewage sludge, dewatered pond sludge, and dissolved air-floatation sludge	953 K for 8–10 min	Limited O_2	Srinivasan et al., 2015
marine macro-algae roots (*Undaria pinnatifida*)	1073 K for 2 h	N_2 atmosphere	Jung et al., 2016
Rice husk and tea waste	973 K for 3 h	Limited O_2	VithanaVithanage et al., 2016
Cherry pulp	673–973 K at rates of 10, 100 and 200° C min^{-1}	N_2 atmosphere	Pehlivan et al., 2017

(Continued)

TABLE 10.1 (Cont.)

Raw material used	Pyrolysis temperature and time	Atmosphere	Reference
bioenergy byproducts (deoiled cakes of *Jatropha carcus* and *Pongamia glabra*), lignocellulose biomass (*Jatropha carcus* seed cover), and a noxious weed (*Parthenium hysterophorus*)	623–923 K, 973 K at rates of 40°C min^{-1}	N_2 atmosphere	Narzari et al., 2017
Rice straw and pig manure	573–973 K for 1.5 h	Anaerobic atmosphere	Liu et al., 2017
De-oiled *Pongamia-pinnata* seed cake	823 K for 30 min	N_2 atmosphere	De et al., 2018
leaf waste biomass	473–673 K for 3 h	O_2 atmosphere	Sahota et al., 2018
Temple flora	623–773 K for 30 min	N_2 atmosphere	Singh et al., 2018

mechanism of biochars is chiefly governed by the type as well as the concentrations of the functional group and charge present on its surface (Yenisoy-Karakaş et al., 2004,). The size and structure of biochar particles also plays a major role in achieving adsorption equilibrium (Zheng et al., 2010). It has been established that smaller biochar particles adsorb pollutants faster as compared to their larger counterparts. The adsorption of organic pollutants onto biochar particles have been anticipated to take place via a number of interactions such as hydrogen bonding, diffusion and partitioning, pore-filling, hydrophobic and electrostatic interactions, π–π interaction or a combination of two or more of these mechanisms (Chen et al., 2015; Zhang et al., 2013). The sorption of organic pollutant on biochar is depicted in Figure 10.4.

Thus, due to the ready and cheap availability of biochar precursor as well as its good adsorptive capacity of organic pollutants, biochar is an environmentally friendly alternative for remediation of aqueous pollutants.

10.3.2 Activated Carbon

Activated carbon (AC) is one of the most commonly used adsorbent in wastewater treatment plants. AC is an efficient carbonaceous adsorbent which can be prepared from any carbon containing material (natural or synthetic) either by chemical or physical activation.

Activated carbon, also known as activated charcoal, does not have characteristic chemical formulae (Cuhadaroglu and Uygun, 2008). However, they can be classified according to their physical properties into granular AC, powder AC, AC pellets, etc. AC often possesses heteroatoms such as hydrogen, oxygen and halogens chemically bonded to the AC structure or as functional groups.

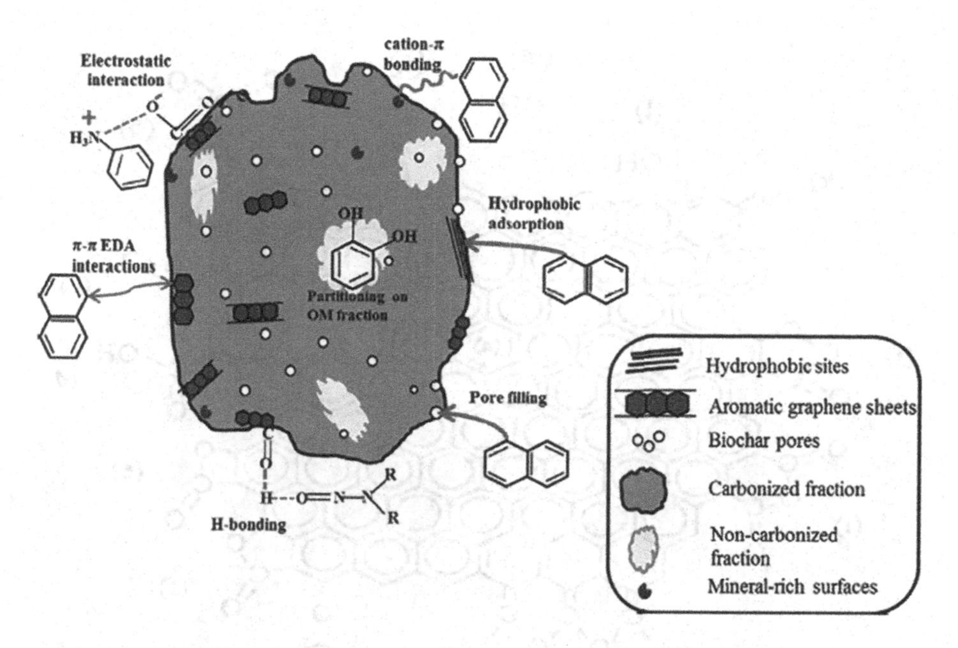

FIGURE 10.4 Mechanisms of organic pollutant interaction with various uptake sites of biochar. The interactions include electrostatic interactions, cation–π bonding, hydrophobic interaction, pore filling, hydrogen bonding and π–π interaction (Adapted from Inyang and Dickenson, 2015).

Oxygen is often the dominant heteroatom found on AC surface as carboxylic, phenolic, carbonyl and other functional groups (Yahya et al., 2015). Figure 10.5 represents the various IR-active substrate functional groups that are observed after oxidation of AC.

AC may also contain different quantity of mineral matter depending on the raw material used as the precursor (Bandosz, 2006). Depending on the surface charge, AC can be classified as H- type and L-type carbon for positive and negatively charged surface, respectively (Nik et al. 2006). The factors influencing the property of AC include (a) AC precursor (b) activating agent (c) precursor: activating agent ratio (d) activation temperature (e) time and (f) inorganic compounds if any. The selection of the raw material for AC preparation is an important step. For sustainable AC prep, the raw material should be easily available and cheap with uninterrupted supply. It is also desirable to have high carbon content, should be readily activated and should not degrade with time (Dias et al., 2007). Production of AC from waste products is a cheap as well as environment friendly method. Green synthesis of AC has an added advantage of waste reduction or converting "waste to wealth" (Hayashi et al., 2002). Commercially available AC is usually prepared from petroleum residues or wood, which have higher yield as compared to agricultural or industrial waste product. However the cost effectiveness of ACs prepared

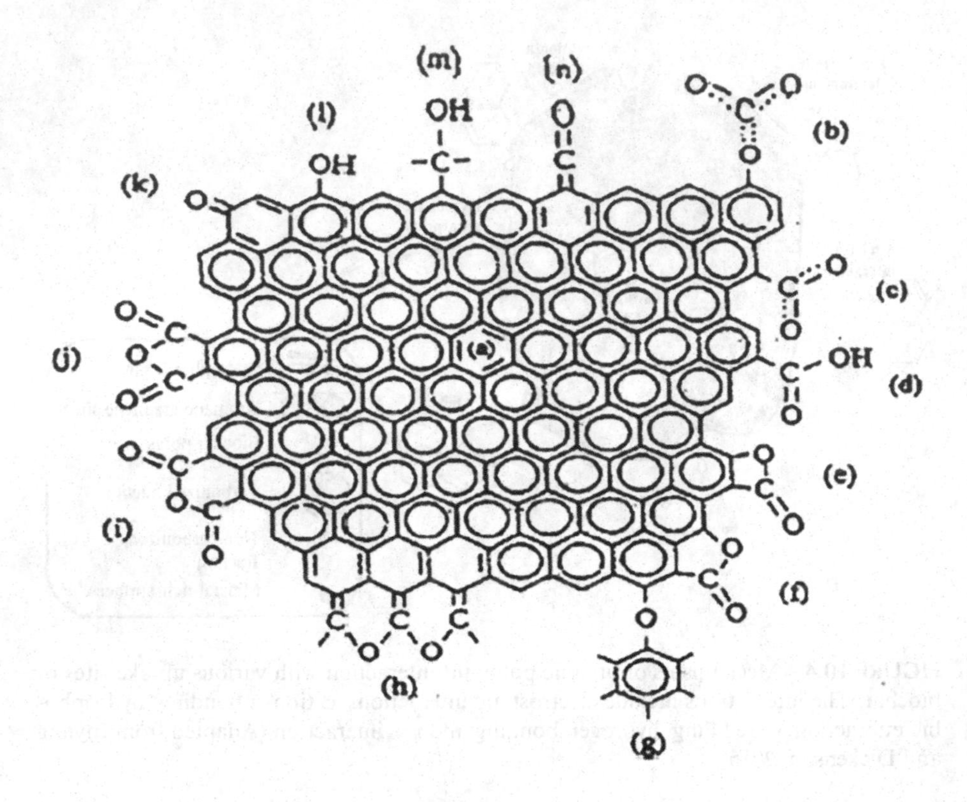

FIGURE 10.5 Functional groups present on activated carbon surface;: (a) aromatic C=C stretching; (b) and (c) carboxyl-carbonates (d) carboxylic acid (e) lactone (4-membered ring) (f) lactone (5-membered ring) (g) ether bridge (h) cyclic ether (i) cyclic anhydride (5-membered ring) (j) cyclic anhydride (6-membered-ring) (k) quinine (l) phenol (m) alcohol and (n) ketene. (Adapted from Yahya et al., 2015).

from waste is higher than that of commercial AC making them significant (Malik et al., 2006). Any cheap material that contains high amount of carbon and low inorganic components can be used for AC preparation.

The activation of the AC precursor, as mentioned earlier, can be done via physical or chemical methods. The physical activation is a two-step process, carbonization of the raw material followed by steam or CO_2 activation. In chemical activation, the precursor material is treated with the chemical activating agent and then subjected to heating under specific conditions. The activating agent assists in crosslink formation in the prepared AC (Örkün et al., 2012). It acts as an oxidant which could catalyzes the process and also boosts yield by preventing formation of ash (Malik et al., 2006). Activating agents also helps in incorporating functional groups on the AC surface. Some common chemical activating agents used in chemical activation include H_3PO_4, KOH, $ZnCl_2$, NaOH, $KMnO_4$, H_2O_2 and K_2CO_3 (Yahya et al., 2015).

TABLE 10.2

Significant work on recycling waste to formulate AC in the last decade

Raw materials	Activating agent	Activating temperature and time	Atmosphere	Reference	Place
Jackfruit peel	H_3PO_4	823 K for 45 min	N_2	Prahas et al., 2008	Surabaya, Indonesia
Olive-waste cakes	H_3PO_4	377 K for 2h	N_2	Baccar et al., 2009	Sfax, Tunisia
Bean pods	K_2CO_3	1223 K for 1.7 h	N_2	Cabal et al., 2009	Oviedo, Spain
Bamboo waste	H_3PO_4	773 K for 2 h	N_2	Ahmad and Hameed, 2010	Penang, Malaysia
waste rubber tire	H_2O_2	1173 K for 2 h	O_2	Gupta et al., 2011	India
Camellia oleifera shell	$ZnCl_2$	773–973 K for 1 h	N_2	Zhang et al., 2012	Hunan, China
Ramulus mori	$(NH_4)_2HPO_4$	973–1123 K for 30 min to 2 h	N_2	Tang et al., 2012	China
Waste rubber tires	Steam	1173 K for 2 h	N_2	Gupta et al., 2013	Dhahran Saudi Arabia
Rice husk	H_3PO_4	673 K for 2 h	O_2	Yakout et al., 2013	Riyadh Saudi Arabia
Rice husk	NaOH and KOH	723–1123 K for 1 h	O_2 and N_2	Muniandy et al., 2014	Penang, Malaysia
Cotton woven waste	H_3PO_4+ $(NH_4)_2HPO_4$	1073 K for 1 h	CO_2	Zheng et al., 2014	Beijing, China
Waste tea	H_3PO_4/K_2CO_3	Microwave for 30 s/ heating at 1073 K for 1 h	N_2	Inal et al., 2015	Ankara, Turkey
Anthracite	KOH + microwaves	700 W for 10 min	N_2	Ge et al., 2015	China
Grape processing wastes	$ZnCl_2$	873 K for 1 h	O_2	Sayğılı and Güzel, 2016	Turkey
Anthracite	KOH,$ZnCl_2$ H_3PO_4 + microwaves	100–1000 W for 4–20 min	Vacuum	Xiao et al., 2015	China
Silk worm cocoon	$(NH_4)_2HPO_4$ pretreatment + KOH	873 K for 2 h	N_2	Li et al., 2015	China

(*Continued*)

TABLE 10.2 (Cont.)

Raw materials	Activating agent	Activating temperature and time	Atmosphere	Reference	Place
Banana peel	H_3PO_4	473 K for 4h followed by reheating at 773 K for 2 h	O_2	Gupta and Gupta, 2016	Uttarakhand, India
Coal	KOH + microwaves+ $Fe(NO_3)_3$+ ultrasound	700 W for 10 m followed by 700 W for 8 min	O_2	Ge et al., 2016	Shihezi, China
Tomato processing waste	$ZnCl_2$	873 K for 1 h	O_2	Sayğılı and Güzel, 2016	Turkey
Coconut shell	Steam	–	O_2	Kaman et al., 2017	Sarawak, Malaysia
lignin	KOH	1073 K for 4 h	N_2	Hao et al., 2017	Stockholm, Sweden
Tea waste	H_3PO_4	723 K for 1h	N_2/O_2/ Steam	Kan et al., 2017	Jinan 250100, China
Denim fabric waste	H_3PO_4	723 K for 2 h	N_2	Silva et al., 2018	Maringá, Paraná, Brazil
Waste polyester textiles	$MgCl_2$	1173 K for 1.5 h	N_2	Yuan et al., 2018	China

It has been established by earlier reports that chemical activation is more advantageous as compared to physical activation as it requires only one step, lower temperatures, and yields more AC with well-developed pores and lower inorganic matter (Hirunpraditkoon et al., 2011; Lillo-Ródenas et al., 2003; Tsai et al., 1998; Zhu et al., 2008).

AC possesses high surface area and well-developed pores and is regarded as an extremely potent adsorbent for removal of different types of pollutants (Huang et al., 2011). In recent times, varied agricultural and household wastes are being used to prepare AC. A range of items, including saw dust, coconut shell and rice husk, have been used for the purpose (Muniandy et al., 2014). This approach also simultaneously reduces waste and thus can be said to be useful in multiple ways. Various raw materials have been used as a precursor of activated carbon.

Table 10.2 represents some significant works in the field of AC preparation from waste products in the last 10 years.

10.3.3 Green Nano-Composites

Although porous carbon is a potent adsorbent for organic pollutants, in recent times increasing interest has developed surrounding carbon-based nano-composites due to their adsorption ability and antimicrobial property (Smith and Rodrigues, 2015). The unique set of properties of these nanomaterials which include large surface area, excellent adsorption capacity for organic and inorganic pollutants, exceptional antimicrobial property, and ease of manipulation and high rate of mass transfer makes them excellent foundation for various research avenues. Huge amount of vacant surface and numerous reactive atoms makes nano particles efficient in wastewater treatment. Additionally, they possess metallic or semi-metallic properties that are available to the target pollutant particle for interaction (Roosta et al., 2014b). Many metal nanoparticles have been produced using plant, bacteria, algae, etc. (Iravani, 2011), making them cheap and easily producible. Copper and copper oxide nanoparticles have been formulated efficiently using *Citrus medica* Linn juice (Shende et al., 2015), leaf extracts of *Euphorbia esula* L. (Nasrollahzadeh et al., 2015) and *Aloe vera* (Kumar et al., 2015), peel extracts of *Punica granatum* (Ghidan et al., 2016), aqueous media of marine algae *Kappaphycus alvarezii* (Khanehzaei et al., 2014,)

Silver nanoparticles have also been extensively biosynthesized using extracts of *Azadirachta indica* plant (Satapathy et al., 2015), seed extracts of *Coffea arabica* (Dhand et al., 2016), leaf extracts of olive plant (Khalil et al., 2014), *Ziziphora tenuior* extracts (Sadeghi and Gholamhoseinpoor, 2015) and freshwater algae *Pithophora oedogonia* (Sinha et al., 2015). Metal nanoparticles of zinc, palladium, gold and lead have also been biosynthesized to some extent (Singh et al., 2016).

Some metal nanoparticles impart magnetic properties to the composites which makes them easy to remove from wastewater. They possess catalytic properties which aid in degradation of organic as well as inorganic pollutants. Carbon-based nano-composites with increased number of functional groups due to the presence of functional nanoparticles on their surface have the property of high specificity for particular pollutants in a mixture of other compounds (Qambrani et al., 2017). Thus, nano composites on carbon backbones are formulated not only to enhance the adsorbent properties but also to achieve novel adsorbents that can combine the advantages of porous carbon and nano particles for greater efficiency. There are also various advantages in the use of porous carbon as a base substrate for these nano-composites as compared to other materials. These include larger surface area available to the nano particles, low energy requirement, presence of reactive functional groups and easy availability of raw materials and lower temperature requirement (Tan et al., 2016).

Both biochar and AC have been used in preparation of nano composites. Nanoparticles-biochar composites have been used to remove heavy metals such as arsenic (Tan et al., 2016), chromium (Gan et al., 2015), copper, platinum (Ghaedi et al., 2011), cadmium and lead (Wang et al., 2015b) from wastewater. Organic chemicals have also been reportedly removed using nanoscale zerovalent ion–biochar composites (Yan et al., 2015) as well as graphite–biochar composites (Pi et al., 2015). On the other hand, various metal nanoparticles AC composites have also been used to treat wastewater. AC has non-localized *n* electrons and reactive centers like OH, NH_2 and COOH, which are capable of attracting nanoparticles via π–π interaction, electrostatic force as well as hydrogen bonding (Ghaedi et al., 2011). Compounds like CuO nanoparticles readily attach to AC surface due to the presence of reactive oxygen centers as well hard–hard or soft–soft interaction with the metal center (Ghaedi and Mosallanejad, 2014).

Silver (Ghaedi et al., 2012a), palladium (Ghaedi et al., 2012b), gold (Roosta et al., 2014a), zinc sulfide (Ghaedi et al., 2015), cobalt ferrite (Mehrabi et al., 2017), iron oxide magnetite (Bagheri et al., 2017), as well as bimetallic nanoparticles (Saleh, 2018) have been loaded on AC for the removal of various dyes, petroleum and other pollutants from water. Aromatic compounds such as benzothiophene, thiophene and dibenzothiophene have been reportedly removed utilizing iron and cerium nanoparticles loaded AC (Danmaliki and Saleh, 2017). It has also been established that several nanoparticles have antimicrobial properties which makes them even more potent for water treatment (Satapathy et al., 2015). Thus, it is clear that the use of porous carbon-based nano composites is a promising avenue for the treatment of aqueous organic as well as inorganic pollutants. The modifications of porous carbon using nano scale materials, with high surface area, enhance the removal percentage and adsorption capacity of adsorbent.

10.4 BACTERIA IMMOBILIZED CARBONACEOUS ADSORBENTS

The process of adsorption is undoubtedly a fast and effective method for the removal of PAHs from water. However, irrespective of positive results, the ultimate detoxification of the pollutants is not achieved via adsorption. Biodegradation on the other hand can break down pollutants in to other non-toxic forms leading to total remediation. Nevertheless, biodegradation can only occur if the conditions are favorable for the biological organisms to grow. High concentration of toxicants, unfavorable pH, temperature or low supply of carbon sources may lead to lower efficiency or total inhibition of biodegradation (Varjani and Upasani, 2017; Venosa and Zhu, 2003; Zapata Acosta et al., 2019). Immobilization of bacterial cells on adsorbents is a technique of combining both adsorption and degradation, which have been reported to enhance the total process of pollutant elimination (El Fantroussi & Agathos, 2005). Immobilization of bacteria on porous surface provides the microorganisms with a protective niche that preserves them against unfavorable conditions (Kwon et al., 2009; Partovinia and Naeimpoor, 2013). This system also enhances biodegradation by forming a biofilm, making the contaminants more bioavailable to the microorganisms and also increasing the available surface area and contact time for biodegradation to take place (Biswas et al., 2015; Chen et al., 2003).

Immobilization of microorganisms on adsorbents have been achieved using methods such as physical adsorption, entrapment, covalent binding, ionic binding and encapsulation (Bayat et al., 2015; Roy et al., 2018). Physical adsorption process of immobilization is a reversible process where interaction between microbes and adsorbate surface mainly occurs via hydrogen bonding, weak van der Waals forces, ionic bonds and hydrophobic bonds (Bayat et al., 2015).

Entrapment method is an irreversible process in which the microbe or enzyme gets embedded in a porous matrix. To prevent enzyme leakage while use, additional chemical covalent bonding is initiated aside from the already present physical barrier (Sheldon and Van Pelt, 2013).Covalent binding process is irreversible and the interaction between adsorbate and cell is chiefly through covalent bonds in presence of a chemical cross linking mediator. Encapsulation is similar to entrapment where the microbes are totally enveloped within the matrix (Bayat et al., 2015).

Various substrates such as fly ash (Roy et al., 2018), chitosan (Muangchinda et al., 2018), polyvinyl alcohol (Ying *et al.*, 2007), bentonite (Duan et al., 2016), sodium alginate beads (Kiran et al., 2018), biochar (Lu et al., 2018) and AC (Zapata Acosta et al., 2019) have been used to immobilize microbes for bioremediation of environmental pollutants. Bacterial strains such as *Pseudomonas mendocina* (Barman et al., 2017), *Alcaligenes denitrificans* (Samanta et al., 2002), *Mycobacterium gilvum* (Xiong et al., 2017), *Acinetobacter* (Shao et al., 2015), *Pseudomonas aeruginosa* (Zhang et al., 2011) and *Pseudomonas putida* (Dutta et al., 2017) have been used to degrade PAHs in previous studies and immobilization of these PAH degrading bacteria onto an adsorbent needs to be done to achieve the desired result.

Both biochar and AC are capable of providing niches to microorganisms which can simultaneously intensify both the techniques as well as eliminate PAHs completely from the environment. Thus, immobilization can be an economically viable and environmental friendly alternative to traditional biodegradation or adsorption.

10.5 CONCLUSION

With increasing problems of waste remediation as well as pollution of natural resources, utilizing waste to formulate adsorbents for pollution control is a promising technology. It is a simultaneous solution to two important environmental issues. Extensive research into the mechanism of formulation of these adsorbents, optimization process parameters for reduced adsorption time, as well as understanding the adsorption kinetics, isotherms and thermodynamics is required. The study of adsorbent surface characteristics and charge is important which helps determine if the adsorbent will efficiently remove a particular pollutant. The reuse and recycling of these adsorbents as well as feasibility for large-scale industrial use, needs to be established. Nanoparticle-loaded carbonaceous adsorbents requires additional attention as it possesses immense potential for pollution remediation and overcoming barriers faced by traditional technologies. With increasing percentage of persistent pollutants in our water resources these nano-material loaded adsorbents can be the answer for the sustainable management of the environment as because the conventional techniques fail to remove these pollutants. Immobilization of bacterial species onto carbon backbone is another environmentally benign technology that needs to be explored as it may potentially have the solution to the problem of final disposal of pollutants. Further research is required in this field to realize the full potential of these adsorbent in cost-effective pollution remediation. Large-scale study of these adsorbents on real effluents is required to determine their effectiveness in real time. There is a requirement to study the effectiveness of these adsorbents by studying the effect of the treated wastewater on plant and animal tissues. The cellular response to the treated and raw wastewater needs to be analyzed and compared to decide if the treated water is potable for reuse. Also the ecological effect of these adsorbent needs to be investigated to eliminate any threat to environmental or human health.

REFERENCES

Adekunle, A. S., Oyekunle, J. A. O., Ojo, O. S., Maxakato, N. W., Olutona, G. O., & Obisesan, O. R. 2017. Determination of polycyclic aromatic hydrocarbon levels of groundwater in Ife north local government area of Osun state, Nigeria. *Toxicology Reports* 4: 39–48.

Adeleye, A. S., Conway, J. R., Garner, K., Huang, Y., Su, Y., & Keller, A. A. 2016. Engineered nanomaterials for water treatment and remediation: costs, benefits, and applicability. *Chemical Engineering Journal* 286: 640–662.

Agah, H., Mehdinia, A., Bastami, K. D., & Rahmanpour, S. 2017. Polycyclic aromatic hydrocarbon pollution in the surface water and sediments of Chabahar Bay, Oman Sea. *Marine Pollution Bulletin* 115(1–2): 515–524.

Ahmad, A. A., & Hameed, B. H. 2010. Fixed-bed adsorption of reactive azo dye onto granular activated carbon prepared from waste. *Journal of Hazardous Materials* 175(1–3): 298–303.

Ahmad, M., Lee, S. S., Dou, X., Mohan, D., Sung, J. K., Yang, J. E., & Ok, Y. S. 2012. Effects of pyrolysis temperature on soybean stover-and peanut shell-derived biochar properties and TCE adsorption in water. *Bioresource Technology* 118: 536–544.

Ali, I., & Gupta, V. K. 2006. Advances in water treatment by adsorption technology. *Nature Protocols* 1(6): 2661.

Al-Wabel, M. I., Al-Omran, A., El-Naggar, A. H., Nadeem, M., & Usman, A. R. 2013. Pyrolysis temperature induced changes in characteristics and chemical composition of biochar produced from conocarpus wastes. *Bioresource Technology* 131: 374–379.

Amaraneni, S. R. 2006. Distribution of pesticides, PAHs and heavy metals in prawn ponds near Kolleru lake wetland, India. *Environment International* 32 (3): 294–302.

ATSDR. 1995. *Chemical and physical information, in toxicological Profile for Polycyclic Aromatic Hydrocarbons (PAHs)*. ATSDR, Atlanta, GA, 209–221.

Attia, H. G. 2012. A comparison between cooking tea-waste and commercial activated carbon for removal of chromium from artificial wastewater. *Journal of Engineering and Sustainable Development* 16(1): 307–325.

Baccar, R., Bouzid, J., Feki, M., & Montiel, A. 2009. Preparation of activated carbon from Tunisian olive-waste cakes and its application for adsorption of heavy metal ions. *Journal of Hazardous Materials* 162(2): 1522–1529.

Bagheri, A. R., Ghaedi, M., Asfaram, A., Bazrafshan, A. A., & Jannesar, R. 2017. Comparative study on ultrasonic assisted adsorption of dyes from single system onto Fe3O4 magnetite nanoparticles loaded on activated carbon: experimental design methodology. *Ultrasonics Sonochemistry* 34: 294–304.

Bamforth, S. M., & Singleton, I. 2005. Bioremediation of polycyclic aromatic hydrocarbons: current knowledge and future directions. *Journal of Chemical Technology & Biotechnology* 80(7): 723–736.

Bandosz, T. J. 2006. *Activated carbon surfaces in environmental remediation* (Vol 7). Academic Press, New York, USA.

Banerjee, P., Barman, S. R., Mukhopadhyay, A., & Das, P. 2017. Ultrasound assisted mixed azo dye adsorption by chitosan–graphene oxide nanocomposite. *Chemical Engineering Research and Design* 117: 43–56.

Barman, S. R., Banerjee, P., Das, P., & Mukhopadhyay, A. 2018. Urban wood waste as precursor of activated carbon and its subsequent application for adsorption of polyaromatic hydrocarbons. *International Journal of Energy and Water Resources* 2(1–4): 1–13.

Barman, S. R., Banerjee, P., Mukhopadhyay, A., & Das, P. 2017. Biodegradation of acenapthene and naphthalene by Pseudomonas mendocina: process optimization, and toxicity evaluation. *Journal of Environmental Chemical Engineering* 5(5): 4803–4812.

Bayat, Z., Hassanshahian, M., & Cappello, S. 2015. Immobilization of microbes for bioremediation of crude oil polluted environments: a mini review. *The Open Microbiology Journal* 9: 48.

Bhatnagar, A., Sillanpää, M., & Witek-Krowiak, A. 2015. Agricultural waste peels as versatile biomass for water purification–a review. *Chemical Engineering Journal* 270: 244–271.

Biswas, B., Sarkar, B., Rusmin, R., & Naidu, R. 2015. Bioremediation of PAHs and VOCs: advances in clay mineral–microbial interaction. *Environment International* 85: 168–181.

Bolong, N., Ismail, A. F., Salim, M. R., & Matsuura, T. 2009. A review of the effects of emerging contaminants in wastewater and options for their removal. *Desalination* 239(1–3): 229–246.

Bomboi, M. T., & Hernandez, A. 1991. Hydrocarbons in urban runoff: their contribution to the wastewaters. *Water Research* 25(5): 557–565.

Boving, T. B., & Zhang, W. 2004. Removal of aqueous-phase polynuclear aromatic hydrocarbons using aspen wood fibers. *Chemosphere* 54(7): 831–839.

Brindha, K., & Elango, L. 2014. PAHs contamination in groundwater from a part of metropolitan city, India: a study based on sampling over a 10-year period. *Environmental Earth Sciences* 71(12): 5113–5120.

Cabal, B., Budinova, T., Ania, C. O., Tsyntsarski, B., Parra, J. B., & Petrova, B. 2009. Adsorption of naphthalene from aqueous solution on activated carbons obtained from bean pods. *Journal of Hazardous Materials* 161(2): 1150–1156.

Cao, X., Ma, L., Gao, B., & Harris, W. 2009. Dairy-manure derived biochar effectively sorbs lead and atrazine. *Environmental Science & Technology* 43(9): 3285–3291.

Cao, X., Ro, K. S., Chappell, M., Li, Y., & Mao, J. 2010. Chemical structures of swine-manure chars produced under different carbonization conditions investigated by advanced solid-state 13C nuclear magnetic resonance (NMR) spectroscopy. *Energy & Fuels* 25(1): 388–397.

Chakraborti, D., Ghorai, S. K., Das, B., Pal, A., Nayak, B., & Shah, B. A. 2009. Arsenic exposure through groundwater to the rural and urban population in the Allahabad-Kanpur track in the upper Ganga plain. *Journal of Environmental Monitoring* 11(8): 1455–1459.

Chen, B., & Chen, Z. 2009. Sorption of naphthalene and 1-naphthol by biochars of orange peels with different pyrolytic temperatures. *Chemosphere* 76(1): 127–133.

Chen, B., Xuan, X., Zhu, L., Wang, J., Gao, Y., Yang, K., Shen, X., & Lou, B. 2004. Distributions of polycyclic aromatic hydrocarbons in surface waters, sediments and soils of Hangzhou City, China. *Water Research* 38(16): 3558–3568.

Chen, B., Zhou, D., & Zhu, L. 2008. Transitional adsorption and partition of nonpolar and polar aromatic contaminants by biochars of pine needles with different pyrolytic temperatures. *Environmental Science & Technology* 42(14): 5137–5143.
Chen, C., Zhou, W., & Lin, D. 2015. Sorption characteristics of N-nitrosodimethylamine onto biochar from aqueous solution. *Bioresource Technology* 17: 359–366.
Chen, K. C., Wu, J. Y., Yang, W. B., & Hwang, S. C. J. 2003. Evaluation of effective diffusion coefficient and intrinsic kinetic parameters on azo dye biodegradation using PVA-immobilized cell beads. *Biotechnology and Bioengineering* 83(7): 821–832.
Cooney, D. O. 1998. *Adsorption design for wastewater treatment*. CRC press, Boca Raton, FL.
Cuhadaroglu, D., & Uygun, O. A. 2008. Production and characterization of activated carbon from a bituminous coal by chemical activation. *African Journal of Biotechnology* 7(20): 3703–3710.
Danmaliki, G. I., & Saleh, T. A. 2017. Effects of bimetallic Ce/Fe nanoparticles on the desulfurization of thiophenes using activated carbon. *Chemical Engineering Journal* 307: 914–927.
De, D., Santosha, S., Aniya, V., Sreeramoju, A., & Satyavathi, B. 2018. Assessing the applicability of an agro-industrial waste to Engineered Bio-char as a dynamic adsorbent for Fluoride Sorption. *Journal of Environmental Chemical Engineering* 6 (2): 2998–3009.
De Gisi, S., Lofrano, G., Grassi, M., & Notarnicola, M. 2016. Characteristics and adsorption capacities of low-cost sorbents for wastewater treatment: a review. *Sustainable Materials and Technologies* 9: 10–40.
Dhand, V., Soumya, L., Bharadwaj, S., Chakra, S., Bhatt, D., & Sreedhar, B. 2016. Green synthesis of silver nanoparticles using Coffea arabica seed extract and its antibacterial activity. *Materials Science and Engineering: C* 58: 36–43.
Dias, J. M., Alvim-Ferraz, M. C., Almeida, M. F., Rivera-Utrilla, J., & Sánchez-Polo, M. 2007. Waste materials for activated carbon preparation and its use in aqueous-phase treatment: a review. *Journal of Environmental Management* 85(4): 833–846.
Duan, L., Wang, H., Sun, Y., & Xie, X. 2016. Biodegradation of phenol from wastewater by microorganism immobilized in bentonite and carboxymethyl cellulose gel. *Chemical Engineering Communications* 203(7): 948–956.
Dutta, K., Shityakov, S., Das, P. P., & Ghosh, C. 2017. Enhanced biodegradation of mixed PAHs by mutated naphthalene 1, 2-dioxygenase encoded by Pseudomonas putida strain KD6 isolated from petroleum refinery waste. *3 Biotech* 7(6): 365.
El Fantroussi, S., & Agathos, S. N. 2005. Is bioaugmentation a feasible strategy for pollutant removal and site remediation? *Current Opinion in Microbiology* 8(3): 268–275.
Gan, C., Liu, Y., Tan, X., Wang, S., Zeng, G., Zheng, B., Li, T., Jiang, Z., & Liu, W. 2015. Effect of porous zinc–biochar nanocomposites on Cr (VI) adsorption from aqueous solution. *RSC Advances* 5(44): 35107–35115.
Gao, B., Yu, J. Z., Li, S. X., Ding, X., He, Q. F., & Wang, X. M. 2011. Roadside and rooftop measurements of polycyclic aromatic hydrocarbons in PM 2.5 in urban Guangzhou: evaluation of vehicular and regional combustion source contributions. *Atmospheric Environment* 45(39): 7184–7191.
Gasperi, J., Garnaud, S., Rocher, V., & Moilleron, R. 2009. Priority pollutants in surface waters and settleable particles within a densely urbanised area: case study of Paris (France). *Science of the Total Environment* 407(8): 2900–2908.
Ge, X., Tian, F., Wu, Z., Yan, Y., Cravotto, G., & Wu, Z. 2015. Adsorption of naphthalene from aqueous solution on coal-based activated carbon modified by microwave induction: microwave power effects. *Chemical Engineering and Processing: Process Intensification* 91: 67–77.

Ge, X., Wu, Z., Wu, Z., Yan, Y., Cravotto, G., & Ye, B. C. 2016. Enhanced PAHs adsorption using iron-modified coal-based activated carbon via microwave radiation. *Journal of the Taiwan Institute of Chemical Engineers* 64: 235–243.
Ghaedi, M., Ansari, A., Bahari, F., Ghaedi, A. M., & Vafaei, A. 2015. A hybrid artificial neural network and particle swarm optimization for prediction of removal of hazardous dye brilliant green from aqueous solution using zinc sulfide nanoparticle loaded on activated carbon. *Spectrochimica Acta Part A: Molecular and Biomolecular Spectroscopy* 137: 1004–1015.
Ghaedi, M., Heidarpour, S., Kokhdan, S. N., Sahraie, R., Daneshfar, A., & Brazesh, B. 2012b. Comparison of silver and palladium nanoparticles loaded on activated carbon for efficient removal of methylene blue: kinetic and isotherm study of removal process. *Powder Technology* 228: 18–25.
Ghaedi, M., & Mosallanejad, N. 2014. Study of competitive adsorption of malachite green and sunset yellow dyes on cadmium hydroxide nanowires loaded on activated carbon. *Journal of Industrial and Engineering Chemistry* 20(3): 1085–1096.
Ghaedi, M., Sadeghian, B., Pebdani, A. A., Sahraei, R., Daneshfar, A., & Duran, C. 2012a. Kinetics, thermodynamics and equilibrium evaluation of direct yellow 12 removal by adsorption onto silver nanoparticles loaded activated carbon. *Chemical Engineering Journal* 187: 133–141.
Ghaedi, M., Tashkhourian, J., Pebdani, A. A., Sadeghian, B., & Ana, F. N. 2011. Equilibrium, kinetic and thermodynamic study of removal of reactive orange 12 on platinum nanoparticle loaded on activated carbon as novel adsorbent. *Korean Journal of Chemical Engineering* 28(12): 2255–2261.
Ghani, W. A. W. A. K., Mohd, A., da Silva, G., Bachmann, R. T., Taufiq-Yap, Y. H., Rashid, U., & Ala'a, H. 2013. Biochar production from waste rubber-wood-sawdust and its potential use in C sequestration: chemical and physical characterization. *Industrial Crops and Products* 44: 18–24.
Ghidan, A. Y., Al-Antary, T. M., & Awwad, A. M. 2016. Green synthesis of copper oxide nanoparticles using Punica granatum peels extract: effect on green peach Aphid. *Environmental Nanotechnology, Monitoring & Management* 6: 95–98.
Goldman, R., Enewold, L., Pellizzari, E., Beach, J. B., Bowman, E. D., Krishnan, S. S., & Shields, P. G. 2001. Smoking increases carcinogenic polycyclic aromatic hydrocarbons in human lung tissue. *Cancer Research* 61(17): 6367–6371.
Gupta, H., & Gupta, B. 2016. Adsorption of polycyclic aromatic hydrocarbons on banana peel activated carbon. *Desalination and Water Treatment* 57(20): 9498–9509.
Gupta, V. K., Ali, I., Saleh, T. A., Siddiqui, M. N., & Agarwal, S. 2013. Chromium removal from water by activated carbon developed from waste rubber tires. *Environmental Science and Pollution Research* 20(3): 1261–1268.
Gupta, V. K., Gupta, B., Rastogi, A., Agarwal, S., & Nayak, A. 2011. A comparative investigation on adsorption performances of mesoporous activated carbon prepared from waste rubber tire and activated carbon for a hazardous azo dye—Acid Blue 113. *Journal of Hazardous Materials* 186(1): 891–901.
Hao, W., Björnerbäck, F., Trushkina, Y., Oregui Bengoechea, M., Salazar-Alvarez, G., Barth, T., & Hedin, N. 2017. High-performance magnetic activated carbon from solid waste from lignin conversion processes. 1. Their use as adsorbents for CO_2. *ACS Sustainable Chemistry & Engineering* 5(4). 3087–3095.
Harvey, D. 2000. *Modern analytical chemistry* (Vol 381). McGraw-Hill, New York.
Hayashi, J. I., Horikawa, T., Takeda, I., Muroyama, K., & Ani, F. N. 2002. Preparing activated carbon from various nutshells by chemical activation with K2CO3. *Carbon* 40(13): 2381–2386.
Hirunpraditkoon, S., Tunthong, N., Ruangchai, A., & Nuithitikul, K. 2011. Adsorption capacities of activated carbons prepared from bamboo by KOH activation.

International Journal of Chemical, Molecular, Nuclear, Materials and Metallurgical Engineering 5(6): 477–481.

Hossain, M. K., Strezov, V., Chan, K. Y., Ziolkowski, A., & Nelson, P. F. 2011. Influence of pyrolysis temperature on production and nutrient properties of wastewater sludge biochar. *Journal of Environmental Management* 92(1): 223–228.

Hu, Y., He, Y., Wang, X., & Wei, C. 2014. Efficient adsorption of phenanthrene by simply synthesized hydrophobic MCM-41 molecular sieves. *Applied Surface Science* 311: 825–830.

Hu, Z., & Vansant, E. F. 1995. Synthesis and characterization of a controlled-micropore-size carbonaceous adsorbent produced from walnut shell. *Microporous Materials* 3(6): 603–612.

Huang, L., Sun, Y., Wang, W., Yue, Q., & Yang, T. 2011. Comparative study on characterization of activated carbons prepared by microwave and conventional heating methods and application in removal of oxytetracycline (OTC). *Chemical Engineering Journal* 171(3): 1446–1453.

Inal, I. I. G., Holmes, S. M., Banford, A., & Aktas, Z. 2015. The performance of supercapacitor electrodes developed from chemically activated carbon produced from waste tea. *Applied Surface Science* 357: 696–703.

Inyang, M., & Dickenson, E. 2015. The potential role of biochar in the removal of organic and microbial contaminants from potable and reuse water: a review. *Chemosphere* 134: 232–240.

Inyang, M., Gao, B., Ding, W., Pullammanappallil, P., Zimmerman, A. R., & Cao, X. 2011. Enhanced lead sorption by biochar derived from anaerobically digested sugarcane bagasse. *Separation Science and Technology* 46(12): 1950–1956.

Iravani, S. 2011. Green synthesis of metal nanoparticles using plants. *Green Chemistry* 13(10): 2638–2650.

Jung, K. W., Kim, K., Jeong, T. U., & Ahn, K. H. 2016. Influence of pyrolysis temperature on characteristics and phosphate adsorption capability of biochar derived from waste-marine macroalgae (*Undaria pinnatifida* roots). *Bioresource Technology* 200: 1024–1028.

Kaman, S. P. D., Tan, I. A. W., & Lim, L. L. P. 2017. Palm oil mill effluent treatment using coconut shell–based activated carbon: adsorption equilibrium and isotherm. In Hasan, A., Khan, A. A., Sutan, Md. N., Othman, Hj. Al-K., Mannan, Md. A., Kabit, M. R., Hipolito, C. N., & Wahab, A. N. (Eds) *MATEC web of conferences* (Vol 87, p. 03009). EDP Sciences, Sarawak, Malaysia.

Kan, Y., Yue, Q., Li, D., Wu, Y., & Gao, B. 2017. Preparation and characterization of activated carbons from waste tea by H_3PO_4 activation in different atmospheres for oxytetracycline removal. *Journal of the Taiwan Institute of Chemical Engineers* 71: 494–500.

Keiluweit, M., Nico, P. S., Johnson, M. G., & Kleber, M. 2010. Dynamic molecular structure of plant biomass-derived black carbon (biochar). *Environmental Science & Technology* 44(4): 1247–1253.

Khalil, M. M., Ismail, E. H., El-Baghdady, K. Z., & Mohamed, D. 2014. Green synthesis of silver nanoparticles using olive leaf extract and its antibacterial activity. *Arabian Journal of Chemistry* 7(6): 1131–1139.

Khanehzaei, H., Ahmad, M. B., Shameli, K., & Ajdari, Z. 2014. Synthesis and characterization of Cu@ Cu_2O core shell nanoparticles prepared in seaweed Kappaphycus alvarezii Media. *International Journal of Electrochemical Science* 9: 8189–8198.

Khattri, S. D., & Singh, M. K. 2009. Removal of malachite green from dye wastewater using neem sawdust by adsorption. *Journal of Hazardous Materials* 167(1–3): 1089–1094.

Kim, K. H., Kim, J. Y., Cho, T. S., & Choi, J. W. 2012. Influence of pyrolysis temperature on physicochemical properties of biochar obtained from the fast pyrolysis of pitch pine (Pinus rigida). *Bioresource Technology* 118: 158–162.

Kiran, M. G., Pakshirajan, K., & Das, G. 2018. Heavy metal removal from aqueous solution using sodium alginate immobilized sulfate reducing bacteria: mechanism and process optimization. *Journal of Environmental Management* 218: 486–496.

Kumar, P. V., Shameem, U., Kollu, P., Kalyani, R. L., & Pammi, S. V. N. 2015. Green synthesis of copper oxide nanoparticles using Aloe vera leaf extract and its antibacterial activity against fish bacterial pathogens. *BioNanoScience* 5(3): 135–139.

Kwak, H. W., Hong, Y., Lee, M. E., & Jin, H. J. 2018. Sericin-derived activated carbon-loaded alginate bead: an effective and recyclable natural polymer-based adsorbent for methylene blue removal. *International Journal of Biological Macromolecules* 120: 906–914.

Kwon, K. H., Jung, K. Y., & Yeom, S. H. 2009. Comparison between entrapment methods for phenol removal and operation of bioreactor packed with co-entrapped activated carbon and Pseudomonas fluorescence KNU417. *Bioprocess and Biosystems Engineering* 32(2): 249–256.

Li, J., Ng, D. H., Song, P., Kong, C., Song, Y., & Yang, P. 2015. Preparation and characterization of high-surface-area activated carbon fibers from silkworm cocoon waste for congo red adsorption. *Biomass and Bioenergy* 75: 189–200.

Lillo-Ródenas, M. A., Cazorla-Amorós, D., & Linares-Solano, A. 2003. Understanding chemical reactions between carbons and NaOH and KOH: an insight into the chemical activation mechanism. *Carbon* 41(2): 267–275.

Liu, K., Han, W., Pan, W. P., & Riley, J. T. 2001. Polycyclic aromatic hydrocarbon (PAH) emissions from a coal-fired pilot FBC system. *Journal of Hazardous Materials* 84 (2–3): 175–188.

Liu, X., Li, Z., Zhang, Y., Feng, R., & Mahmood, I. B. 2014. Characterization of human manure-derived biochar and energy-balance analysis of slow pyrolysis process. *Waste Management* 34(9): 1619–1626.

Liu, Y., Yao, S., Wang, Y., Lu, H., Brar, S. K., & Yang, S. 2017. Bio-and hydrochars from rice straw and pig manure: inter-comparison. *Bioresource Technology* 235: 332–337.

Lu, L., Li, A., Ji, X., Yang, C., & He, S. 2018. Removal of acenaphthene from water by Triton X-100-facilitated biochar-immobilized Pseudomonas aeruginosa. *RSC Advances 8*(41): 23426–23432.

Luo, Y., Guo, W., Ngo, H. H., Nghiem, L. D., Hai, F. I., Zhang, J., Liang, S., & Wang, X. C. 2014. A review on the occurrence of micropollutants in the aquatic environment and their fate and removal during wastewater treatment. *Science of the Total Environment* 473: 619–641.

Maletić, S. P., Dalmacija, B. D., Rončević, S. D., Agbaba, J. R., & Perović, S. D. U. 2011. Impact of hydrocarbon type, concentration and weathering on its biodegradability in soil. *Journal of Environmental Science and Health, Part A* 46(10): 1042–1049.

Malik, A., Verma, P., Singh, A. K., & Singh, K. P. 2011. Distribution of polycyclic aromatic hydrocarbons in water and bed sediments of the Gomti River, India. *Environmental Monitoring and Assessment* 172(1–4): 529–545.

Malik, R., Ramteke, D. S., & Wate, S. R. 2006. Physico-chemical and surface characterization of adsorbent prepared from groundnut shell by $ZnCl_2$ activation and its ability to adsorb colour. *Indian Journal of Chemical Technology* 13: 319–328.

Manoli, E., & Samara, C. 1999. Polycyclic aromatic hydrocarbons in natural waters: sources, occurrence and analysis. *TrAC Trends in Analytical Chemistry* 18(6): 417–428.

Martin, M. J., Serra, E., Ros, A., Balaguer, M. D., & Rigola, M. 2004. Carbonaceous adsorbents from sewage sludge and their application in a combined activated sludge-powdered activated carbon (AS-PAC) treatment. *Carbon* 42(7): 1389–1394.

Masoud, M. S., El-Saraf, W. M., Abdel-Halim, A. M., Ali, A. E., Mohamed, E. A., & Hasan, H. M. 2016. Rice husk and activated carbon for waste water treatment of El-Mex Bay, Alexandria Coast, Egypt. *Arabian Journal of Chemistry* 9: S1590–S1596.

Mehrabi, F., Vafaei, A., Ghaedi, M., Ghaedi, A. M., Dil, E. A., & Asfaram, A. 2017. Ultrasound assisted extraction of Maxilon Red GRL dye from water samples using cobalt ferrite nanoparticles loaded on activated carbon as sorbent: optimization and modeling. *Ultrasonics Sonochemistry* 38: 672–680.

Mohanta, D., & Ahmaruzzaman, M. 2018. Bio-inspired adsorption of arsenite and fluoride from aqueous solutions using activated carbon@ SnO_2 nanocomposites: isotherms, kinetics, thermodynamics, cost estimation and regeneration studies. *Journal of Environmental Chemical Engineering* 6(1): 356–366.

Mrozik, A., Piotrowska-Seget, Z., & Labuzek, S. 2003. Bacterial degradation and bioremediation of polycyclic aromatic hydrocarbons. *Polish Journal of Environmental Studies* 12(1): 15.

Muangchinda, C., Chamcheun, C., Sawatsing, R., & Pinyakong, O. 2018. Diesel oil removal by Serratia sp. W4-01 immobilized in chitosan-activated carbon beads. *Environmental Science and Pollution Research* 25(27): 26927–26938.

Muniandy, L., Adam, F., Mohamed, A. R., & Ng, E. P. 2014. The synthesis and characterization of high purity mixed microporous/mesoporous activated carbon from rice husk using chemical activation with NaOH and KOH. *Microporous and Mesoporous Materials* 197: 316–323.

Narzari, R., Bordoloi, N., Sarma, B., Gogoi, L., Gogoi, N., Borkotoki, B., & Kataki, R. 2017. Fabrication of biochars obtained from valorization of biowaste and evaluation of its physicochemical properties. *Bioresource Technology* 242: 324–328.

Nasrollahzadeh, M., Sajadi, S. M., Rostami-Vartooni, A., Bagherzadeh, M., & Safari, R. 2015. Immobilization of copper nanoparticles on perlite: green synthesis, characterization and catalytic activity on aqueous reduction of 4-nitrophenol. *Journal of Molecular Catalysis A: Chemical* 400: 22–30.

Nik, W. W., Rahman, M. M., Yusof, A. M., Ani, F. N., & Adnan, C. 2006. Production of activated carbon from palm oil shell waste and its adsorption characteristics. In *1st international conference on natural resources engineering and technology*, Putrajaya, Malaysia (pp. 646–654).

Oleszczuk, P., Hale, S. E., Lehmann, J., & Cornelissen, G. 2012. Activated carbon and biochar amendments decrease pore-water concentrations of polycyclic aromatic hydrocarbons (PAHs) in sewage sludge. *Bioresource Technology* 111: 84–91.

Örkün, Y., Karatepe, N., & Yavuz, R. 2012. Influence of temperature and impregnation ratio of H_3PO_4 on the production of activated carbon from hazelnut shell. *Acta Physica Polonica-Series A General Physics* 121(1): 277.

Partovinia, A., & Naeimpoor, F. 2013. Phenanthrene biodegradation by immobilized microbial consortium in polyvinyl alcohol cryogel beads. *International Biodeterioration & Biodegradation* 85: 337–344.

Pavoni, B., Caliceti, M., Sperni, L., & Sfriso, A. 2003. Organic micropollutants (PAHs, PCBs, pesticides) in seaweeds of the lagoon of Venice. *Oceanologica Acta* 26(5–6): 585–596.

Pehlivan, E., Özbay, N., Yargıç, A. S., & Şahin, R. Z. 2017. Production and characterization of chars from cherry pulp via pyrolysis. *Journal of Environmental Management* 203: 1017–1025.

Pi, L., Jiang, R., Zhou, W., Zhu, H., Xiao, W., Wang, D., & Mao, X. 2015. g-C3N4 modified biochar as an adsorptive and photocatalytic material for decontamination of aqueous organic pollutants. *Applied Surface Science* 358: 231–239.

Pitarch, E., Medina, C., Portolés, T., López, F. J., & Hernández, F. 2007. Determination of priority organic micro-pollutants in water by gas chromatography coupled to triple quadrupole mass spectrometry. *Analytica Chimica Acta* 583(2): 246–258.

Prahas, D., Kartika, Y., Indraswati, N., & Ismadji, S. 2008. Activated carbon from jackfruit peel waste by H 3 PO 4 chemical activation: pore structure and surface chemistry characterization. *Chemical Engineering Journal* 140(1): 32–42.

Priac, A., Morin-Crini, N., Druart, C., Gavoille, S., Bradu, C., Lagarrigue, C., Torri, G., Winterton, P., & Crini, G. 2017. Alkylphenol and alkylphenol polyethoxylates in water and wastewater: a review of options for their elimination. *Arabian Journal of Chemistry* 10: S3749–S3773.

Qambrani, N. A., Rahman, M. M., Won, S., Shim, S., & Ra, C. 2017. Biochar properties and eco-friendly applications for climate change mitigation, waste management, and wastewater treatment: a review. *Renewable and Sustainable Energy Reviews* 79: 255–273.

Rajasekhar, B., Nambi, I. M., & Govindarajan, S. K. 2018. Human health risk assessment of ground water contaminated with petroleum PAHs using Monte Carlo simulations: a case study of an Indian metropolitan city. *Journal of Environmental Management* 205: 183–191.

Reddy, M. S., Basha, S., Joshi, H. V., & Ramachandraiah, G. 2005. Seasonal distribution and contamination levels of total PHCs, PAHs and heavy metals in coastal waters of the Alang–Sosiya ship scrapping yard, Gulf of Cambay, India. *Chemosphere* 61(11): 1587–1593.

Roosta, M., Ghaedi, M., Daneshfar, A., Sahraei, R., & Asghari, A. 2014a. Optimization of the ultrasonic assisted removal of methylene blue by gold nanoparticles loaded on activated carbon using experimental design methodology. *Ultrasonics Sonochemistry* 21(1): 242–252.

Roosta, M. A., Ghaedi, M. A., Shokri, N., Daneshfar, A., Sahraei, R., & Asghari, A. 2014b. Optimization of the combined ultrasonic assisted/adsorption method for the removal of malachite green by gold nanoparticles loaded on activated carbon: experimental design. *Spectrochimica Acta Part A: Molecular and Biomolecular Spectroscopy* 118: 55–65.

Roy, U., Sengupta, S., Das, P., Bhowal, A., & Datta, S. 2018. Integral approach of sorption coupled with biodegradation for treatment of azo dye using Pseudomonas sp.: batch, toxicity, and artificial neural network. *3 Biotech* 8(4): 192.

Saad, M. E. K., Khiari, R., Elaloui, E., & Moussaoui, Y. 2014. Adsorption of anthracene using activated carbon and Posidonia oceanica. *Arabian Journal of Chemistry* 7(1): 109–113.

Sadeghi, B., & Gholamhoseinpoor, F. 2015. A study on the stability and green synthesis of silver nanoparticles using Ziziphora tenuior (Zt) extract at room temperature. *Spectrochimica Acta Part A: Molecular and Biomolecular Spectroscopy* 134: 310–315.

Sahota, S., Vijay, V. K., Subbarao, P. M. V., Chandra, R., Ghosh, P., Shah, G., Kapoor, R., Vijay, V., Koutu, V., & Thakur, I. S. 2018. Characterization of leaf waste based biochar for cost effective hydrogen sulphide removal from biogas. *Bioresource Technology* 250: 635–641.

Saleh, T. A. 2018. Simultaneous adsorptive desulfurization of diesel fuel over bimetallic nanoparticles loaded on activated carbon. *Journal of Cleaner Production* 172. 2123–2132.

Samanta, S. K., Singh, O. V., & Jain, R. K. 2002. Polycyclic aromatic hydrocarbons: environmental pollution and bioremediation. *TRENDS in Biotechnology* 20(6): 243–248.

Satapathy, M. K., Banerjee, P., & Das, P. 2015. Plant-mediated synthesis of silver-nanocomposite as novel effective azo dye adsorbent. *Applied Nanoscience* 5 (1): 1–9.

Sayğılı, H., & Güzel, F. 2016. High surface area mesoporous activated carbon from tomato processing solid waste by zinc chloride activation: process optimization, characterization and dyes adsorption. *Journal of Cleaner Production* 113: 995–1004.

Schwarzenbach, R. P., Escher, B. I., Fenner, K., Hofstetter, T. B., Johnson, C. A., Von Gunten, U., & Wehrli, B. 2006. The challenge of micropollutants in aquatic systems. *Science* 313(5790): 1072–1077.

Şen, A., Pereira, H., Olivella, M. A., & Villaescusa, I. 2015. Heavy metals removal in aqueous environments using bark as a biosorbent. *International Journal of Environmental Science and Technology* 12(1): 391–404.

Seth, P. K. 2014. Chemical contaminants in water and associated health hazards. In Singh, P. & Sharma, V. (Eds) *Water and Health*, 375–384. Springer, New Delhi.

Shao, Y., Wang, Y., Wu, X., Xu, X., Kong, S., Tong, L., Jiang, Z., & Li, B. 2015. Biodegradation of PAHs by Acinetobacter isolated from karst groundwater in a coal-mining area. *Environmental Earth Sciences* 73(11): 7479–7488.

Sheldon, R. A., & Van Pelt, S. 2013. Enzyme immobilisation in biocatalysis: why, what and how. *Chemical Society Reviews* 42(15): 6223–6235.

Shende, S., Ingle, A. P., Gade, A., & Rai, M. 2015. Green synthesis of copper nanoparticles by Citrus medica Linn. (Idilimbu) juice and its antimicrobial activity. *World Journal of Microbiology and Biotechnology* 31(6): 865–873.

Silva, T. L., Cazetta, A. L., Souza, P. S., Zhang, T., Asefa, T., & Almeida, V. C. 2018. Mesoporous activated carbon fibers synthesized from denim fabric waste: efficient adsorbents for removal of textile dye from aqueous solutions. *Journal of Cleaner Production* 171: 482–490.

Singh, P., Kim, Y. J., Zhang, D., & Yang, D. C. 2016. Biological synthesis of nanoparticles from plants and microorganisms. *Trends in Biotechnology* 34(7): 588–599.

Singh, P., Singh, R., Borthakur, A., Madhav, S., Singh, V. K., Tiwary, D., Srivastava, V. C., & Mishra, P. K. 2018. Exploring temple floral refuse for biochar production as a closed loop perspective for environmental management. *Waste Management* 77: 78–86.

Sinha, S. N., Paul, D., Halder, N., Sengupta, D., & Patra, S. K. 2015. Green synthesis of silver nanoparticles using fresh water green alga Pithophora oedogonia (Mont.) Wittrock and evaluation of their antibacterial activity. *Applied Nanoscience* 5(6): 703–709.

Smith, S. C., & Rodrigues, D. F. 2015. Carbon-based nanomaterials for removal of chemical and biological contaminants from water: a review of mechanisms and applications. *Carbon* 91: 122–143.

Srinivasan, P., Sarmah, A. K., Smernik, R., Das, O., Farid, M., & Gao, W. 2015. A feasibility study of agricultural and sewage biomass as biochar, bioenergy and biocomposite feedstock: production, characterization and potential applications. *Science of the Total Environment* 512: 495–505.

Sulyman, M., Namiesnik, J., & Gierak, A. 2017. Low-cost adsorbents derived from agricultural by-products/wastes for enhancing contaminant uptakes from wastewater: a review. *Polish Journal of Environmental Studies* 26(3): 479–510.

Sun, L., Wan, S., & Luo, W. 2013. Biochars prepared from anaerobic digestion residue, palm bark, and eucalyptus for adsorption of cationic methylene blue dye: characterization, equilibrium, and kinetic studies. *Bioresource Technology* 140: 406–413.

Tan, X. F., Liu, Y. G., Gu, Y. L., Xu, Y., Zeng, G. M., Hu, X. J., Liu, S. B., Wang, X., Liu, S. M., & Li, J. 2016. Biochar-based nano-composites for the decontamination of wastewater: a review. *Bioresource Technology* 212: 318–333.

Tang, Y. B., Liu, Q., & Chen, F. Y. 2012. Preparation and characterization of activated carbon from waste ramulus mori. *Chemical Engineering Journal* 203: 19–24.

Tong, D. S., Zhou, C. H. C., Lu, Y., Yu, H., Zhang, G. F., & Yu, W. H. 2010. Adsorption of acid red G dye on octadecyl trimethylammonium montmorillonite. *Applied Clay Science* 50(3): 427–431.

Touray, N., Tsai, W. T., Chen, H. R., & Liu, S. C. 2014. Thermochemical and pore properties of goat-manure-derived biochars prepared from different pyrolysis temperatures. *Journal of Analytical and Applied Pyrolysis* 109: 116–122.

Tsai, W. T., Chang, C. Y., & Lee, S. L. 1998. A low cost adsorbent from agricultural waste corn cob by zinc chloride activation. *Bioresource Technology* 64(3): 211–217.

US EPA. 1984. List of sixteen pahs with highest carcinogenic effect. *IEA Coal Research, London.*

US EPA, 2016. Poly cyclic organic matter. Available at: www3.epa.gov/ttn/atw/hlthef/polycycl.html.

Usman, A. R., Abduljabbar, A., Vithanage, M., Ok, Y. S., Ahmad, M., Ahmad, M., & Al-Wabel, M. I. 2015. Biochar production from date palm waste: charring temperature induced changes in composition and surface chemistry. *Journal of Analytical and Applied Pyrolysis* 115: 392–400.

Van Zwieten, L., Kimber, S., Morris, S., Chan, K. Y., Downie, A., Rust, J., & Cowie, A. 2010. Effects of biochar from slow pyrolysis of papermill waste on agronomic performance and soil fertility. *Plant and Soil* 327(1–2): 235–246.

Varjani, S. J., & Upasani, V. N. 2017. A new look on factors affecting microbial degradation of petroleum hydrocarbon pollutants. *International Biodeterioration & Biodegradation* 120: 71–83.

Vasudevan, S., & Oturan, M. A. 2014. Electrochemistry: as cause and cure in water pollution—an overview. *Environmental Chemistry Letters* 12(1): 97–108.

Venosa, A. D., & Zhu, X. 2003. Biodegradation of crude oil contaminating marine shorelines and freshwater wetlands. *Spill Science & Technology Bulletin 8*(2): 163–178.

Vilanova, R. M., Fernández, P., Martínez, C., & Grimalt, J. O. 2001. Polycyclic aromatic hydrocarbons in remote mountain lake waters. *Water Research* 35(16): 3916–3926.

Vithanage, M., Mayakaduwa, S. S., Herath, I., Ok, Y. S., & Mohan, D. 2016. Kinetics, thermodynamics and mechanistic studies of carbofuran removal using biochars from tea waste and rice husks. *Chemosphere* 150: 781–789.

Wang, C., Wang, T., Li, W., Yan, J., Li, Z., Ahmad, R., & Zhu, N. 2014. Adsorption of deoxyribonucleic acid (DNA) by willow wood biochars produced at different pyrolysis temperatures. *Biology and Fertility of Soils* 50(1): 87–94.

Wang, H., Gao, B., Wang, S., Fang, J., Xue, Y., & Yang, K. 2015b. Removal of Pb (II), Cu (II), and Cd (II) from aqueous solutions by biochar derived from KMnO 4 treated hickory wood. *Bioresource Technology* 197: 356–362.

Wang, J., Wang, C., Huang, Q., Ding, F., & He, X. 2015a. Adsorption of PAHs on the sediments from the yellow river delta as a function of particle size and salinity. *Soil and Sediment Contamination: An International Journal* 24(2): 103–115.

Wang, X., Sato, T., & Xing, B. 2006. Competitive sorption of pyrene on wood chars. *Environmental Science & Technology* 40(10): 3267–3272.

Wenzl, T., Simon, R., Anklam, E., & Kleiner, J. 2006. Analytical methods for polycyclic aromatic hydrocarbons (PAHs) in food and the environment needed for new food legislation in the European Union. *TrAC Trends in Analytical Chemistry 25*(7): 716–725.

WHO (World Health Organization). 2015. Drinking-water: fact sheet No. 391. Available at: www.who.int/mediacentre/factsheets/fs391/en/.

Wong, K. T., Eu, N. C., Ibrahim, S., Kim, H., Yoon, Y., & Jang, M. 2016. Recyclable magnetite-loaded palm shell-waste based activated carbon for the effective removal

of methylene blue from aqueous solution. *Journal of Cleaner Production* 115: 337–342.
World Health Organization. 2017. Guidelines for drinking-water quality: incorporating first addendum.
Xiao, X., Liu, D., Yan, Y., Wu, Z., Wu, Z., & Cravotto, G. 2015. Preparation of activated carbon from Xinjiang region coal by microwave activation and its application in naphthalene, phenanthrene, and pyrene adsorption. *Journal of the Taiwan Institute of Chemical Engineers* 53: 160–167.
Xiong, B., Zhang, Y., Hou, Y., Arp, H. P. H., Reid, B. J., & Cai, C. 2017. Enhanced biodegradation of PAHs in historically contaminated soil by M. gilvum inoculated biochar. *Chemosphere* 182: 316–324.
Yahya, M. A., Al-Qodah, Z., & Ngah, C. Z. 2015. Agricultural bio-waste materials as potential sustainable precursors used for activated carbon production: a review. *Renewable and Sustainable Energy Reviews* 46: 218–235.
Yakout, S. M., Daifullah, A. A. M., & El-Reefy, S. A. 2013. Adsorption of naphthalene, phenanthrene and pyrene from aqueous solution using low-cost activated carbon derived from agricultural wastes. *Adsorption Science & Technology* 31(4): 293–302.
Yan, J., Han, L., Gao, W., Xue, S., & Chen, M. 2015. Biochar supported nanoscale zerovalent iron composite used as persulfate activator for removing trichloroethylene. *Bioresource Technology* 175: 269–274.
Yan, J., Wang, L., Fu, P. P., & Yu, H. 2004. Photomutagenicity of 16 polycyclic aromatic hydrocarbons from the US EPA priority pollutant list. *Mutation Research/Genetic Toxicology and Environmental Mutagenesis* 557(1): 99–108.
Yenisoy-Karakaş, S., Aygün, A., Güneş, M., & Tahtasakal, E. 2004. Physical and chemical characteristics of polymer-based spherical activated carbon and its ability to adsorb organics. *Carbon* 42(3): 477–484.
Ying, W., Ye, T., Bin, H., Zhao, H. B., Bi, J. N., & Cai, B. L. 2007. Biodegradation of phenol by free and immobilized Acinetobacter sp. strain PD12. *Journal of Environmental Sciences 19*(2): 222–225.
Young, L. Y., & Cerniglia, C. 1995. *Microbial transformation and degradation of toxic organic chemicals* (Vol 15*)*. Wiley-Liss, New York.
Yuan, J. H., Xu, R. K., & Zhang, H. 2011. The forms of alkalis in the biochar produced from crop residues at different temperatures. *Bioresource Technology* 102(3): 3488–3497.
Yuan, M., Tong, S., Zhao, S., & Jia, C. Q. 2010. Adsorption of polycyclic aromatic hydrocarbons from water using petroleum coke-derived porous carbon. *Journal of Hazardous Materials* 181(1): 1115–1120.
Yuan, Z., Xu, Z., Zhang, D., Chen, W., Zhang, T., Huang, Y., & Tian, D. 2018. Box-Behnken design approach towards optimization of activated carbon synthesized by co-pyrolysis of waste polyester textiles and MgCl2. *Applied Surface Science* 427: 340–348.
Zapata Acosta, K., Carrasco-Marin, F., Cortés, F. B., Franco, C. A., Lopera, S. H., & Rojano, B. A. 2019. Immobilization of P. stutzeri on activated carbons for degradation of hydrocarbons from oil-in-saltwater emulsions. *Nanomaterials* 9(4): 500.
Zeledon-Toruno, Z. C., Lao-Luque, C., de Las Heras, F. X. C., & Sole-Sardans, M. 2007. Removal of PAHs from water using an immature coal (leonardite). *Chemosphere* 67(3): 505–512.
Zhang, J., Gong, L., Sun, K., Jiang, J., & Zhang, X. 2012. Preparation of activated carbon from waste Camellia oleifera shell for supercapacitor application. *Journal of Solid State Electrochemistry* 16(6): 2179–2186.

Zhang, P., Sun, H., Yu, L., & Sun, T. 2013. Adsorption and catalytic hydrolysis of carbaryl and atrazine on pig manure-derived biochars: impact of structural properties of biochars. *Journal of Hazardous Materials* 244: 217–224.

Zhang, Z., Hou, Z., Yang, C., Ma, C., Tao, F., & Xu, P. 2011. Degradation of n-alkanes and polycyclic aromatic hydrocarbons in petroleum by a newly isolated Pseudomonas aeruginosa DQ8. *Bioresource Technology* 102(5): 4111–4116.

Zheng, J., Zhao, Q., & Ye, Z. 2014. Preparation and characterization of activated carbon fiber (ACF) from cotton woven waste. *Applied Surface Science* 299: 86–91.

Zheng, W., Guo, M., Chow, T., Bennett, D. N., & Rajagopalan, N. 2010. Sorption properties of greenwaste biochar for two triazine pesticides. *Journal of Hazardous Materials* 181(1–3): 121–126.

Zhu, Z. L., Li, A. M., Xia, M. F., Wan, J. N., & Zhang, Q. X. 2008. Preparation and characterization of polymer-based spherical activated carbons. *Chinese Journal of Polymer Science* 26(5): 645–651.

Index

Printed in the United States
by Baker & Taylor Publisher Services

Printed in the United States
by Baker & Taylor Publisher Services